国家示范性中等职业技术教育精品教材

智能手机APP软件开发：Android

主　编　李小军

副主编　欧阳元东

编　委　苏伟斌　周清流　赵英姿

华南理工大学出版社

SOUTH CHINA UNIVERSITY OF TECHNOLOGY PRESS

·广州·

内容简介

本课程是中等职业学校计算机专业的重点专业课程之一。课程紧密结合国家工信部通信行业技能鉴定中心要求，以项目和任务的方式将课程内容的各种实际操作“项目化”，通过13个项目逐一介绍Android的系统架构、程序设计、用户界面设计、四大组件、存储本地数据、多媒体、网络编程、游戏开发、实习安全管理平台开发等知识。从基础知识到完整游戏再到管理程序APP实现，学生可轻松掌握Android下应用程序设计的流程以及方法。可作为Android开发类教材以及Android开发人员开发参考书。

图书在版编目(CIP)数据

智能手机APP软件开发:Android/李小军主编.—广州:华南理工大学出版社,2015.1（2020.1重印）
国家示范性中等职业技术教育精品教材
ISBN 978-7-5623-4539-8

Ⅰ.①智… Ⅱ.①李… Ⅲ.①移动电话机-应用程序-程序设计-中等专业学校-教材
Ⅳ.TN929.53

中国版本图书馆CIP数据核字(2015)第021488号

ZHINENG SHOUJI APP RUANJIAN KAIFA:Android
智能手机APP软件开发:Android
李小军 主编

出 版 人：卢家明
出版发行：华南理工大学出版社
（广州五山华南理工大学17号楼，邮编510640）
http://www.scutpress.com.cn E-mail:scutc13@scut.edu.cn
营销部电话：020-87113487 87111048（传真）
策划编辑：何丽云
责任编辑：何丽云
印 刷 者：虎彩印艺股份有限公司
开 本：787mm×1092mm 1/16 **印张：**21 **字数：**521千
版 次：2015年1月第1版 2020年1月第3次印刷
定 价：46.00元

前言

本课程是中等职业学校计算机专业的重点专业课程之一。课程以培养技能型人才为导向，注重理论与案例相结合的教学。同时遵循中等职业院校学生的认知规律，紧密结合 Android 手机实际开发，以具体 Android 开发项目为驱动，配备具体的任务制作和开发实例分析，以培养和提高学生分析问题和解决问题的能力。整个课程围绕 Android 的实际开发技术，以任务驱动的方式培养和提高中等职业院校学生的 Android 实际开发能力。

本书在编写过程前对中等职业院校学生的实际情况做过摸底调查，编写过程中也吸收企业技术人员参与教材编写，紧密结合工作岗位，与职业岗位对接；选取的任务贴近生活、贴近开发实际；将创新理念贯彻到内容、任务选取等方面，培养学生 Android 开发能力、创新能力和自学能力。

本书坚持中职技能型人才培养的定位，在编写时努力贯彻教学改革的有关精神，努力体现以下特色：

1. 以项目任务为驱动，强化知识与技能的整合。

本课程采用项目驱动模式，以实施理论实践一体化、教学做一体化教学。教材编写以每个项目中的具体任务为切入点，以实际开发 APP 应用任务为导向，既重视学生实际操作技能的提高，又保证学生对基本理论知识的学习和掌握；又根据中等职业学校学生的认知特点，以项目和任务的方式强化学生的实际学习效果，努力让学生做到“做中学，学中悟”，尽量将不同知识点与技能点以实际的开发任务的方式有机连贯起来，以降低 Android 开发的学习难度，满足学生学习的需求。

2. 以学生的实际水平为中心，培养学生创新能力和自学能力。

中等职业院校学生在本课程方面的起点是比较低的，而且同其他计算机专业课程相比，本课程的学习范围和难度都相对要大。因此，本课程设计了多个实际的 Android 开发具体任务，每个任务都基本覆盖了该部分的 Android 开发的知识点，使原本抽象、难懂、难学的教学内容变得直观、易懂和容易掌握，提高了老师的教学效率和学生的学习效率。每个任务后基本都留有与该任务相关的习题，以检测和考核学生的学习情况，培养学生的创新能力和自学能力。

本书主要由 13 个项目组成，每个项目都将介绍与 Android 开发相关的基础知识，并重点讲解典型的开发技术。

项目一为初探 Android，重点介绍了 Android 的运行环境的安装和配置，以及 Android 开发流程。

项目二为 UI 界面布局，主要介绍了用户界面的概念以及创建，并对 Android 中常用的四种布局管理器做了重点介绍。

项目三为基本控件及常用事件，重点介绍了 Android 常用基本控件及其用法。

项目四为 Android 中的 Activity，重点介绍了 Activity 的概念及其生命周期。

项目五为 Android 中的 Intent，重点介绍了 Intent 的基本知识，以及 Intent 的开发和应用。

项目六为 Android 中的 Service，重点介绍了 Service 组件以及开发流程和方法。

项目七为 Android 中的数据存储，重点介绍了 Share Preferereences 类、File 存储、SQLite 数据库的操作、ContentProvider 类及它们的使用方法。

项目八为 Android 多媒体方面，重点介绍了 Android 多媒体与娱乐开发的基本知识以及实际开发。

项目九为 Android 绘画与动画，重点介绍了 Android 绘图的基本知识以及几种 Android 简单动画的制作。

项目十为 Android 地图服务，重点介绍了 Android 在百度地图方面的开发。

项目十一为 Android 网络编程，重点介绍了 Android 网络编程方面的基本知识和实际开发。

项目十二为 Android 游戏开发，重点介绍了常见的游戏开发框架。

项目十三以一综合实例——学生实习安全管理平台，重点介绍布局文件编写、常用控件的应用、Intent 组件 、Service 组件、百度地图编程、网络编程技术的应用。

本教材由李小军主编，欧阳元东任副主编，其中项目一、项目七、项目十三由欧阳元东编写，项目二、项目十二由苏伟斌编写，项目三、项目九由赵英姿编写，项目四、项目十一由周清流编写，项目五、项目六、项目八、项目十由李小军编写，李小军、欧阳元东对全书进行统稿和修改。

由于编者水平有限，书中难免有错误和不妥之处，恳请广大读者批评指正。

编　者

目录

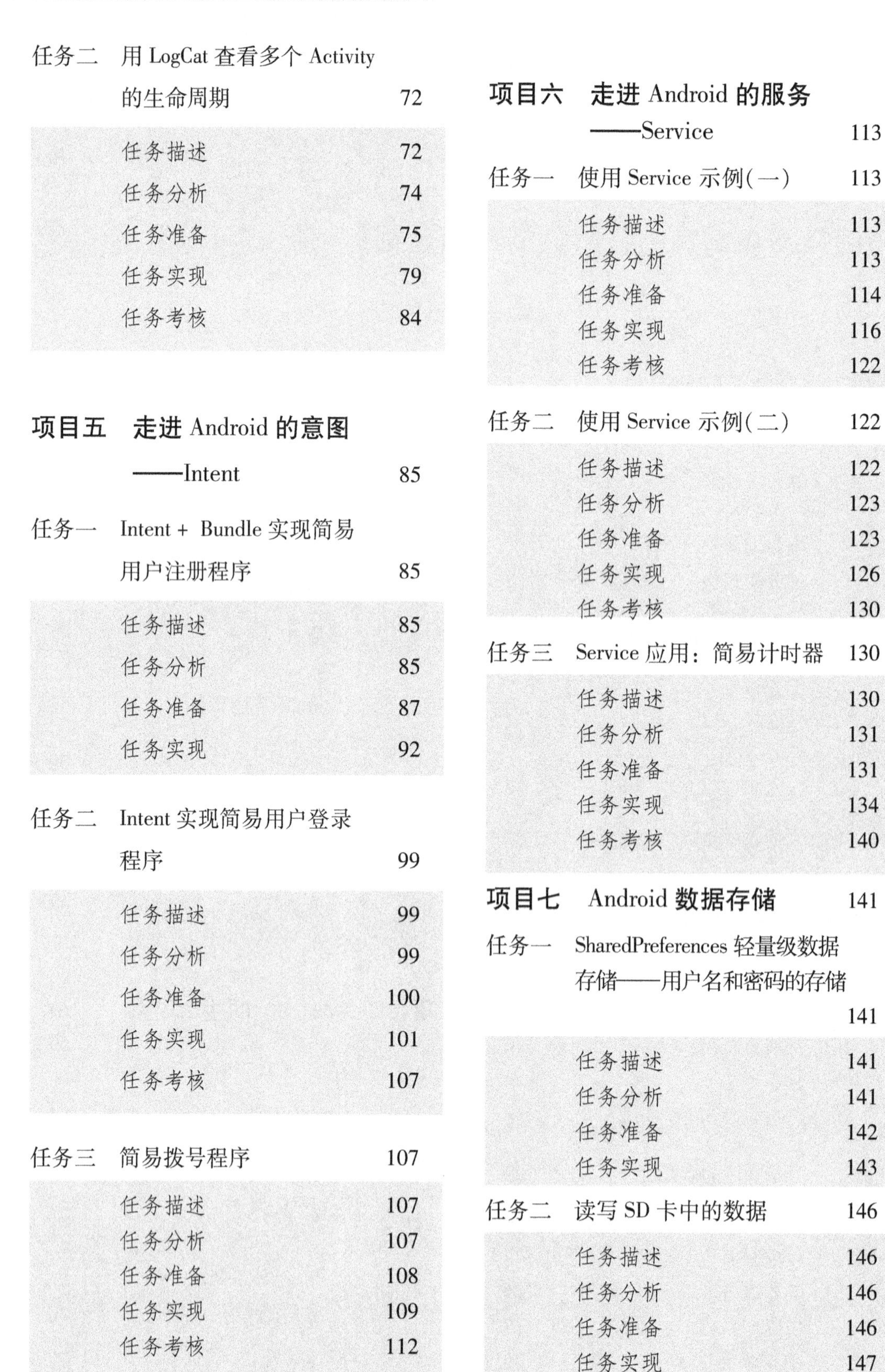

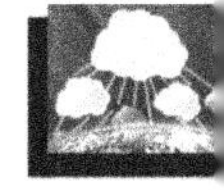

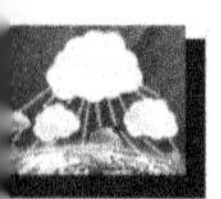

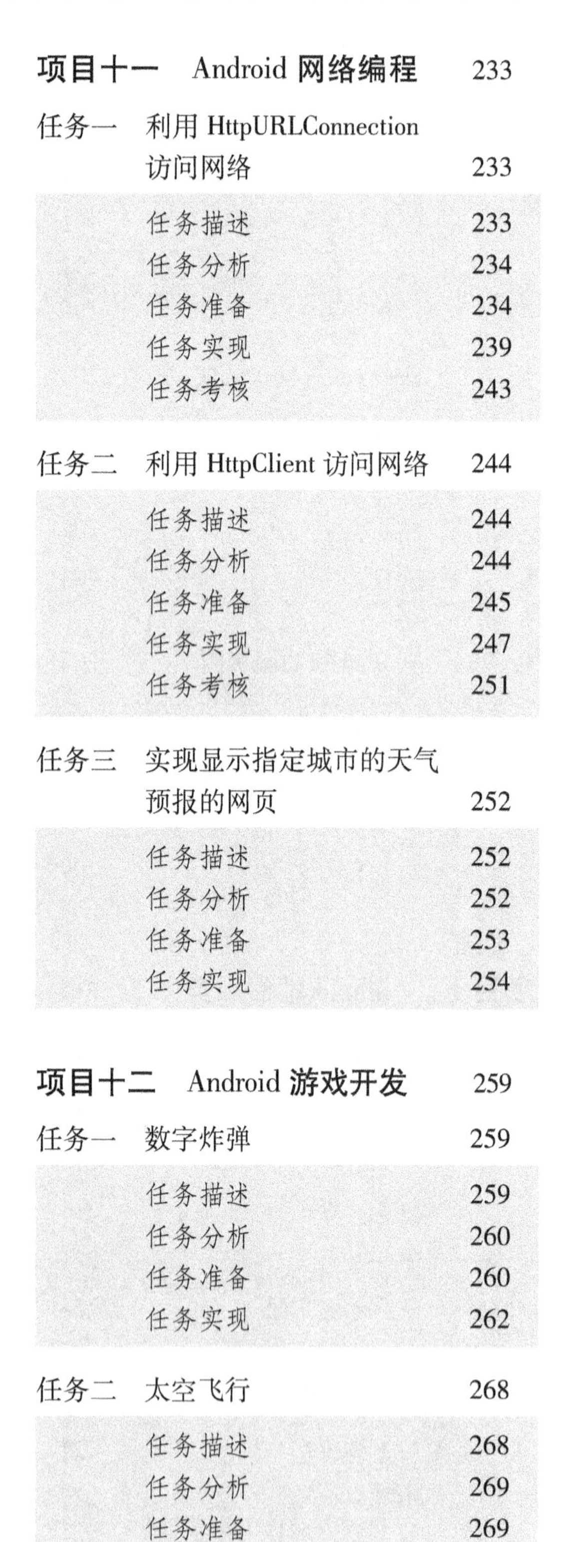

项目一

初探Android

Android是Google于2007年11月5日发布的基于Linux内核的移动开发平台，主要用于便携式设备。该平台由操作系统、中间件、用户界面和应用软件组成，是一个真正的移动开发平台。本项目的任务一详细讲授Android开发平台的搭建，任务二是开发一个简单的Android程序，并通过对该程序进行简单的分析，带领大家迈进Android开发世界的殿堂。

了解 Android 的架构和特点
掌握 Android 开发平台的搭建
学会创建 Android 的工程项目
学会创建模拟器
理解 Android 工程文件夹的内涵

任务一　Android 开发环境的搭建

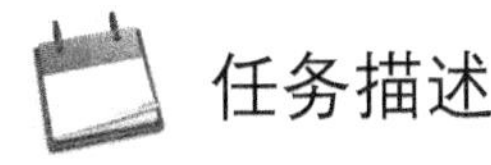

任务描述

Android 操作系统是一个由 Google 和开发手机联盟共同开发的移动设备操作系统，其最早的一个发布版本开始于 2007 年 11 月的 Android1.0 beta，并且已经发布了多个更新版本。2011 年第一季度，Android 在全球的市场份额首次超过塞班系统，跃居全球第一。2012 年 2 月数据显示 Android 占据全球智能手机操作系统市场 52.5% 的份额，中国市场占有率为 68.4%。

Android 的一个重要特点就是它的应用框架和 GUI 库都是用 Java 语言实现。Android 内部有一个叫做 Dalvik 的 Java 虚拟机，Java 程序由这个虚拟机解释运行。所以，Android 平台的应用程序必须用 Java 语言开发。

本任务重点是搭建 Android 的开发平台。本平台在 Eclipse 中构建，需要安装开发工具 JDK、SDK 和 ADT，同时配置 JDK 和 SDK 的环境变量。

任务分析

完成本任务，我们需解决以下步骤：

第一步：安装 JDK，因为 Android 开发是基于 Java 语言的。

第二步：安装 Eclipse，本书源代码是在 Eclipse 环境下调试。

第三步：安装 Android SDK 工具包。

第四步：安装 Eclipse 的 ADT 插件。

第五步：创建 AVD。

第六步：设置 Android 系统语言。

任务准备

CPU 主频在 1.5 以上，1G 内存，显存 128M 以上，可以正常运行 Windows XP 或者 Windows7 的计算机。

任务实现

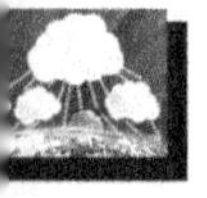

1. 下载 JDK 安装包

在浏览器的地址栏中输入网址“http：//www. oracle. com”，在网站中找到链接下载 JDK 安装包，笔者使用的是 jdk-8u11-windows-i586。创建文件夹 d:\Android，把 JDK 安装包放入此文件夹中。JDK 下载完后，双击安装程序图标，使用安装向导一步一步进行下去，注意安装的目标文件夹。操作过程如图 1.1 所示。

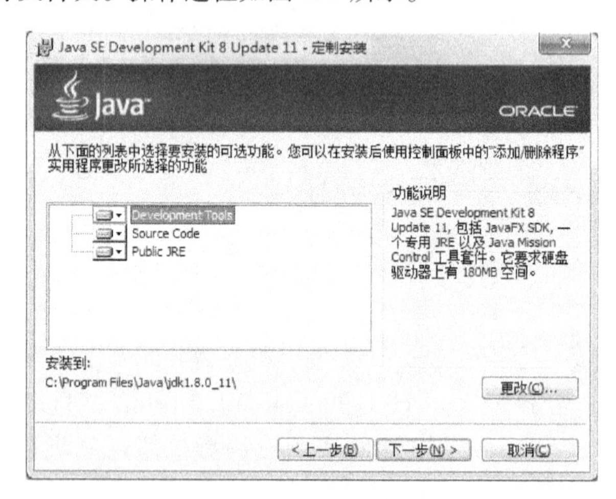

图 1.1　JDK 安装

2. JDK 配置

JDK 安装完毕后，需要进行环境变量 JAVA_HOME 和 CLASSPATH 的配置，如图 1.2 和图 1.3 所示。

图 1.2 JAVA_HOME 环境变量的配置

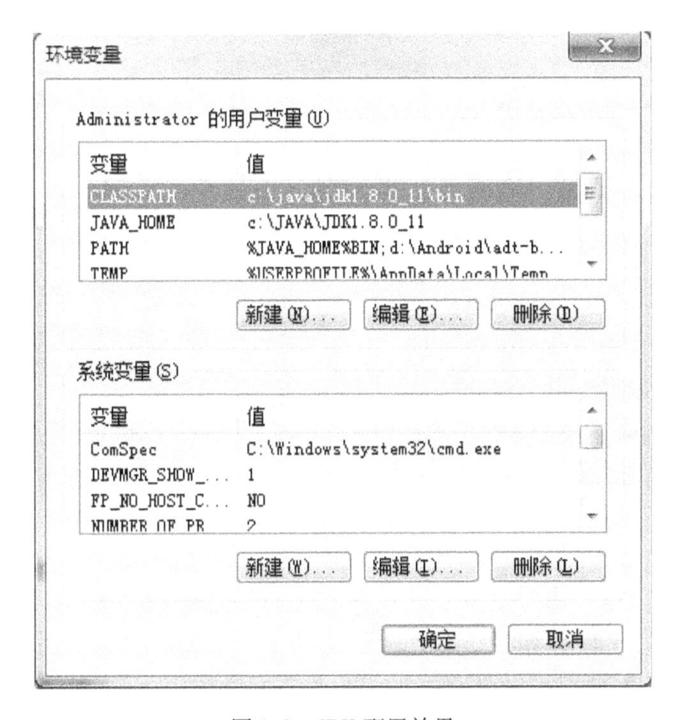

图 1.3 JDK 配置效果

3. 下载安装 Eclipse

在浏览器的地址栏中输入官方网址“http://www.eclipse.org”，进入 Eclipse 安装包的下载网页。下载的安装包是 eclipse-standard-luna-R-win32.zip。将下载的安装包解压到 d:\Android 文件夹中。解压后，文件夹中的 Eclipse.exe 可以直接运行。

4. 安装并配置 Android SDK

Android SDK 为 Android 开发者提供了库文件和开发工具，它们是整个开发中使用的工具包。

在浏览器的地址栏内输入“http://developer.android.om”，找到要下载的 SDK 安装包，下载的是 Android_SDK_windows.zip。

将下载的 SDK 安装包解压到 d:\Android 文件夹中，解压后得到如下的文件结构。

Add-ons：该文件夹下存放第三方公司为 Android 平台开发的附加功能工具，刚解压时该文件夹是空的。

Platforms：该文件夹下存放不同版本的 Android 系统。刚解压时该文件夹为空。

Tools：该文件夹下存放大量的 Android 开发和调试工具。

AVD Manager. exe：它是 Android 虚拟设备的管理器。

SDK Manager. exe：该程序是 Android SDK 管理器。启动它，结果如图 1.4，可以在线安装 SDK 及相关工具。

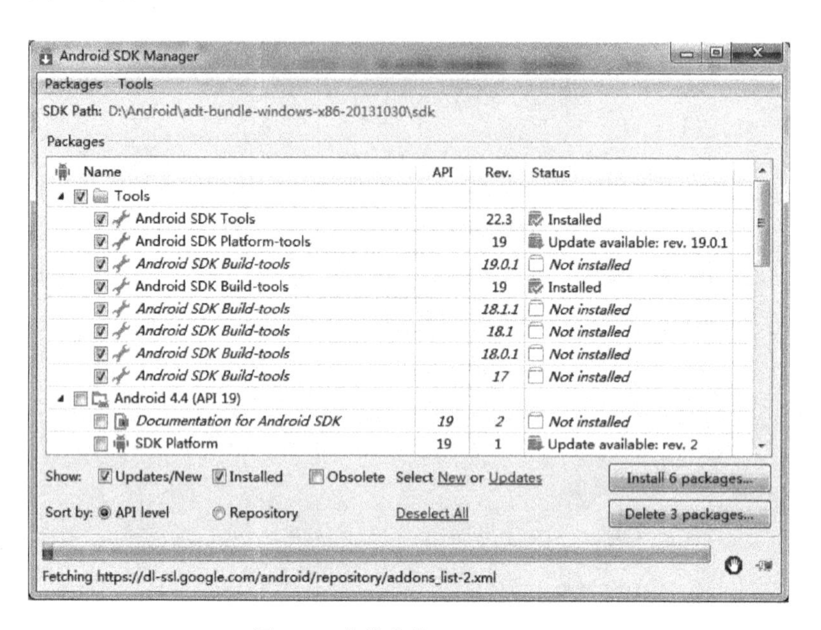

图 1.4　在线安装 Android SDK

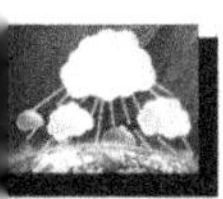

如果在线更新失败，请用记事本打开本机的 c:\Windows\System32\drivers\etc 下面的 hosts 文件，在最后一行添加：74.125.237.1 dll-ssl.goole.com，重新使用 Android SDK Manager 进行更新，问题就可以解决。

安装完成后可以看到文件结构增加了下列文件夹。

Docs：该文件夹下存放了 Android SDK 开发文件和 API 文档。

Extras：该文件夹下存放了 Google 提供的 USB 驱动、Intel 提供的硬件加速等附加工具包。

Platform - tools：该文件夹存放了 Android 平台相关工具。

Samples：该文件夹下存放了不同的 Android 平台的示例。

Sources：该文件下存放了平台最高版本的源代码。

配置 Android SDK 的作用是通知系统 SDK 存放在哪里，设置 PATH 的变量值为 SDK 的路径即可。

PATH ="d:\Android\adt-bundle-windows-x86-20131030/sdk/tools"，如图 1.5 所示。

5. 安装 Android ADT

在 Eclipse 编译环境中，ADT 为 Android 开发提供了开发工具的更新和升级。

运行 Eclipse，启动后依次点击【Finish】、【Help】、【Install New Software】选项，打开"Install"对话框。点击【Add】按钮，进入"Add　Repository"对话框，如图 1.6 所示。

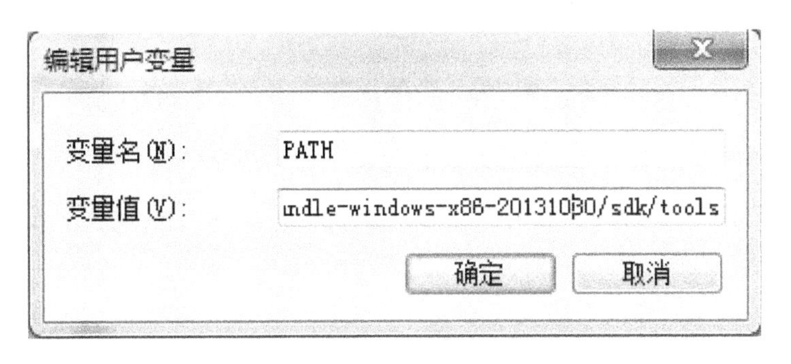

图 1.5　SDK 环境变量的配置

图 1.6　Add Repository 配置(在线安装)

在“Name”中输入任意名字，在“Location”输入 ADT 的下载地址：“http：//dl-ssl.google.com/android/repository/”。值得注意的是，我们可以下载 ADT 的安装包，在本地安装 ADT。具体如图 1.7 所示。

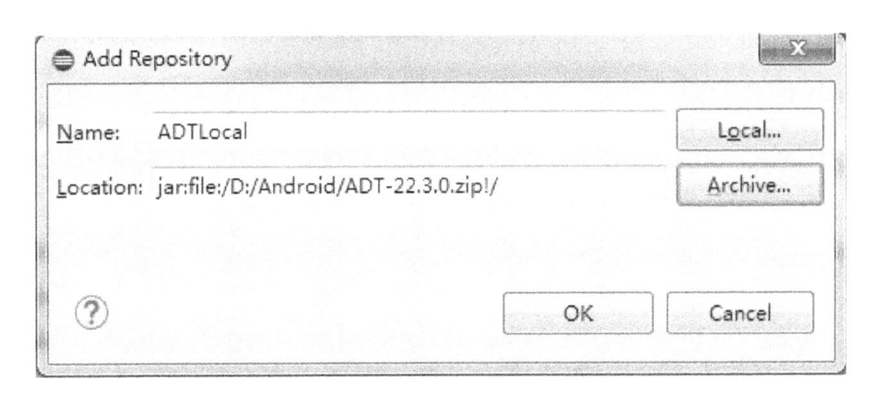

图 1.7　Add Repository 配置(本地安装)

6. 指定 SDK 的位置

ADT 安装完毕后，需要指定 SDK 的位置。操作顺序是重启 Eclipse，依次选择【Window】、【Preferences】选项，打开“Preferences”对话框，选取“Android”栏目，在“SDK Location”文本框内输入 SDK 的位置，如图 1.8 所示。注意 SDK 路径里不要包含汉字，否则Android平台可能不能正常启动。

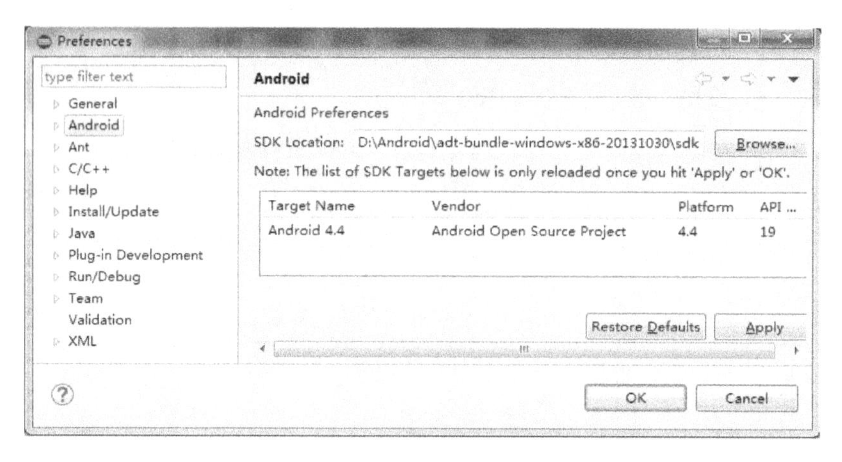

图1.8　指定SDK的位置

7. 创建和管理模拟器

运行Eclipse.exe，点击常用工具栏的可以打开“Android Virtual Device Manager”窗口。该窗口是管理模拟器的。【New】是定义新的模拟器，【Edit】是编辑模拟器信息，【Delete】可以删除模拟器，【Details】查看模拟器信息，【Start】启动模拟器，【Refresh】是刷新选定的模拟器的。

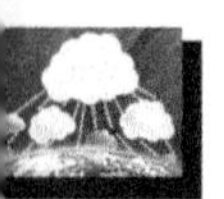

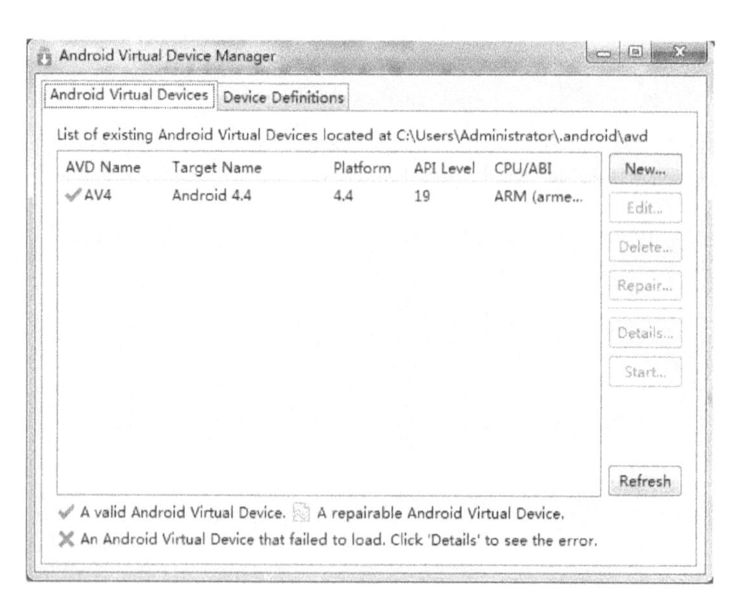

图1.9　模拟器管理窗口

点击图1.9中的【New】按钮，弹出“Create new Android Virtual Device”窗口，它可以定义一个新模拟器，如图1.10所示。

各项目描述如下：

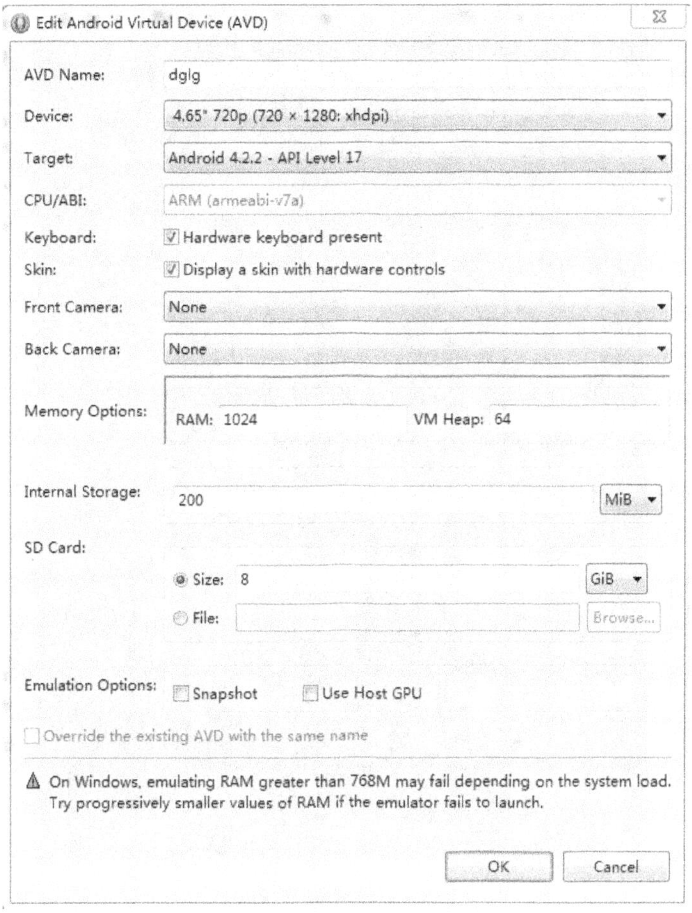

图 1.10 创建模拟器窗口

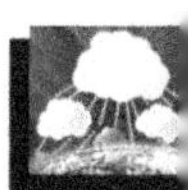

AVD Name：用来输入模拟器名称。

Device：选择屏幕分辨率。

Target：选择 Android 平台版本。

Front Camera 和 Back Camera：分别用来启用前后置摄像头。

Memory Options：其中 RAM 表示设置手机内存多大，VM Heap 可以设置模拟器的堆内存多大。

SD Card：设置 SD 卡的大小。

点击【OK】就可以创建 AVD 模拟器。

选中已创建的 AVD 模拟器，点击【Start】就可以启动模拟器，如图 1.11 所示。

8. 设置 Android 系统语言

在已启动的模拟器中，可以设置 Android 的系统语言，操作是点击图 1.11 的【Dev Settings】，然后依次点击【Language&Input】、【中文(简体)】就可以了。

到此，Android 开发平台搭建成功！

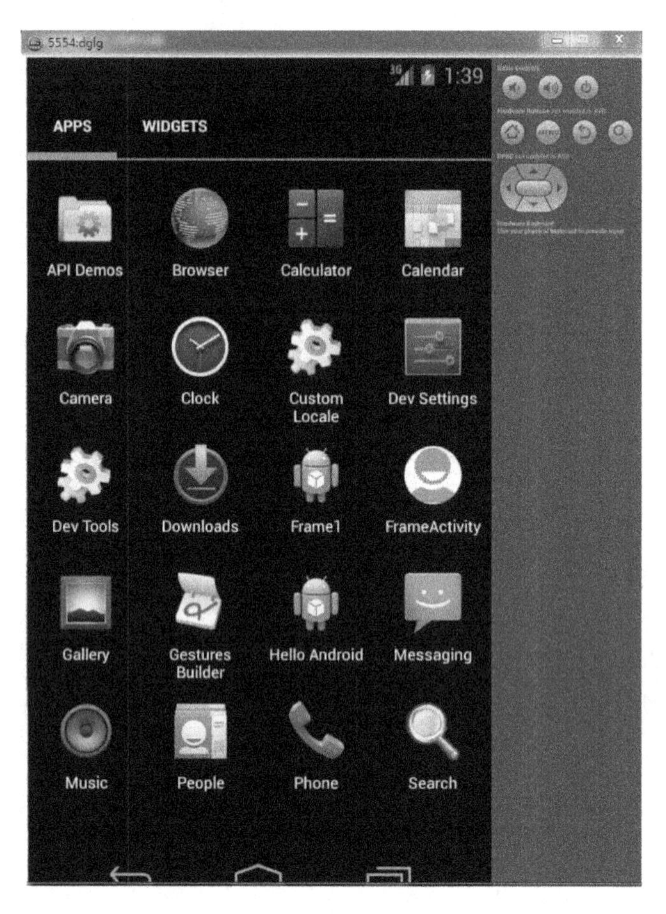

图 1.11　模拟器界面

任务二　第一个 Android 工程

任务描述

本任务完成第一个 Android 工程，界面只有一个按钮和一个文本框，点击按钮“试试点击我”，文本框的标题会变为“Hello Android”。任务虽然简单，但对后续章节的学习很重要，因此讲解创建工程方法外，还比较详细地介绍了相关知识。

任务分析

完成本任务，我们需完成以下工作：

① 创建一个 Android 工程项目。

② 在 XML 布局文件中定义应用程序的界面。

③ 在 Java 中编写逻辑代码实现业务。

④ 调试运行项目。

任务准备

准备安装好了 Android 开发环境，能正常启动 Eclipse 的计算机。

任务实现

1. 创建 Android 工程

打开 Eclipse，依次点击菜单的【File】【New】【Project】，选择【Android Application Project】栏目，如图 1.12 所示。

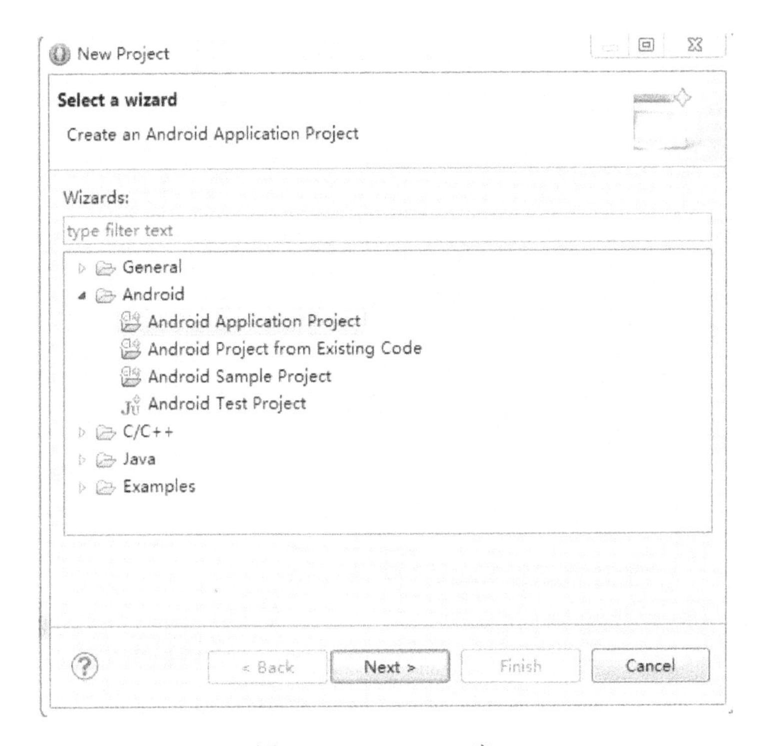

图 1.12 New Project 窗口

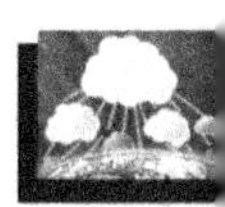

继续点击图 1.12 的【Next】窗口，出现图 1.13。

在图 1.13 所示窗口中输入 Android 应用的项目名、应用程序名和包名，选择 Android 应用的版本。点击【Next】按钮。当出现图 1.14 时，是告诉我们是否创建 activity 和选择创建什么类型的 Activity。

如果取消“Create Activity”选项的，点击【Finish】，Eclipse 成功地创建了一个 Android 项目。勾选了“Create Activity”选项后，点击【Next】按钮继续往下。出现图 1.15 所示。

在 Activity Name 中输入 Activity 的名称。Layout Name 布局文件名称。点击【Finish】按钮，宣告一个 Android 项目创建成功！

2. 创建布局文件

Android 项目创建成功后，在 layout 文件夹下有一个 hello_android. xml 文件，该文件就是应用程序界面，文件类型是 XML，代码如下。

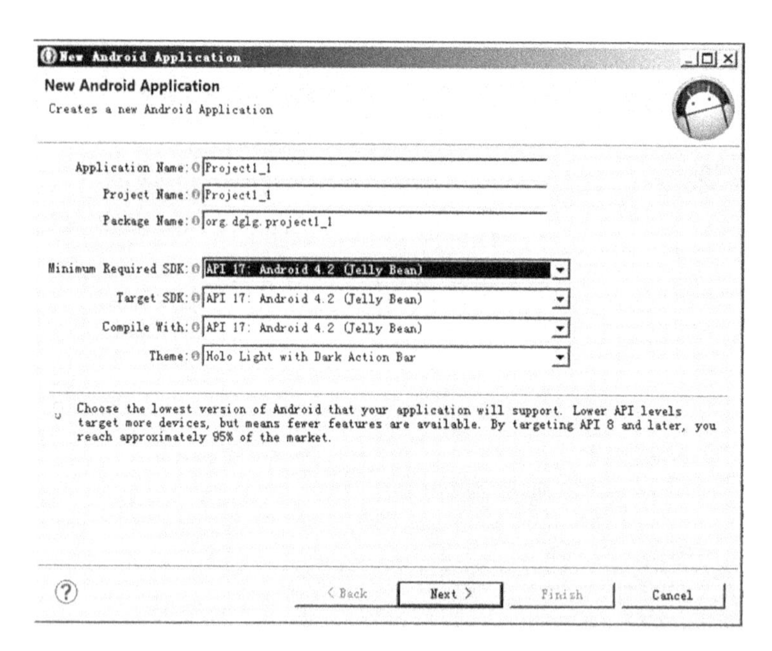

图 1.13　创建 Android Application 窗口

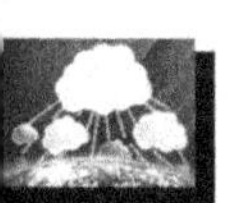

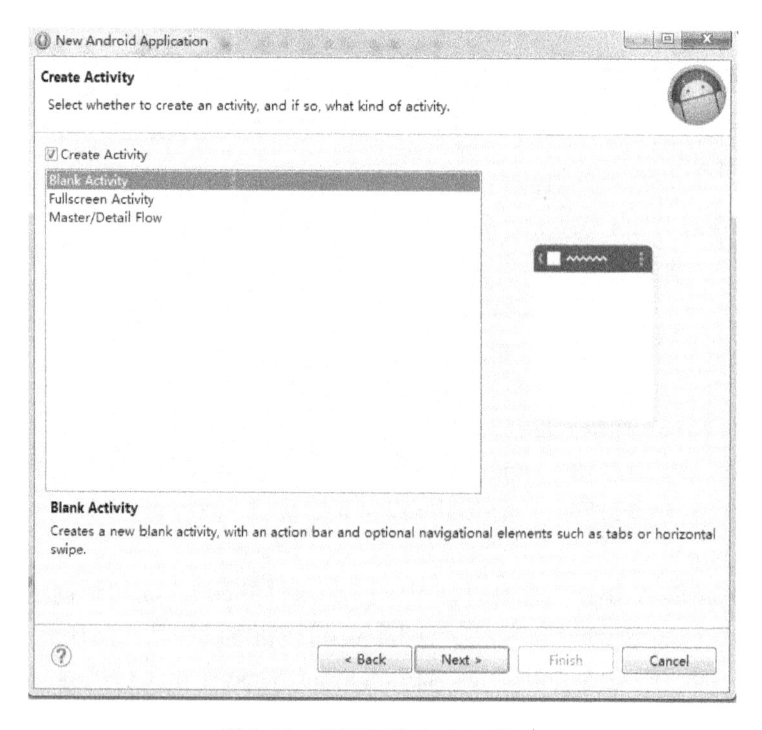

图 1.14　是否创建 Activity 窗口

New Android Application

Blank Activity

Creates a new blank activity, with an action bar and optional navigational elements such as tabs or horizontal swipe.

Activity Name HelloAndroidActivity

Layout Name hello_android

Navigation Type None

The name of the layout to create for the activity

< Back Next > Finish Cancel

图 1.15　为创建 Activity 设置信息

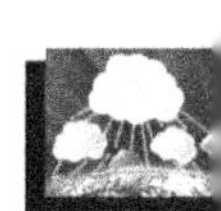

```
<RelativeLayout xmlns：android ="http：//schemas.android.com/apk/res/android"
    xmlns：tools ="http：//schemas.android.com/tools"
    android：layout_width ="match_parent"
    android：layout_height ="match_parent"
    tools：context =".HelloAndroidActivity" >
    <TextView
        android：id ="@ +id/textView1"
        android：layout_width ="wrap_content"
        android：layout_height ="wrap_content"
        android：textSize ="30sp"
        android：text ="你好，安卓" />
    <Button
        android：id ="@ +id/button1"
        android：layout_width ="200dp"
        android：layout_height ="60dp"
        android：layout_below ="@ +id/textView1"
        android：layout_marginTop ="30dp"
```

```
        android: layout_toRightOf = "@ +id/textView1"
        android: text = "试试点击我" />
</RelativeLayout>
```

该文件是XML文件，其中RelativeLayout是根元素，它代表了一个相对布局。该布局文件中，有两个控件，一个是TextView，代表一个文本框。另一个是Button，代表一个普通按钮。<TextView … />定义了文本框控件的名称，控件大小位置等。<Button … />定义了按钮控件的名称，控件大小位置等。开发者可以在界面上添加需要的控件。

3. 编写业务逻辑

在已经创建的project1_1项目的src文件夹下存放的是项目的源代码文件，该文件叫HelloAndroidActivity.java。如果开发者要完成某些动作，就可以在该文件里添加代码。

HelloAndroidActivity.java代码清单：

```
public class HelloAndroidActivity extends Activity {
Button btBtn;
  TextView tvTxtView;
  protected void onCreate(Bundle savedInstanceState) {
  super.onCreate(savedInstanceState);
  setContentView(R.layout.hello_android);
  tvTxtView = (TextView) findViewById(R.id.textView1);
  btBtn = (Button) findViewById(R.id.button1);
      btBtn.setOnClickListener(new OnClickListener() {
  public void onClick(View arg0) {
      tvTxtView.setText("Hello Android");
      }
      });
    }
  }
}
```

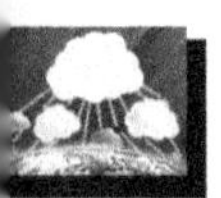

上面代码中粗体部分就是添加上去的代码，完成一个点击按钮监听事件，事件的响应是在控件TextView中显示"Hello Android"文本。

4. 调试并运行Android项目

项目创建和业务逻辑编写完成后，可在【Package Explorer】窗口中看到项目的目录结构。右击项目名project1_1，在弹出的快捷菜单中依次点击【Run As】【Android Application】选项，启动Android模拟器，就可以看到运行结果(如图1.16所示)。

图 1.16　project1_1 运行结果

5. 项目的目录结构概述

图 1.17 是 project1_1 项目在 Eclipse 中的目录结构。

图 1.17　project1_1 目录结构

项目的根目录有以下文件夹：

src/：专门存放编写的 Java 源代码的包。

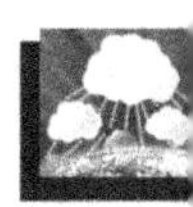

android4.4：存放 Android 自身的 jar 包。

gen：该目录不用开发人员维护，是用来存放 Android 开发工具所生成的目录。该目录下的 R.java 文件非常重要。

assets：该目录用来存放应用中使用到的如视频文件、mp3 媒体文件。

res：res 是 resource 的缩写。该目录称为资源目录，可以存放一些图标、界面文件、应用中用到的文字信息。

AndroidManifest.xml：该文件是功能清单文件，列出了应用中所用到的所有组件，包括 activity、广播接收者、服务等组件。

该项目详细讲解了 Android 开发平台的搭建步骤，开发了第一个 Android 程序，简单地说明了各个目录和文件的作用。通过本项目的学习，读者应该掌握环境变量的配置和 AVD 模拟器的创建，并对 Android 平台应用程序的开发步骤有一定的了解。

任务考核

1. 搭建一个 Android 平台，并在 Eclipse 中创建 AVD(模拟器)，设置名称为“Frist AVD”，“SDCard”为 5G。

2. 创建一个 Android 项目，一个 Activity，一个文本框，在文本框里显示“你好，安卓”。

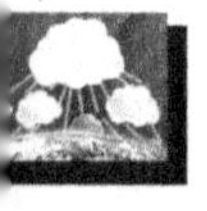

项目二

走进Android UI布局

用户界面（User Interface，简称UI）是Android程序与用户进行交互的接口，好的用户界面应该是美观的、简洁的和易用的。

在Andriod程序中，Activity相当于一个窗口，我们在这个窗口中放置文本、图像和按钮等元素就构成了我们看到的用户界面。

一直以来，Android系统的用户界面没有得到重视，无法与苹果的IOS相比。从Android 4.0开始，Google开始重视用户界面的设计，又为开发者设立了网站Google Design（http://developer.android.com/design/index.html），这为用户带来良好的使用体验。如今用户界面设计已经成为一个专业性很强的工作，下面我们用实例介绍简单的用户界面设计和布局管理。

理解 Android UI 的基本知识
了解制作 UI 的三种方法
学会使用 Eclipse 可视化地制作 UI
理解 Android 布局管理器的使用
学会合理选择布局管理器

任务一　简单用户登录界面制作

任务描述

本任务利用 Eclipse 创建一个简单的用户登录界面，用户能在此界面输入用户姓名和密码(密码隐藏显示)，并通过一个按钮提交登录信息，如图 2.1 所示。

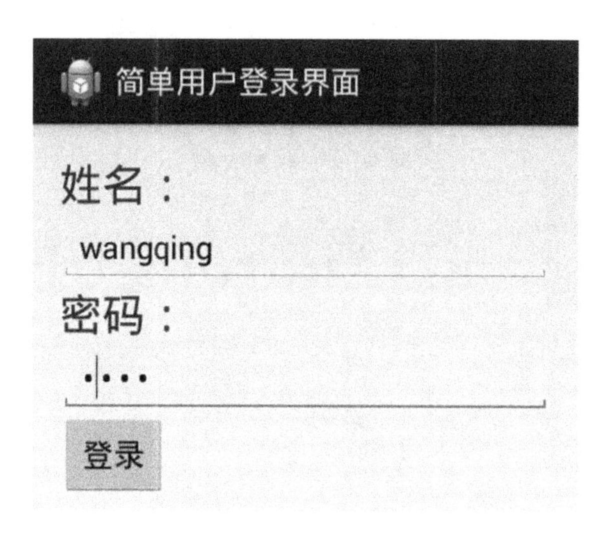

图 2.1　用户登录界面效果图

任务分析

我们可以看到，该用户界面从上到下分别有标题、姓名标签、姓名编辑框、密码标签、密码编辑框和登录按钮这些元素。通常来说，更改标题栏的文字我们可以通过修改 app_name 的值来实现；显示文字的标签可以用 TextView 控件；涉及文字输入的就用 EditText 控件；普通按钮就用 Button 控件。制作用户界面有三种常用的方法：

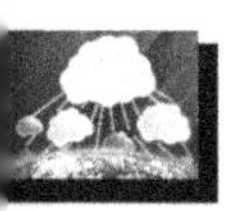

1. 使用 Eclipse 等可视化工具来创建，这种方法的优点是可视化操作，基本不需要编写代码，所见即所得。

2. 编辑 XML 代码实现，这种方法优点是能精确定位各界面元素，方便制作复杂的用户界面。

3. 用 Java 代码实现，该方法能实现动态创建、修改用户界面，可是代码的可读性比较差，需要开发者很好的 Java 编程基础。

本任务的重点和难点如下：

(1)初步了解用户界面
(2)熟悉 Eclipse 的使用方法
(3)初步了解控件的属性
(4)初步了解三种制作用户界面的方法

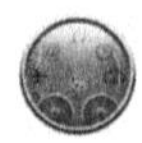

任务准备

1. Eclipse 可视化制作用户界面简介

如今，所有应用都开始重视用户界面，Eclipse 为我们提供了可视化制作用户界面的功能，让我们简单地把需要的控件用鼠标拖动到视图容器中，用户无须通过编写代码就可以创建自己的界面而且实现了所见即所得的效果。尽管用户没有自己亲自编写代码，可是

Eclipse 已经自动为每一个用户界面生成了 XML 格式的代码文件。

尽管通过 Eclipse 可视化地创建用户界面很容易，可是这种方法是有局限的，它只能创建简单的、静态的界面，而对复杂的、动态的用户界面就无能为力了。所以对于初学者，我们有必要了解这种可视化制作界面的方法，但随着学习的深入，我们更应该学习通过 XML 或者 Java 代码来创建、编辑和修改用户界面。

2. 通过 XML 代码制作用户界面简介

XML(Extensible Markup Language)又称可扩展标记语言，Android 允许用户直接通过编写 XML 代码来制作用户界面，这种方法允许开发者精确定位界面上的所有元素，方便制作和控制复杂的用户界面。

3. 通过 Java 代码制作用户界面简介

Android 还允许用户直接通过 Java 代码来动态地制作用户界面，由于这种方法创建的用户界面不易维护，所以通常用在不方便使用 XML 代码制作的情况下使用。

任务实现

1. 利用 Eclipse 可视化制作

1)新建 Android 项目

在 Eclipse 中，新建一名为 Porject2_1 的 Android 项目，具体方法可参考项目一。

在默认情况下将得到图 2.2 所示的模拟手机界面窗口。如果窗口没有显示出来，可以在 res/layout 文件夹下双击 activity_main. xml 文件，再单击屏幕中间的【Graphical Layout】。

图 2.2　模拟手机界面窗口

2)调整布局类型

默认情况下，Eclipse 创建的 Android 项目使用的都是 RelativeLayout(相对布局)的，在本任务中，我们需要把布局的类型改变为 LinearLayout(线性布局)，在【Outlines】窗口中右击【RelativeLayout】，在弹出如图 2.3 所示的菜单中选择【Change Layout】，在图 2.4 所示的“Change Layout”对话框中选择【LinearLayout(Vertical)】纵向线性布局。

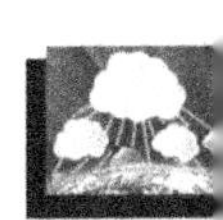

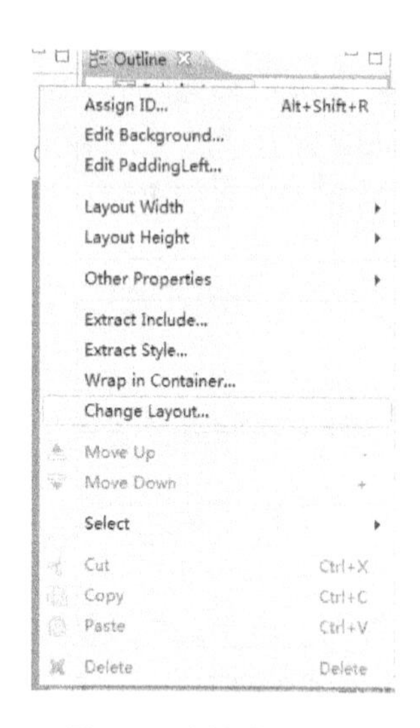

图 2.3 右键弹出菜单

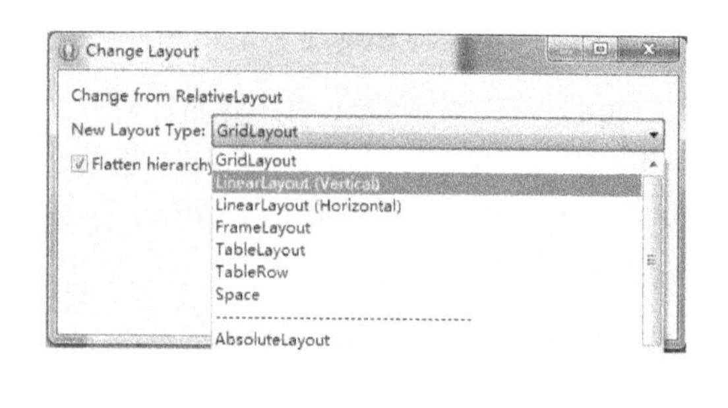

图 2.4 "Change Layout"对话框

3)添加界面元素

Eclipse 所有可视化的组件都在【Palette】面板中，这些组件根据不同类型被分到不同的文件夹中，例如：From Widgets(控件类)、Text Fields(文本类)和 Time&Date(时间日期类)等等。本任务中我们需要添加两个 TextView 标签、一个 Button 按钮和两个 EditText 文本框控件，各控件的具体功能和用法会在以后章节详细介绍。

(1)添加姓名标签

单击并打开【Palette】面板中的【From Widgets】得到如图 2.5 所示面板，拖动一个 TextView 到手机屏幕上，该 TextView 会自动对齐左上角。

单击手机屏幕上的 TextView 就能在【Properties】(属性)面板中显示和修改 TextView 的属性。单击并修改【Text】属性为"姓名:"，【Text Size】属性为"25SP"效果如图 2.6 所示。其中 Text 属性就是要显示的文本，Text Size 属性就是显示的文本字体大小。

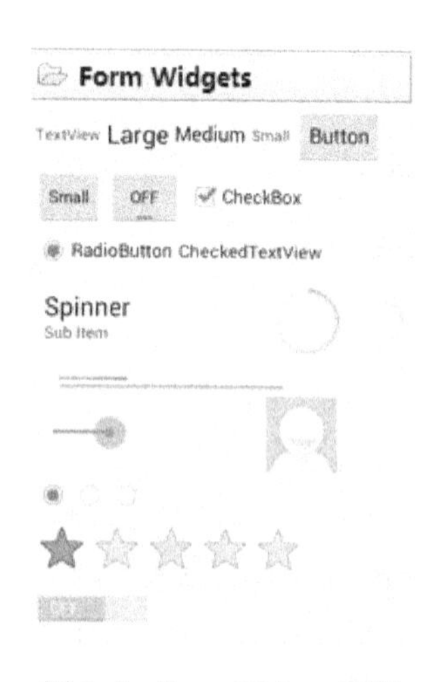

图 2.5 From Widgets 面板

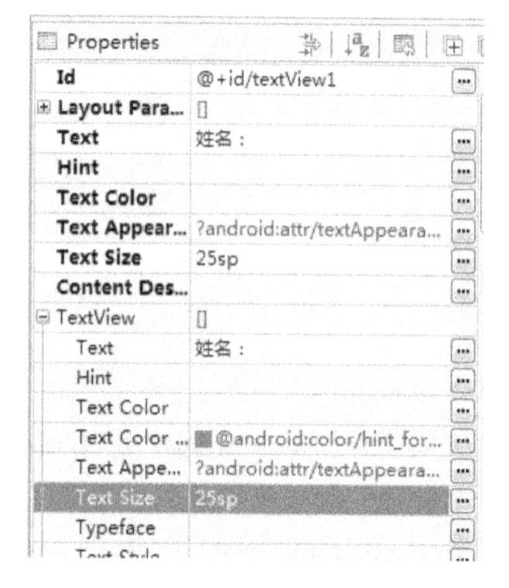

图 2.6 属性窗口

知识拓展

Palette 面板中的组件都提供不同大小的预览功能，可以在这些组件中单击右键，然后选择 Show Previews（大图预览）、Show SamllPreviews（小图预览）、Show Tiny Previews（微小图预览）等。

（2）添加姓名编辑框

单击并打开【Palette】面板中的【Text Fields】得到图 2.7 所示面板，拖动一个 Plain Text 到“姓名”标签下面，效果如图 2.8 所示。

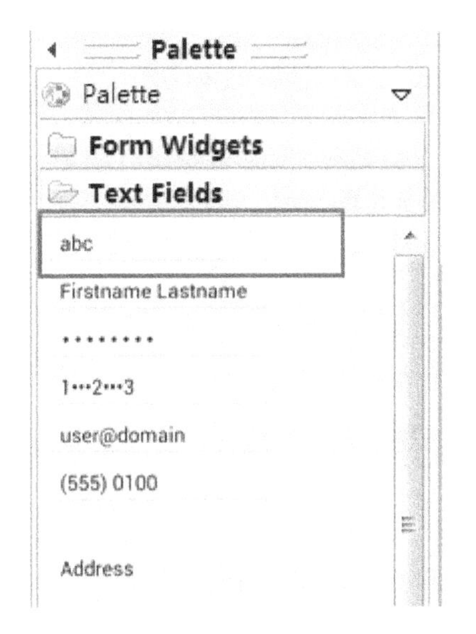

图 2.7 Text Fields 面板

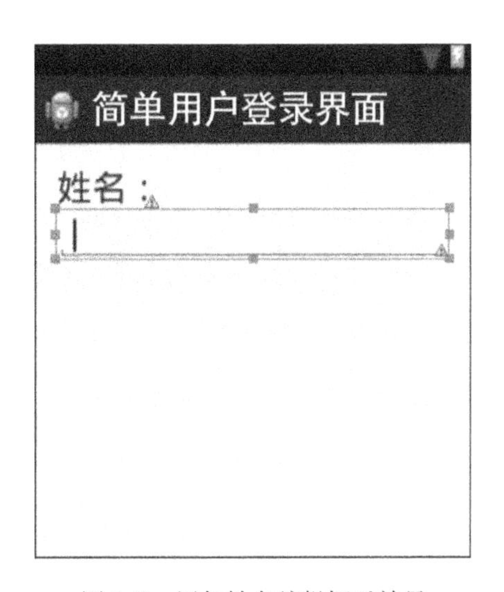

图 2.8 添加姓名编辑框后效果

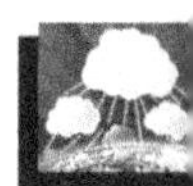

知识拓展

你可能会注意到在“姓名”两个字的右下角有一个小小的黄色叹号，这只是一个警告，并不代表你的代码有错误，一般不会影响程序的运行。要解决这个问题可以把文字存入 strings. xml 中，然后进行引用。

（3）添加密码标签

请参考第一步“添加姓名标签”，效果和属性设置如图 2.9 所示。

（4）添加密码编辑框

单击并打开【Palette】面板中的【Text Fields】（参考图 2.7），拖动一个 Password 控件到“密码”标签下面，效果和属性设置如图 2.10 所示。

（5）添加登录按钮

单击并打开【Palette】面板中的【From Widgets】（参考图 2.5），拖动一个 Button 到手机屏幕上，在【Properties】面板中修改【Text】属性为“登录”，效果和属性如图 2.11 所示。

运行程序后将得到图 2.1 效果的登录窗口。到此，简单用户登录界面就完成了。

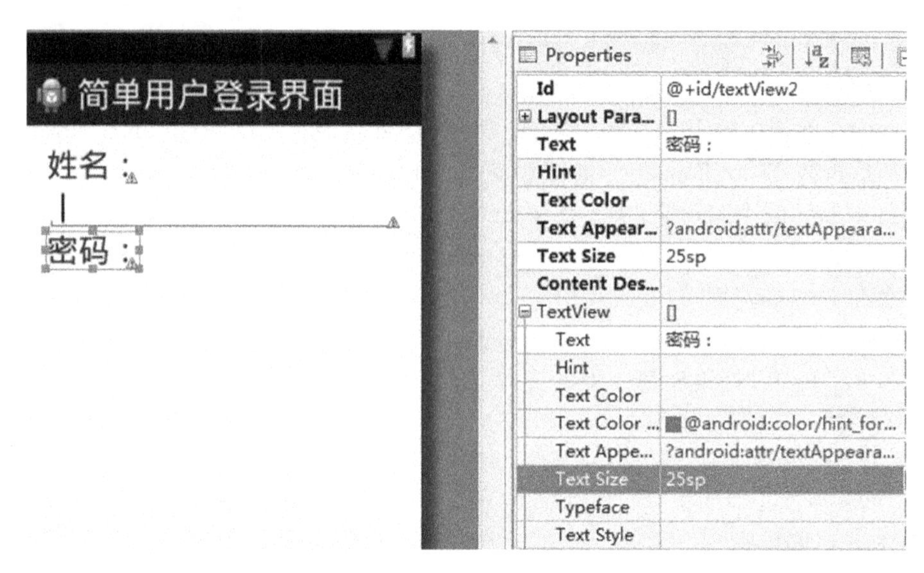

图 2.9　添加密码标签效果与密码标签属性

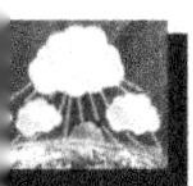

图 2.10　添加密码编辑框效果与密码编辑框属性

图 2.11　添加登录按钮效果

2. 编写 XML 代码制作

正如前面所说，我们使用 Eclipse 可视化制作用户界面时，基本不需要自己编写布局代码而系统会自动生成。单击【activity_main. xml】可以看到如图 2. 12 所示的 XML 布局文件，这是用户界面的布局代码。通常来说，熟识了 Android 用户界面开发以后，都会采用直接编写 XML 布局代码的方法来制作用户界面。

```
activity_main.xml

<TextView
    android:id="@+id/textView1"
    android:layout_width="wrap_content"
    android:layout_height="wrap_content"
    android:text="姓名: "
    android:textSize="25sp" />

<EditText
    android:id="@+id/editText1"
    android:layout_width="match_parent"
    android:layout_height="wrap_content"
    android:ems="10" >

    <requestFocus />
</EditText>

<TextView
    android:id="@+id/textView2"
    android:layout_width="wrap_content"
    android:layout_height="wrap_content"
    android:text="密码: "

Graphical Layout | activity_main.xml
```

图 2. 12　activity_main. xml 窗口

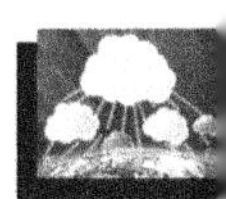

新建一个项目，打开 res/layout/activity_main. xml 布局文件并输入如下代码。

```
<! - -声明一个线性布局 - - >
<LinearLayout xmlns: android = http: //schemas. android. com/apk/res/android
<! —“@ + id/LinearLayout1”表示如果系统没有 LinearLayout1 这个 id 的话，就增
加一个名字为 LinearLayout1 的 id - - >
    android: id = "@ + id/LinearLayout1"
<! - - : layout_width、layout_height 分别表示布局的宽度和高度，match_parent
指与容器匹配 - - >
    android: layout_width = "match_parent"
    android: layout_height = "match_parent"
<! —orientation 表示线性布局方向，vertical 为纵向 - - >
    android: orientation = "vertical"  >
    <TextView
        android: id = "@ + id/textView1"
        android: layout_width = "wrap_content"
        android: layout_height = "wrap_content"
        android: text = "姓名:"
```

```
<! —文字大小为25sp - - >
        android: textSize = "25sp" />
    <EditText
        android: id = "@ +id/editText1"
        android: layout_width = "match_parent"
<! —wrap_content 表示文本框的高度为包裹内容 - - >
        android: layout_height = "wrap_content"
        android: ems = "10"  >
         <requestFocus />
    </EditText>
    <TextView
        android: id = "@ +id/textView2"
        android: layout_width = "wrap_content"
        android: layout_height = "wrap_content"
        android: text = "密码:"
        android: textSize = "25sp" />
    <EditText
        android: id = "@ +id/editText2"
        android: layout_width = "match_parent"
        android: layout_height = "wrap_content"
        android: ems = "10"
<! —文本框的类型为 textPassword , 文本密码 - - >
        android: inputType = "textPassword" />
    <Button
        android: id = "@ +id/button1"
        android: layout_width = "wrap_content"
        android: layout_height = "wrap_content"
        android: text = "登录" />
</LinearLayout>
```

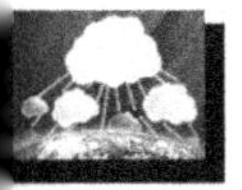

运行后我们同样可以得到一模一样的登录界面。

小提示

编写 Android 布局文件使用的是 XML 语言，其中 <! —注释 - - > 符号表示注释，程序运行时不参与编译。

3. 编写 Java 代码制作

编写 Java 代码制作用户界面是一种直接通过 Java 代码实现动态生成用户界面的方法。在 Eclipse 中新建一个项目，然后在 MainActivity. java 中改写 OnCreate 函数，代码如下：

```
protected void onCreate( Bundle savedInstanceState) {
            super. onCreate( savedInstanceState) ;
            // 创建 LinearLayout 名字为 ll
            LinearLayout ll = new LinearLayout( getApplicationContext( ) ) ;
            // 指定 LinearLayout 的布局方式(宽度和高度)
            ll. setLayoutParams ( new LinearLayout. LayoutParams ( LayoutParams. FILL
_PARENT, LayoutParams. FILL_PARENT) ) ;
        // 设置布局方向为纵向
        ll. setOrientation( LinearLayout. VERTICAL) ;
        // 创建姓名标签
        TextView tvName = new TextView( this) ;
        // 设置姓名标签
        tvName. setText( "姓名:" ) ;
        //创建姓名编辑框
        EditText etName = new EditText( this) ;
        //创建密码标签
        TextView tvPsw = new TextView( this) ;
        //设置密码标签
        tvPsw. setText( "密码:" ) ;
        //创建密码编辑框
        EditText etPsw = new EditText( this) ;
        //登录按钮
        Button btnLog = new Button( this) ;
        //设置登录按钮
            btnLog. setText( "登录" ) ;
    //把以上五个元素添加到 ll 中，并设置好各元素的宽度和高度
    ll. addView( tvName, new LinearLayout. LayoutParams
                ( LayoutParams. WRAP _ CONTENT, LayoutParams. WRAP _
CONTENT) ) ;
        ll. addView( etName, new LinearLayout. LayoutParams
            ( LayoutParams. FILL_PARENT, LayoutParams. WRAP_CONTENT) ) ;
        ll. addView( tvPsw, new LinearLayout. LayoutParams
                ( LayoutParams. WRAP _ CONTENT, LayoutParams. WRAP _
CONTENT) ) ;
        ll. addView( etPsw, new LinearLayout. LayoutParams
            ( LayoutParams. FILL_PARENT, LayoutParams. WRAP_CONTENT) ) ;
        ll. addView( btnLog, new LinearLayout. LayoutParams
```

```
                            ( LayoutParams. WRAP _ CONTENT,  LayoutParams. WRAP _
CONTENT));
    // 把这个布局添加到 activity 中
    setContentView(ll);
        }
```

运行后我们同样可以得到一模一样的登录界面。

再深入一点学习就会发现 Activity 是 Android 程序与用户交互的窗口，这个窗口本来是空白的，它自己本身并不显示在屏幕上，能看到的界面元素都是使用 View(视图)或 ViewGroup(视图组)绘制出来的。程序运行时，在 Activity 类的 onCreate()函数中调用 setContentView()方法来加载 XML 文件并经过编译后生成所见的用户界面。可以打开 MainActivity. java 看到如下这段代码。

```
protected void onCreate(Bundle savedInstanceState) {
        super.onCreate(savedInstanceState);
        //加载用户界面
        setContentView(R.layout.activity_main);
    }
```

在本任务中，创建的用户界面已经使用了 LinearLayout 线性布局，关于 Android 布局将在下一任务中继续深入探讨。

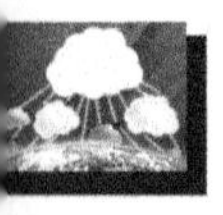

任务二　Android 布局实例

任务描述

为了方便地让开发者管理自己的用户界面，Android 为我们提供了布局管理器(以下简称布局)。本次任务通过五个实例介绍了 Android 开发中常用的五种布局，分别是 LinearLayout(线性布局)、RelativeLayout(相对布局)、AbsoluteLayout(绝对布局)、FrameLayout(帧布局)和 TableLayout(表格布局)。

任务分析

经过上一个任务，已经大概知道如何在 Eclipse 中创建自己的用户界面，在 Eclipse 中除了提供许多可视化的基本界面元素，如 Botton(按钮)、TextView(文本标签)等，还提供了一组 ViewGroup 作为管理这些界面控件的容器，这就是布局。每个布局实现了一种管理其中控件或子布局的大小、位置的特定策略。

本任务的重点和难点如下：

(1)理解 View 和 ViewGroup 类；
(2)熟悉各种布局的属性与方法；
(3)区别各种布局特点，学会为自己的程序选择合适的布局。

任务准备

1. View 和 ViewGroup 类概述

在上一任务中已经知道 Android 程序中的界面都是由 View 类绘制出来的，也就是说 View 类是所有可视化控件的基类，看见的所有屏幕控件都是 View 类的对象。

ViewGroup 类是 View 类的子类，它提供了一组如何布局应用程序屏幕控件的属性和方法，而将要学习的四种常用布局都是 ViewGroup 类的子类，具体关系可以参考图 2. 13。

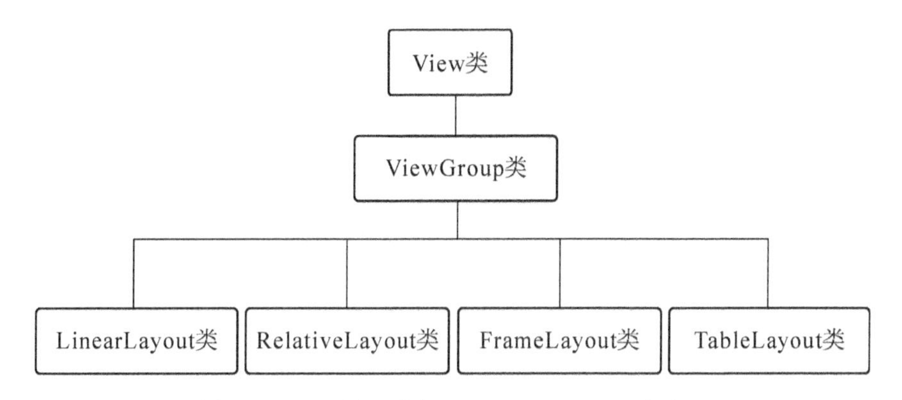

图 2. 13　五种布局类与 View 和 ViewGroup 类关系

每一个 View 对象都有一组公共的属性，常用的属性介绍如表 2. 1。

表 2. 1　View 类的常用属性

属性名称	描　　述
android：layout_width	设置宽度
android：layout_height	设置高度
android：layout_x	设置位置的 x 坐标
android：layout_y	设置位置的 y 坐标
android：layout_gravity	设置与父元素的相对位置
android：layout_weight	设置空间布局元素所占空间的权重
android：background	设置背景
android：clickable	设置是否响应点击事件
android：visibility	设置可见性
android：focusable	设置是否可以获取焦点
android：id	设置标识符

续表 2.1

属性名称	描　述
android：longClickable	设置是否响应长点击事件
android：saveEnabled	如果未作设置，当 View 被冻结时将不会保存其状态

凡是继承于 View 类的子类都会继承它的属性，可这些属性并不是在每一个子类中都起作用，例如：layout_weight 和 layout_gravity 属性在 LinearLayout 或 TableLayout 布局中起作用。

2. LinearLayout

LinearLayout 是 Android 界面开发常用的布局，它允许开发者以横向或者纵向的方式来组织自己的界面元素。当设置成 horizontal(横向)的时候，元素会默认以屏幕的左上角开始，从左到右一个接一个排列在屏幕上方；当设置成 vertical(纵向)的时候，元素会默认以屏幕的左上角开始，从上到下一个接一个排列在屏幕的左侧。LinearLayout 布局的属性可以在 Eclipse 的可视化界面编辑器中修改，也可以在 XML 文件中修改或者用 Java 代码设置和修改。

表 2.2　LinearLayout 布局的常用属性

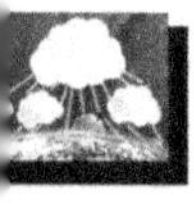

属性名称	属性描述
android：orientation	设置 LinearLayout 的朝向，可取 horizontal(横向) 和 vertical(纵向) 两种排列方式
android：gravity	设置 LinearLayout 的内部元素的布局方式

这里要特别说明一下 android：gravity 属性，该属性用来设置控件在容器内的对齐方式，从表 2.3 可以看出它的取值与说明，该属性可以同时取两个或以上的值，这时需要用“|”把不同的值分开。如：android：gravity = "bottom | right" 表示元素在容器内的对齐方式为底部右对齐。

表 2.3　gravity 属性取值

属性值	属性说明
fill_vertical	元素纵向拉伸，以填满容器
fill_horizontal	元素横向拉伸，以填满容器
fill	如果允许的话拉伸元素以填满容器
top	元素顶端对齐，不改变其大小
bottom	元素底部对齐，不改变其大小
left	元素左侧对齐，不改变其大小
right	元素右侧对齐，不改变其大小
center_vertical	元素纵向居中对齐，不改变其大小

续表 2.3

属性值	属性说明
center_horizontal	元素横向居中对齐，不改变其大小
center	元素中部剧中对齐，不改变其大小

当然，上表的 android：gravity 取值不仅在 LinearLayout 布局中适用，在其他类型布局中同样适用。

3. RelativeLayout

相对布局可以使布局在容器中的界面元素指定相对于它所在的布局容器或者兄弟元素的位置，这种布局方式灵活方便，越来越受到开发者的欢迎。在 RelativeLayout 类中很多属性都是用来设置元素相对位置的，常用的如表 2.4 所示。

表 2.4　取值为 true 或 false 的属性

属性名称	属性说明
android：layout_alignParentLeft	贴紧父容器的左边缘
android：layout_alignParentTop	贴紧父容器的上边缘
android：layout_alignParentRight	贴紧父容器的右边缘
android：layout_alignParentBottom	贴紧父容器的下边缘
android：layout_centerInparent	相对于父容器中部居中(垂直、水平居中)
android：layout_centerHrizontal	相对于父容器水平居中
android：layout_centerVertical	相对于父容器垂直居中
android：layout_alignWithParentIfMissing	如果对应的兄弟元素找不到的话就以父容器做参照物
android：layout_alignParentStart	贴紧父容器的结束位置开始
android：layout_alignParentEnd	贴紧父容器的结束位置结束

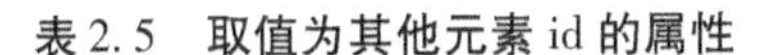

表 2.5　取值为其他元素 id 的属性

属性名称	属性说明
android：layout_below	在所引用 id 元素的下方
android：layout_above	在所引用 id 元素的上方
android：layout_toLeftOf	在所引用 id 元素的左边
android：layout_toRightOf	在所引用 id 元素的右边
android：layout_alignTop	本元素的上边缘和所引用 id 元素的上边缘对齐
android：layout_alignLeft	本元素的左边缘和所引用 id 元素的左边缘对齐
android：layout_alignBottom	本元素的下边缘和所引用 id 元素的下边缘对齐
android：layout_alignRight	本元素的右边缘和所引用 id 元素的右边缘对齐
android：layout_alignBaseline	本元素的基准线和所引用 id 元素的基准线对齐
android：layout_toStartOf	本元素从所引用 id 元素开始

续表 2.5

属性名称	属性说明
android：layout_toEndOf	本元素从所引用 id 元素结束
android：layout_alignStart	本元素与开始的父元素对齐
android：layout_alignEnd	本元素与结束的父元素对齐

表 2.6　取值为具体像素的属性

属性名称	属性说明
android：layout_margin	本元素离上下左右间的距离
android：layout_marginBottom	本元素离相对元素的底边缘的距离
android：layout_marginLeft	本元素离相对元素的左边缘的距离
android：layout_marginRight	本元素离相对元素的右边缘的距离
android：layout_marginTop	本元素离相对元素的上边缘的距离
android：layout_marginStart	本元素离开始的位置的距离
android：layout_marginEnd	本元素离结束位置的距离

4. FrameLayout

与以上两种布局不同，FrameLayout 帧布局允许开发人员把布局元素以层叠的方式添加到屏幕上，在同一个容器中，后面元素的会覆盖在前面元素上面，也就是说后面元素在前面元素的上层，顺序如堆栈。所以如果后面的元素比前面元素大就有可能完全遮挡住前面的元素，这种布局方式最大的好处是空间利用率高，属性如表 2.7 所示。

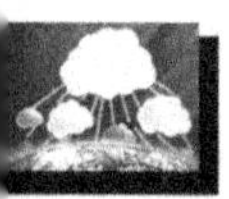

表 2.7　FrameLayout 布局常用属性

属性名称	属性说明
android：foreground	设置该布局的前景
android：foregroundGravity	设置该布局前景的对齐方式
android：measureAllChildren	在切换图片时是否测量所有组件大小

5. TableLayout

TableLayout 表格布局通常使用在需要做出表格效果的界面上。在实际使用过程中，TableLayout 与 TableRow 配合使用，TableRow 表示表格里的一行，所有其他的布局元素都放在 <TableRow> </TableRow>标签中，属性如表 2.8 所示。

表 2.8　TableLayout 布局常用属性

属性名称	属性描述
android：collapseColumns	设置指定列为隐藏，列号从 0 开始计算
android：shrinkColumns	设置指定列根据内容自动延伸，列号从 0 开始计算
android：stretchColumns	把指定列空白部分填满，列号从 0 开始计算

小提示

当 TableRow 里面的控件还没有布满布局时，shrinkColumns 属性不起作用。

 任务实现

1. LinearLayout 实例

制作如图 2.14 所示界面，界面由三个按钮元素组成，按顺序从上到下垂直靠左排列在屏幕上。

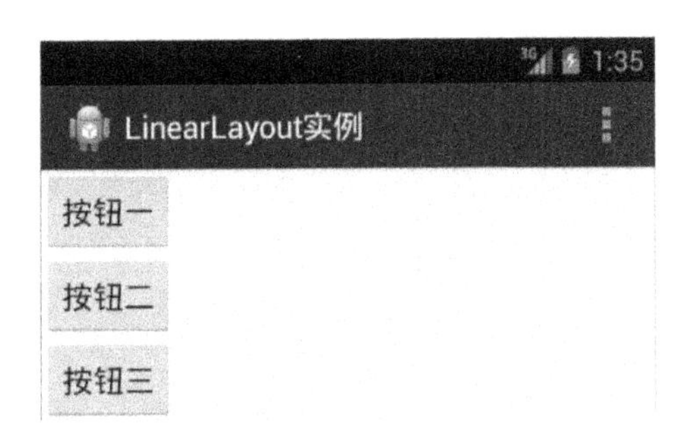

图 2.14　LinearLayout 效果

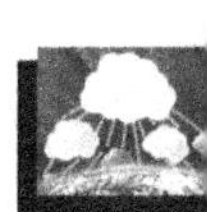

新建一个名为 Project2_2_1 的 Android 项目。

打开 res/values 文件夹中的 string. xml 文件，输入如下代码：

```
<?xml version="1.0" encoding="utf-8"?>
<resources>
    <string name="app_name">Project2_2_1</string>
    <string name="action_settings">Settings</string>
    <string name="hello_world">Hello world!</string>
    <string name="text1">按钮一</string>
    <string name="text2">按钮二</string>
    <string name="text3">按钮三</string>
</resources>
```

string. xml 文件定义了布局中用到的字符串的名称和值，下面标签：

```
<string name="text1">按钮一</string>
```

表示定义一个字符串 id 为“text1”，值为“按钮一”。这样定义的好处是显而易见的，当需要修改“text1”的值的时候只需要在 string. xml 文件里修改而不必查找程序代码。在布局中引用“text1”的值可以采用以下方法：

```
android: text="@string/text1"
```

"@ string/text1"表示在 string. xml 文件中 id 为"text1"的值，注意，"@"不可以漏写。

打开 res/layout 文件夹中的 activity_main. xml 文件，输入如下代码：

```
<LinearLayout xmlns: android = "http: //schemas. android. com/apk/res/android"
    android: id = "@ +id/LinearLayout1"
    android: layout_width = "match_parent"
    android: layout_height = "match_parent"
    android: orientation = " vertical"  >
    <Button
        android: id = "@ +id/button1"
        android: layout_width = "wrap_content"
        android: layout_height = "wrap_content"
        android: text = "@ string/text1"  />
    <Button
        android: id = "@ +id/button2"
        android: layout_width = "wrap_content"
        android: layout_height = "wrap_content"
        android: text = "@ string/text2"  />
    <Button
        android: id = "@ +id/button3"
        android: layout_width = "wrap_content"
        android: layout_height = "wrap_content"
        android: text = "@ string/text3"  />
</LinearLayout>
```

运行后得到图 2. 14 所示的界面。本界面采用 LinearLayout 布局，在屏幕中布置了三个按钮，并声明布局方向为垂直。可以看到三个按钮会依次从上到下垂直排列在屏幕上。如果把以上代码中的 android: orientation = "vertical" 改为 android: orientation = "horizontal"，即表示把 LinearLayout 布局方式改为水平，运行后将得到图 2. 15 所示界面，三个按钮从左到右水平排列在屏幕上。

如果在第二个按钮的 XML 代码中添加一行 android: layout_gravity = "right"，表示把第二个按钮的对齐方式设置为相对于其父容器靠右对齐，在第三个按钮 XML 代码中添加一行 android: layout_weight = "1. 0"，表示分配给按钮三的权重是 1. 0，将得到如图 2. 16 所示界面。

图 2.15 LinearLayout 水平布局

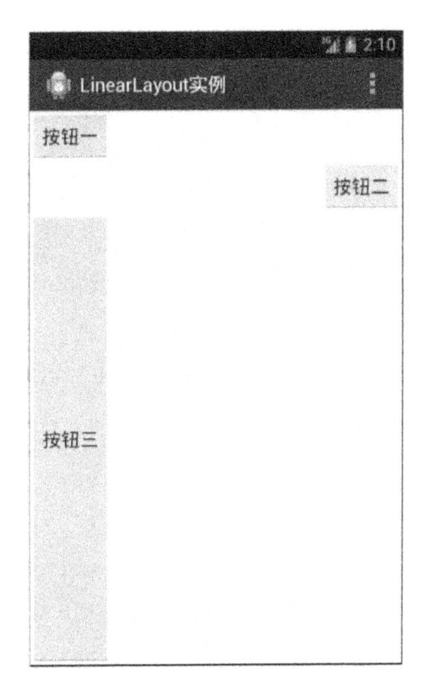

图 2.16 修改后的布局

知识拓展

android：gravity 与 android：layout_gravity 两个属性非常相似，可是它们的应用范围不同。android：gravity 属性作用在元素本身，如 Button 内的文字右对齐就用 android：gravity = "right"；android：layout_gravity 属性是对元素在其容器里面的对齐方式进行设置，如上例子中的“按钮二”设置在 LinearLayout1 内右对齐就用 android：layout_gravity = "right"。

在 LinearLayout 布局中，元素垂直分布时占一列，水平分布时占一行。特别要注意的是，水平排列时如果超过一行，则不会自动换行，超出屏幕的元素将不会被显示。同样的，垂直排列时如果超过一列也不会自动换列，超出屏幕的元素也将不会被显示。

在上面修改后的 LinearLayout 布局中，为“按钮三”设置了 android：layout_weight 属性，表示要给“按钮三”分配 1.0 权重的屏幕空间。android：layout_weight 属性默认值为 0，表示按需要分配屏幕空间，而如果多个控件元素设置了大于 0 的值，表示按所拥有的权重的比例分配屏幕空间。例如按钮一的 layout_weight 设置为 1，按钮二的 layout_weight 设置为 2，按钮三的 layout_weight 也设置为 2，则系统会分配给按钮一 1/5 的屏幕空间，按钮二 2/5，按钮三 2/5，效果如图 2.17 所示。

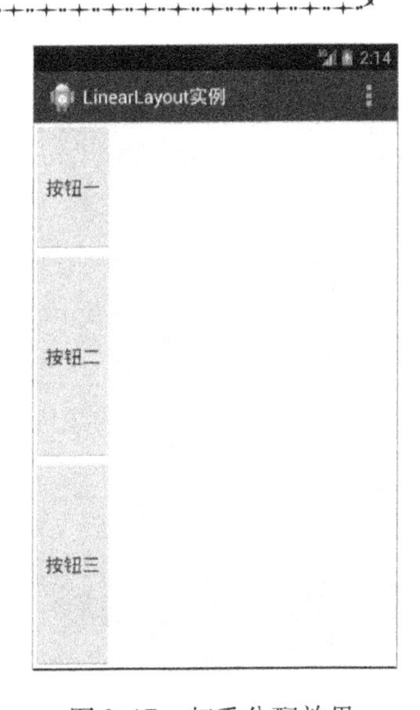

图 2.17 权重分配效果

2. RelativeLayout 实例

制作如图2.18所示的界面，该界面由五个按钮元素组成，其中一个位于屏幕的正中间其余四个分别位于中间按钮的左上角、右上角、左下角和右下角。

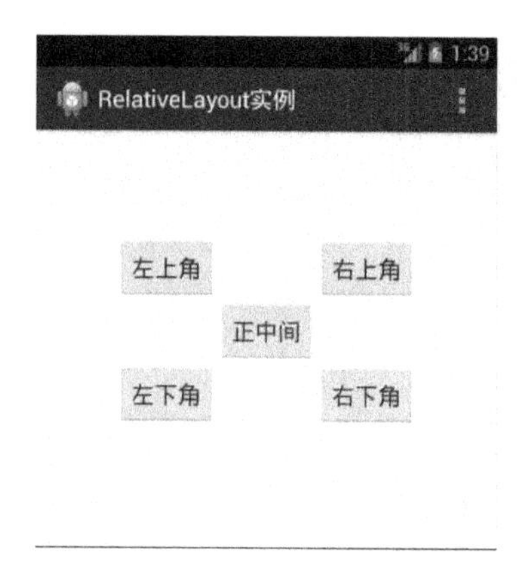

图2.18 RelativeLayout 效果

新建一个名为 Project2_2_2 的 Android 项目。

打开 res/values 文件夹中的 string. xml 文件，添加如下代码：

```
<string name = "btn_ce" >正中间 </string >
    <string name = "btn_la" >左上角 </string >
    <string name = "btn_ra" >右上角 </string >
    <string name = "btn_lb" >左下角 </string >
    <string name = "btn_rb" >右下角 </string >
```

打开 res/layout 文件夹的 activity_main. xml 文件，输入如下代码：

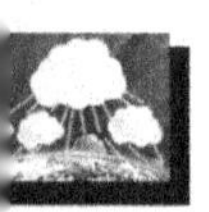

```
<RelativeLayout xmlns: android = "http: //schemas. android. com/apk/res/android"
    xmlns: tools = "http: //schemas. android. com/tools"
    android: id = "@ +id/container"
    android: layout_width = "match_parent"
    android: layout_height = "match_parent"
    tools: context = "com. example. project2_2_2. MainActivity"
    tools: ignore = "MergeRootFrame"  >
<Button
        android: id = "@ +id/button1"
        android: layout_width = "wrap_content"
        android: layout_height = "wrap_content"
<! —第一个 Button 位于屏幕的正中间 - - >
        android: layout_centerInParent = "true"
        android: text = "@ string/btn_ce"  / >
<Button
        android: id = "@ +id/button2"
        android: layout_width = "wrap_content"
        android: layout_height = "wrap_content"
<! —第二个按钮位于 button1 的左上方 - - >
        android: layout_above = "@ +id/button1"
```

```
            android：layout_toLeftOf = "@ +id/button1"
            android：text = "@ string/btn_la" / >
    < Button
            android：id = "@ +id/button3"
            android：layout_width = "wrap_content"
            android：layout_height = "wrap_content"
  <！—第三个按钮位于 button1 的右上方 - - >
            android：layout_above = "@ +id/button1"
            android：layout_toRightOf = "@ +id/button1"
            android：text = "@ string/btn_ra" / >
    < Button
            android：id = "@ +id/button4"
            android：layout_width = "wrap_content"
            android：layout_height = "wrap_content"
  <！—第四个按钮位于 button1 的左下方 - - >
            android：layout_below = "@ +id/button1"
            android：layout_toLeftOf = "@ +id/button1"
            android：text = "@ string/btn_lb" / >
    < Button
            android：id = "@ +id/button5"
            android：layout_width = "wrap_content"
            android：layout_height = "wrap_content"
  <！—第五个按钮位于 button1 的右下方 - - >
            android：layout_below = "@ +id/button1"
            android：layout_toRightOf = "@ +id/button1"
            android：text = "@ string/btn_rb" / >
  </RelativeLayout >
```

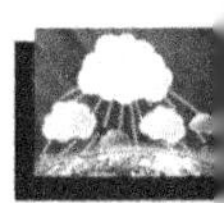

从上述代码我们可以看出，在相对布局中每一个布局元素都会以另外一个布局元素的位置来确定的，所以在编写代码的时候同一个元素同样的位置可以有多种不同的写法，换个相对位置元素就可以了。在相对布局中如果用来确定位置的元素位置发生变化，那么相对它的元素的位置也会跟着变化，所以如果要想改变相对布局里的元素位置就要查清楚有没有与之位置相对的其他元素也同时被改变了。

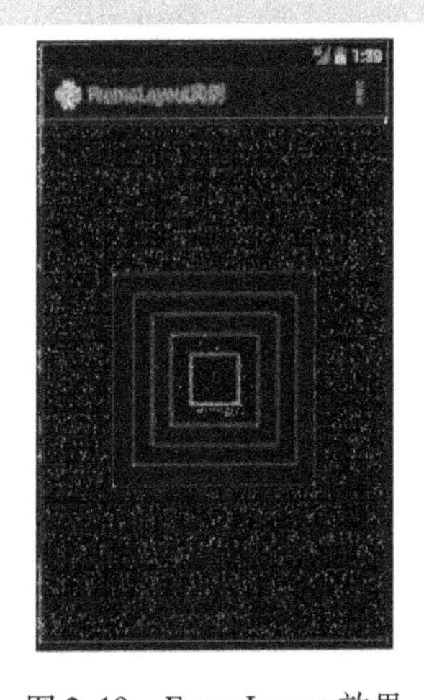

图 2. 19　FrameLayout 效果

3. FrameLayout 实例

制作如图 2. 19 所示的界面，该界面由一组不同大小和颜色的正方形组成，最大的正方形在最底下，而最小的在最上面。显然，这种界面用 FrameLayout 布局最方便。

新建一个名为 Project2_2_3 的 Android 项目。

打开 res/layout 文件夹的 activity_main. xml 文件，输入如下代码：

```
<FrameLayout xmlns: android = "http: //schemas. android. com/apk/res/android"
    xmlns: tools = "http: //schemas. android. com/tools"
    android: id = "@ +id/container"
    android: layout_width = "match_parent"
    android: layout_height = "match_parent"
        >
    <TextView
        android: id = "@ +id/textView1"
        android: layout_width = "210dp"
        android: layout_height = "210dp"
        <! —设置标签居中对齐 - - >
        android: layout_gravity = "center"
        <! —设置标签背景颜色 - - >
        android: background = "#0000FF"/>
    <TextView
        android: id = "@ +id/textView2"
        android: layout_width = "170dp"
        android: layout_height = "170dp"
        android: layout_gravity = "center"
        android: background = "#014efd"/>
    <TextView
        android: id = "@ +id/textView3"
        android: layout_width = "130dp"
        android: layout_height = "130dp"
        android: layout_gravity = "center"
        android: background = "#0195fd"/>
    <TextView
        android: id = "@ +id/textView4"
        android: layout_width = "90dp"
        android: layout_height = "90dp"
        android: layout_gravity = "center"
        android: background = "#01d1fd"/>
      <TextView
        android: id = "@ +id/textView5"
        android: layout_width = "50dp"
```

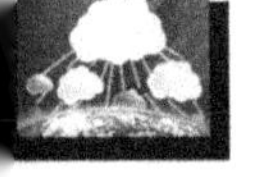

```
        android: layout_height = "50dp"
        android: layout_gravity = "center"
        android: background = "#01fddc"/>
</FrameLayout>
```

帧布局是比较少用的布局方式，可只要能利用好也可以做出很漂亮的用户界面。

4. TableLayout 实例

制作如图 2.20 所示的计算器键盘界面，可以看出，该界面的框架是一个 4 行 4 列的表格，表格的每个单元格内放置了一个按钮，如果选用相对布局也可以把这个界面做出来，可是代码编写比较麻烦，效率不高，所以选用 TableLayout 布局来完成这个任务。

图 2.20　TableLayout 效果

新建一个名为 Project2_2_3 的 Android 项目。

打开 res/layout 文件夹的 activity_main. xml 文件，输入如下代码：

```
<TableLayout xmlns: android = "http: //schemas. android. com/apk/res/android"
    xmlns: tools = "http: //schemas. android. com/tools"
    android: id = "@ +id/TableLayout1"
    android: layout_width = "match_parent"
    android: layout_height = "match_parent"
    android: stretchColumns = "0, 1, 2, 3" >
        <TableRow >
```

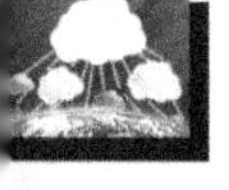

```
            <Button android: id = "@ +id/btn1" android: text = "1" />
            <Button android: id = "@ +id/btn2" android: text = "2" />
            <Button android: id = "@ +id/btn3" android: text = "3" />
            <Button android: id = "@ +id/btnPlus" android: text = " + " />
        </TableRow>
        <TableRow>
            <Button android: id = "@ +id/btn4" android: text = "4" />
            <Button android: id = "@ +id/btn5" android: text = "5" />
            <Button android: id = "@ +id/btn6" android: text = "6" />
            <Button android: id = "@ +id/btnSub"  android: text = " - " />
        </TableRow>
        <TableRow>
            <Button android: id = "@ +id/btn7" android: text = "7" />
            <Button android: id = "@ +id/btn8" android: text = "8" />
            <Button android: id = "@ +id/btn9"  android: text = "9" />
            <Button android: id = "@ +id/btnMul" android: text = " * " />
        </TableRow>
        <TableRow>
            <Button android: id = "@ +id/btnDot" android: text = ". " />
            <Button android: id = "@ +id/btn0" android: text = "0" />
            <Button android: id = "@ +id/btnEqu" android: text = " = " />
            <Button android: id = "@ +id/btnDiv" android: text = "/" />
        </TableRow>
    </TableLayout>
```

从上述代码可以看出，在 TableLayout 布局中每一行的内容都会包裹在 <TableRow> 和 </TableRow> 标签中，这与 HTML 语言非常像。另外，在表格布局中每一列的宽度由该列最宽的单元格的宽度决定，并且允许单元格跨行显示。

至此，已经利用四个实例介绍了 Android 中常用的四种布局，在设计用户界面的时候，选择正确的布局可以让工作效率更高，设计效果更好。

 任务考核

请根据前面所学知识，制作一个如图 2.21 所示的密码输入界面，答案可参考网络资源。

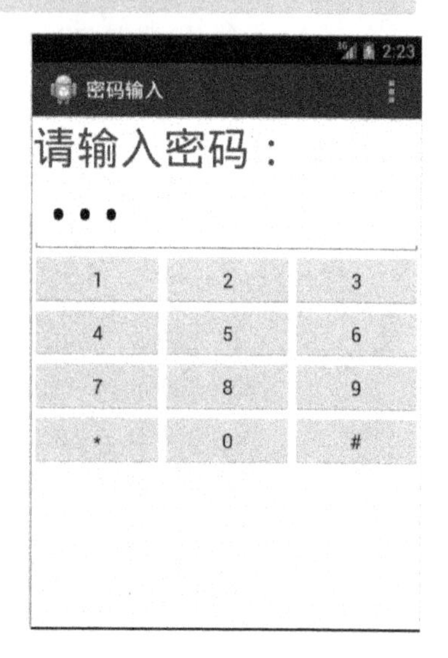

图 2.21　密码输入界面

项目三

基本控件及常用事件

Android 中提供了丰富多彩的控件，开发人员只需简单的几句调用或者参数设置的语句就可以用其构建完整的用户界面。Android 中常用控件有很多，项目三列举了一些常用控件的使用方法。

掌握 Android UI 层次体系
掌握 Android 常用基本控件
掌握 Android 常用事件

任务一　Android 常用控件使用

任务描述

本任务主要是学习一些基本的 Android 常用控件，先来了解 Android 常用控件的一些基本属性及用法，然后通过一个简单例子来学习控件的基本用法。

任务分析

在本任务中，一共要学习文本显示框(TextView)、文本编辑框(EditView)、按钮(Button)、图像(ImageView、ImageButton)、单选按钮(RadioButton)、复选框(CheckBox)、列表(ListView)等，重点和难点如下：

(1)掌握多个常用控件的创建及常用属性的设置；
(2)熟悉常用控件的布局方法。

任务准备

1. 文本显示框(TextView)

文本显示框(TextView)是用来显示文本的控件，文本显示框一般只有显示功能，不能编辑。它的常用属性 android：id 用来设置唯一标识；android ：layout_ width、android：layout_height 设置文本所占宽、高；android：text 设置文本内容等。

Android 推荐，文本显示框内容即 android：text 属性值，应用 string 索引方式存贮，不建议直接赋值，以便以后进行文字修改及应用多语言版本开发。具体用法是在 res/values/strings. xml 文件中新建 string 标签，将文字内容写在这里，如下所示：

```
<string name = "TxtSex" >性别</string >
```

上述语句定义了名为 TxtSex 值为“性别”的字符型常量，在布局文件引用方法：

```
android：text  = "@ string/ TxtSex "
```

上述语句中的 TxtSex 就是 strings. xml 中定义好的 TxtSex 常量，其他控件上的文字内容都可以用这种方式处理。

布局文件中的配置。如图 3. 1 和图 3. 2 所示。

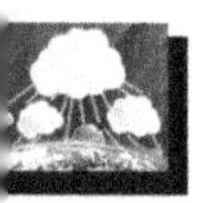

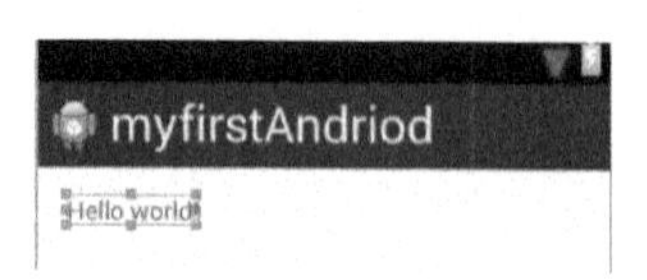

图 3. 1　文本显示框

图 3. 2　文本显示框

以下各控件的布局代码，大家都可在布局文件中尝试使用并了解布局后各控件的外观，以熟悉各控件的布局方法。

```
<TextView                                    //文本显示框标签
    android：id = "@ + id/textView1"   //设置该控件的唯一标识
    android：layout_width = "fill_parent"   //设置控件宽度为充满上级控件
    android：layout_height = "wrap_content"   //设置控件高度适应文本内容
    android：text = "@ string/text"   //设置文本的内容
/>
```

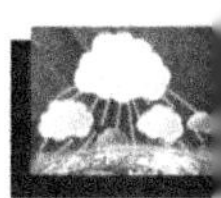

知识拓展

android：id = "@ + id/textView1"语句中的"@"符号是将这个控件的 id 自动记载在 R. java 文件中，"+"号表示向 R. java 文件中的内部类 id 中添加一个变量，名字叫 textView1，"+"相当于代码中的 new，"+id"就表示产生一个新的 id，如果没有"+"，而是"@ ID"的话，就是引用其他地方已经定义过的 id 了。

2. 文本编辑框(EditView)

文本编辑框(EditView)是 TextView 的子类，它和 TextView 的主要区别就是 EditView 可以编辑，EditText 与 TextView 继承关系是 View→TextView→EditText。文本编辑框作为一个可编辑的文本框，在控件中也占有重要地位，常用属性如表 3.1 所示。

表 3.1 文本编辑框常用属性

属 性	作 用
android：textSize	输入的字体大小
android：textColor	输入的字体颜色
android：singleLine	设置单行输入，一旦设置为 true，则文字不会自动换行
android：hint	设置显示在控件上的提示信息
android：inputType	设置文本框的输入类型。如设置为 textPassword 表示输入类型为密码形式
android：password	设置替代显示文本的符号。如以小点"."显示文本
android：text	设置显示文本的内容
android：height	设置文本区域的高度
android：width	设置文本区域的宽度
android：background	设置文本区域的背景颜色

文本编辑框的外观及属性如图 3.3 及图 3.4 所示。

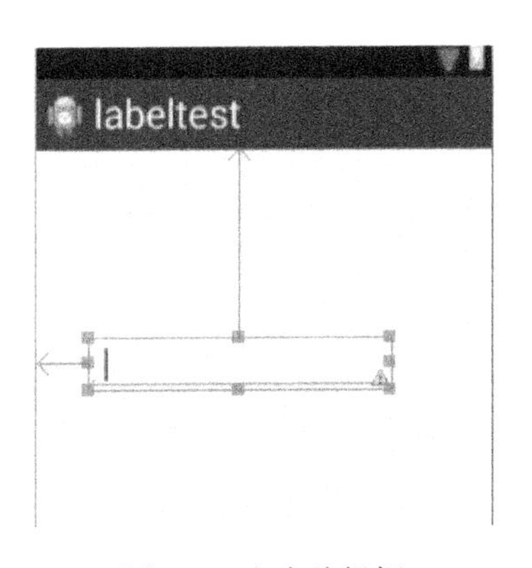

图 3.3 文本编辑框

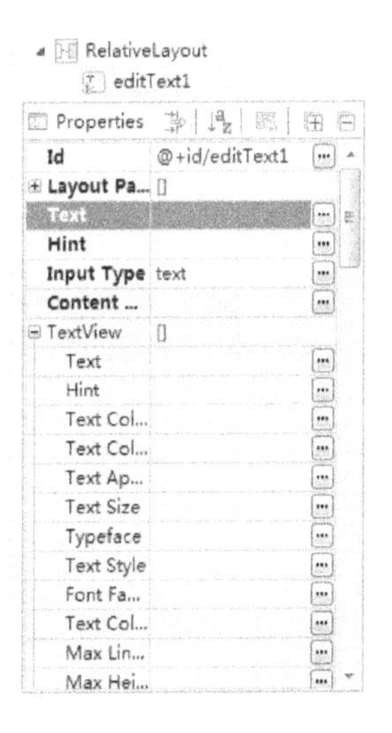

图 3.4 文本编辑框属性

以下为EditText在布局文件中的设置举例：

```
<EditText                                      //文本编辑框标签
        android: id = "@ +id/HeightEdit"
        android: layout_width = "123px"
        android: layout_height = "wrap_content"
        android: text = ""
        android: textSize = "18sp"
        android: layout_x = "124px"
        android: layout_y = "160px" >
    </EditText>
```

知识拓展

上例中，在android: layout_width = "123px"语句中将控件宽度设为了123px，即123像素。Android一般不推荐给出控件的绝对大小，为了能适应更多的移动设备，Android推荐使用dp。dp和px的转换公式为：px = dp * (density/160) dp = (px * 160)/density。其中density是指设备屏幕密度，即每英寸的显示点数目。

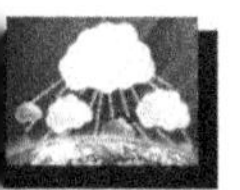

3. 按钮(Button)

按钮(Button)在Android中是一个非常常用的控件，按钮的用法也和其他面向对象的语言中非常类似。图3.5及图3.6就是按钮控件的外观及其属性。

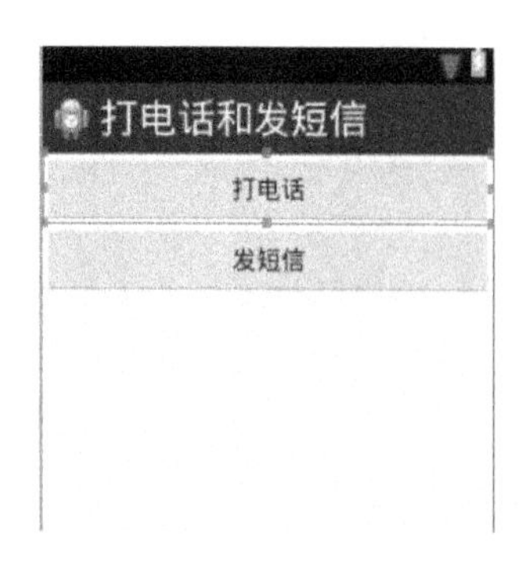

图3.5 按钮编辑框

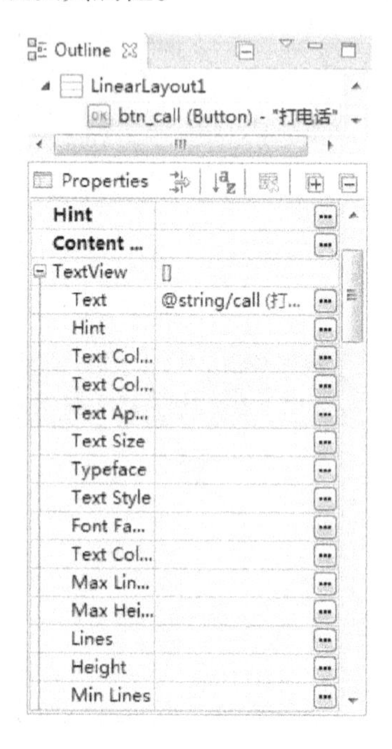

图3.6 按钮属性

布局文件中的配置如下：

```
<Button
        android: id = "@ +id/button3"   //设置该控件的唯一标识
        android: layout_width = "match_parent"   //设置文本所占的宽度
        android: layout_height = "wrap_content"   //设置文本所占的高度
        android: text = "@ string/TxtButton"   //设置按钮显示文本
        android: onClick = "clickHandler"   //设置按钮点击事件
        />
```

其中，Button 的 OnClick 属性为 clickHandler。

```
public void clickHandler( View v) {
    // Do something
}
```

4. 其他

图像框和图像按钮(ImageView、ImageButton)是两个非常相似的控件。从名称上就可以看出，控件 ImageView 的功能是显示图片；而控件 ImageButton 的功能是带有图片的按钮，既可显示图片，又具有按钮的功能。

```
<ImageView                   //图像框标签
    android: id = "@ +id/img"   //设置该控件的唯一标识
    android: layout_width = "fill_parent"   //设置控件所占的宽度
    android: layout_height = "wrap_content"   //设置控件所占的高度
    android: contentDescription = "ImageView Demo"   //对图片的一个简要说明
    android: src = "@ drawable/ic_launcher"   //调用放置在 drawable 文件夹内的图片
ic_launcher
    />
```

提示：

使用图像框前先要把所要用到的一系列图片复制到 res/drawable 文件夹内，在程序中只需通过@ drawable/来调用图片资源即可。放置在 drawable 文件夹内的图片，会在 R. java 中自动注册。

单选按钮(RadioButton)在 Android 中使用得也非常多，经常在做一些唯一性的选择项目的时候，就会用到单选按钮。实现单选按钮一般由两部分组成，也就是 RadioButton 和单选组合框(RadioGroup) 配合使用。单选组合框，用于将 RadioButton 框起来；在没有 RadioGroup 框起来的情况下，多个 RadioButton 可以全部都被选中；当多个 RadioButton 被 RadioGroup 包含起来的情况下，RadioButton 只可以做唯一性选择。

RadioButton 的使用方法及属性也和其他面向对象的程序设计非常相似，表 3. 2 中列出了一些它的重要属性。

表 3.2　单选按钮的常用属性

属　　性	作　　用
GroupName	组名。注意同一组单选按钮必须有相同的组名
Checked	是否为选中状态。值为(True 或 False)

```
<RadioGroup android: layout_width = "wrap_content"          //单选按钮组标签
            android: layout_height = "wrap_content"
            android: orientation = "vertical"              //指定分布方式为垂直分布
            android: id = "@ +id/radioGroup" >
   //以下为 female, male 两个单选按钮
   <RadioButton android: layout_width = "wrap_content"    //单选按钮标签
                android: layout_height = "wrap_content"
                android: id = "@ +id/radio1"
                android: text = "@ string/female"/ >
   <RadioButton android: layout_width = "wrap_content"
                android: layout_height = "wrap_content"
                android: id = "@ +id/radio2"
                android: text = "@ string/male"/ >
</RadioGroup >
```

上例中，建立了一个单选按钮组。此单选按钮组中有两个单选按钮，按垂直方式分布。单选按钮中显示的文本为 string. xml 中将要定义的常量 famale 和 male。图 3.7 就是按上述代码布局后的界面(注意此时常量 famale 和 male 暂时还未定义)，图 3.8 是单选按钮属性。

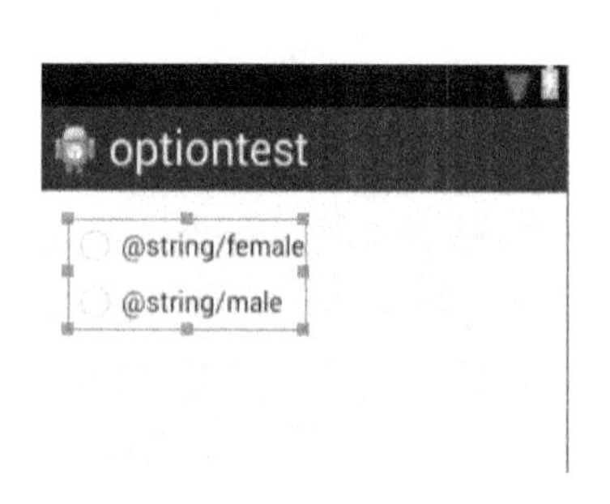

图 3.7　单选按钮

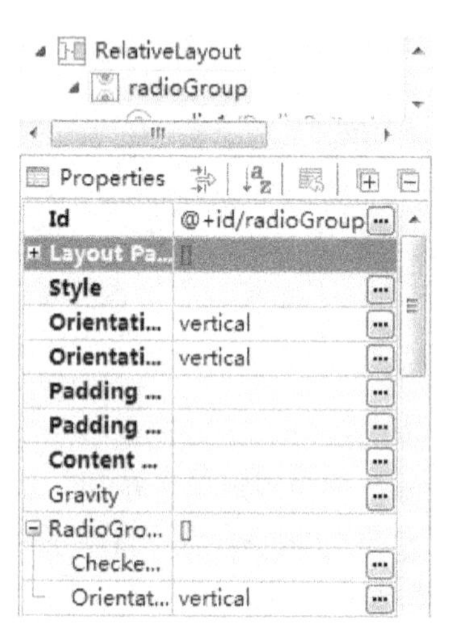

图 3.8　单选按钮属性

复选框(CheckBox)与单选按钮非常类似，只是它可以同时选中多个选项。

```
<LinearLayout,
        android: layout_width = "wrap_content"
        android: layout_height = "wrap_content"
        android: layout_gravity = "center_horizontal"
        android: orientation = "vertical"  >
    <CheckBox
        android: layout_width = "wrap_content"
        android: layout_height = "wrap_content"
        android: checked = "true"
        android: text = "红色" />
    <CheckBox
        android: layout_width = "wrap_content"
        android: layout_height = "wrap_content"
        android: text = "蓝色" />
    <CheckBox
        android: layout_width = "wrap_content"
        android: layout_height = "wrap_content"
        android: text = "绿色" />
    </LinearLayout>
```

上例中，建立了一组复选框。此例中有三个复选框，按垂直方式分布，复选框中显示的文本分别为红色、蓝色、绿色。为便于大家理解，在这里采用了直接赋值。注意：Android 不推荐类似 android：text = "绿色"这样的显示赋值。一般应该按前面文本显示框内容中讲过的 string 索引的方式来赋值。图 3.9 就是布局后的界面，图 3.10 是复选框属性。

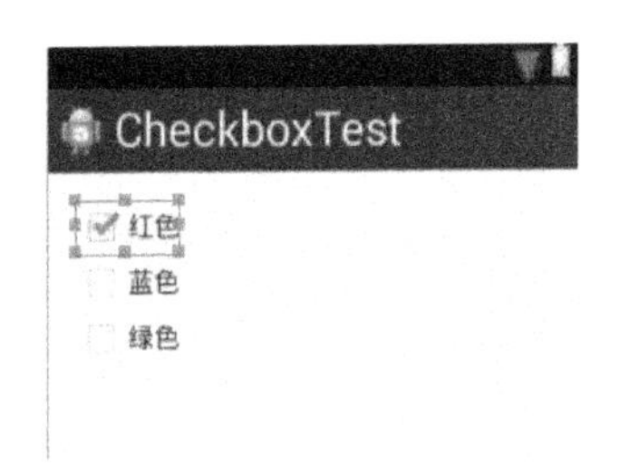

图 3.9 复选框

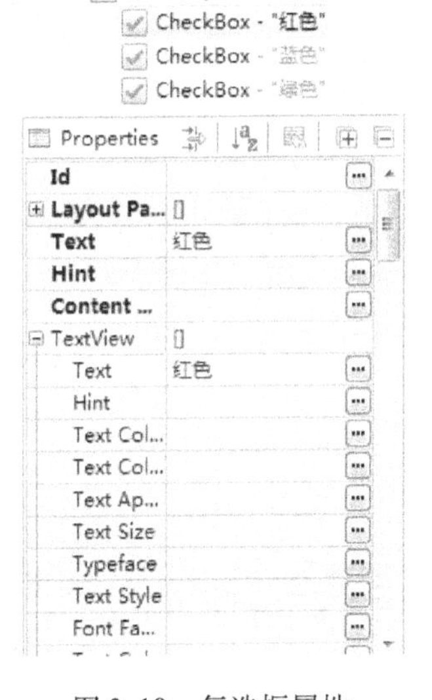

图 3.10 复选框属性

列表（ListView）也是比较常用的控件之一，它以列表的形式显示具体内容，并且能够根据数据的长度自适应显示。在ListView中可以根据需要显示自定义的列表内容，包括文字（TextView）、图片（ImageView）、按钮（Button）等，以此构成图文并茂的显示效果，常用属性如表3.3所示。

表3.3　ListView的常用属性

属　　性	作　　用
android：divider	设置ListView中每一项中间的分割线
android：dividerHeight	分割线的高度
android：background	指定ListView的背景图片或者颜色
android：cacheColorHint	值设为透明（#00000000）防止滑动变化
android：fadeScrollbars	设置这个属性为true可实现滚动条自动隐藏和显示

ListView布局文件如下：

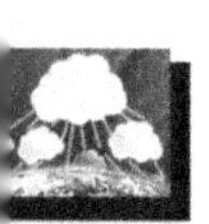

```
< ListView                                    //ListView 的标签
    android: id = "@ +id/listview"
    android: layout_width = "match_parent"
    android: layout_height = " wrap_content"
    android: divider = "@ drawable/dividerimage"   //用图片 dividerimage 做分隔样式
>
```

效果如图3.11与图3.12所示。

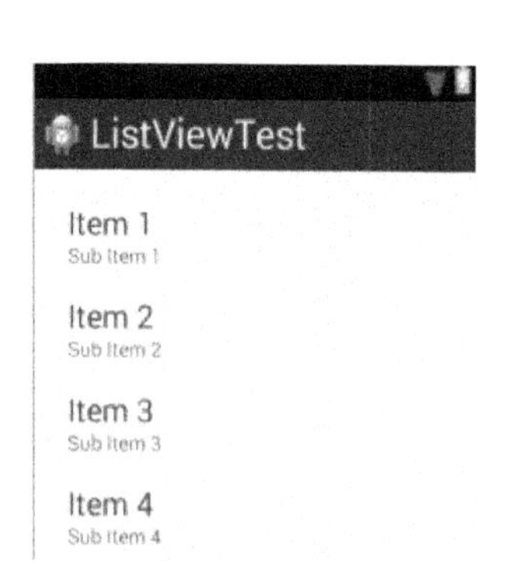

图3.11　列表

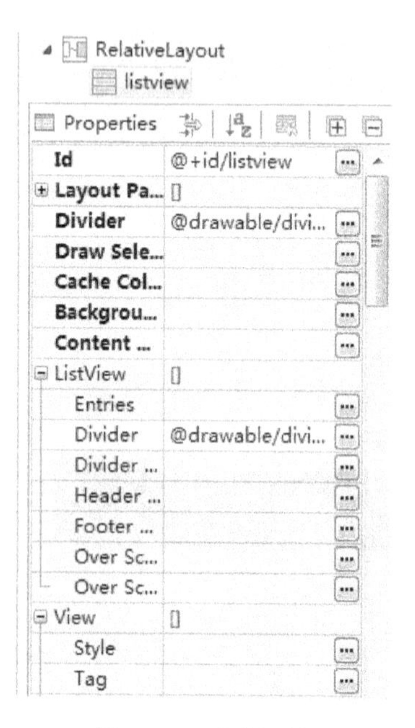

图3.12　列表属性

任务实现

1. 新建 Android 项目

此项目的功能为根据用户输入的性别和身高来计算标准体重，标准体重信息在 Toast 中弹出显示。

标准体重计算公式如下：

男性标准体重 =（身高 cm − 80）×70 %　女性标准体重 =（身高 cm − 70）×60 %

在 Eclipse 中，新建一 Android 应用，并命名为 project3_1，方法可参考项目一。新建一个名为 BMIActivity. java 的 Activity，其相应布局文件为 Activity_bmi. xml。

2. 布局

此项目中所用的控件共有三个文本显示框 TextView 和两个单选按钮 RadioButton。两个单选按钮要放置在一个单选按钮组中，一个文本编辑框 EditText，一个按钮 Button。注意各控件的 id 应该事先在/res/values/目录下的 string. xml 文件中进行定义，不要直接赋值。string. xml 文件内容在后序内容中会列出。

各控件的命名及布局情况如表 3. 4 所示。

表 3. 4　任务一所用的控件及其命名

控件种类	控件的唯一标识(id)
TextView	ShowText、TextSex、TextHeight
RadioButton	SexMan、SexWoman
Button	ButtonOk
EditText	HeightEdit

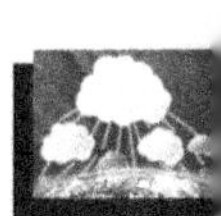

各控件的命名及布局情况如图 3. 13 及图 3. 14 所示。

图 3. 13　界面设计

图 3. 14　各控件命名

3. 代码

对 BMIActivity 布局的布局文件 Activity_bmi. xml 如下所示，各个控件的具体布局方式

见如下具体代码。(为让大家多了解并练习使用控件的多种属性，本实例采用了绝对布局，目前 Android 一般不建议用绝对布局，大家可根据学过的布局知识将自行将本例改为相对布局。)

```
<AbsoluteLayout  xmlns: android = "http: //schemas. android. com/apk/res/android"
    xmlns: tools = "http: //schemas. android. com/tools"
    android: layout_width = "match_parent"
    android: layout_height = "match_parent"
    tools: context = ". BMIActivity"  >
    <TextView                     //以下对三个文本显示框进行布局
        android: id = "@ +id/ShowText"
        android: layout_width = "wrap_content"
        android: layout_height = "20px"
        android: text = "@ string/TxtCount"
        android: textSize = "18px"
        android: layout_x = "65px"
        android: layout_y = "21px" >
    </TextView>
    <TextView
        android: id = "@ +id/TextSex"
        android: layout_width = "wrap_content"
        android: layout_height = "wrap_content"
        android: text = "@ string/TxtSex"
        android: layout_x = "71px"
        android: layout_y = "103px" >
    </TextView>
    <TextView
        android: id = "@ +id/TextHeight"
        android: layout_width = "wrap_content"
        android: layout_height = "wrap_content"
        android: text = "@ string/TxtHeight"
        android: layout_x = "72px"
        android: layout_y = "169px" >
    </TextView>
    <RadioGroup     //以下将两个单选按钮布局入同一个单选按钮组中
        android: id = "@ +id/RadioGroup"
        android: layout_width = "wrap_content"
        android: layout_height = "37px"
        android: orientation = "horizontal"
```

```
        android: layout_x = "124px"    //控件的 x 坐标值
        android: layout_y = "101px" >    //控件的 y 坐标值
    <RadioButton
        android: id = "@ +id/SexMan"
        android: layout_width = "wrap_content"
        android: layout_height = "wrap_content"
        android: text = "@ string/TxtMale" >
    </RadioButton>
    <RadioButton
        android: id = "@ +id/SexWoman"
        android: layout_width = "wrap_content"
        android: layout_height = "wrap_content"
        android: text = "@ string/TxtFemale" >
    </RadioButton>
    </RadioGroup>
    //以对文本编辑框进行布局
    <EditText
        android: id = "@ +id/HeightEdit"
        android: layout_width = "123px"
        android: layout_height = "wrap_content"
        android: text = ""
        android: textSize = "18sp"    //设置字体大小
        android: layout_x = "124px"
        android: layout_y = "160px" >
    </EditText>
    <Button                          //以下对按钮进行布局
        android: id = "@ +id/ButtonOK"
        android: layout_width = "100px"
        android: layout_height = "wrap_content"
        android: text = "@ string/BtCount"
        android: layout_x = "125px"
        android: layout_y = "263px" >
    </Button>
</AbsoluteLayout>
```

BMIActivity. java 文件内容如下所示：

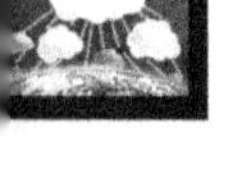

```
package org.dglg.project3_1;
import java.text.DecimalFormat;
import java.text.NumberFormat;
import android.app.Activity;
import android.os.Bundle;
import android.view.View;
import android.view.View.OnClickListener;
import android.widget.Button;
import android.widget.EditText;
import android.widget.RadioButton;
import android.widget.Toast;
public class BMIActivity extends Activity {
    private Button countButton;
    private EditText heighText;
    private RadioButton maleBtn, femaleBtn;
    String sex = "";
    double height;
    @Override
    public void onCreate(Bundle savedInstanceState) {
        super.onCreate(savedInstanceState);
        setContentView(R.layout.activity_bmi);
        creadView();    //调用创建视图函数
        sexChoose();    //调用性别选择函数
        setListener();    //调用 Button 注册监听器函数
  }
     private void setListener() {                    //响应 Button 事件函数
        countButton.setOnClickListener(countListner);
    }
    private OnClickListener countListner = new OnClickListener() {
            @Override
        public void onClick(View v) {    //以下利用 Toast 在屏幕上输出信息
            Toast.makeText(BMIActivity.this, "你是一位" + sexChoose() + "\n"
 + "你的身高为" + Double.parseDouble(heighText.getText( ).toString()) + "cm"
 + "\n 你的标准体重为" + getWeight(sexChoose(), height) + "kg", Toast.LENGTH_
LONG).show();
            }
    };
    private String sexChoose() {      //性别选择函数
```

```
            if (maleBtn.isChecked()) {
                sex = "男性";
            }
            else if(femaleBtn.isChecked()) {
                sex = "女性";
            }
            return sex;
        }
    public void creadView() {                               //创建视图函数
            //txt = (TextView)findViewById(R.id.txt);
            countButton = (Button)findViewById(R.id.ButtonOK);
            heighText = (EditText)findViewById(R.id.HeightEdit);
            maleBtn = (RadioButton)findViewById(R.id.SexMan);
            femaleBtn = (RadioButton)findViewById(R.id.SexWoman);
            //txt.setBackgroundResource(R.drawable.bg);
        }
    private String format(double num) {                 //标准体重格式化输出函数
            NumberFormat formatter = new DecimalFormat("0.00");
            String str = formatter.format(num);
            return str;
            }
    private String getWeight(String sex, double height) {       //计算标准体重的函数
            height = Double.parseDouble(heighText.getText().toString());
            String weight = "";
            if (sex.equals("男性")) {
                    weight =format((height - 80) * 0.7);
            }
            else {
                    weight = format((height - 70) * 0.6);
            }
            return weight;
        }
}
```

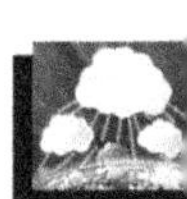

此外，还需事先在 string.xml 文件中添加如下常量定义：

```
<string name = "TxtCount" >计算您的标准体重 </string >
```

```
<string name = "TxtMale" >男 </string >
<string name = "TxtFemale" >女 </string >
<string name = "TxtSex" >性别 </string >
<string name = "TxtHeight" >身高 </string >
<string name = "BtCount" >计算 </string >
```

运行结果如图3.15所示。

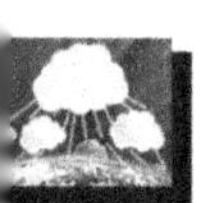

图3.15　项目project3_1运行界面

任务二　Android常用事件

任务描述

本任务主要是学习一些基本的Android常用事件。先了解Android常用事件的触发机制及一些基本用法，然后通过一个简单例子来熟悉并且掌握事件的基本用法。

任务分析

在本任务中，首先要学习事件处理的概念及机制，然后分别介绍：按钮单击事件(onClick)、长按事件(onLongClick)、上下文菜单事件(onCreateContextMenu)、焦点事件

(onFocusChange)、触屏事件(onTouchEvent)、键盘或遥控事件(onKeyUp、onKeyDown)等，重点和难点如下：

(1)了解 Android 事件的基础知识；
(2)掌握多个常用事件的处理方式。

任务准备

1. 事件处理的概念

基于用户图形界面的任何应用最终都是要面对用户的，经常处理的是用户的动作，也就是说要基于用户图形界面的应用程序是要为用户的动作提供响应，这种为用户动作提供响应的机制就是事件处理。Android 的事件与其他面向对象的程序设计的事件概念是一致的。

2. 几种常用的事件

下面是常见的一些 Android 的常用事件，具体名称及触发机制如表 3.5 所示。

表 3.5　Android 常用事件名称及触发机制

事　　件	事件名称	触发机制
onClick	按钮单击事件	在单击按钮时被触发
onLongClick	长按事件	在长按控件时被触发
onCreateContextMenu	上下文菜单事件	在上下文菜单应该显示的时候被触发
onFocusChange	焦点事件	焦点发生变化时被触发
onTouchEvent	触屏事件	用户触摸或手指在屏幕上移动时被触发
onKeyUp、onKeyDown	键盘或遥控事件	在用户按下任何键盘键(包括系统按钮，如箭头键和功能键)时被触发

3. Android 常用事件的常用处理方法

1)onClick：按钮单击事件

实现步骤：

(1)用 findViewById 方法通过控件 ID 获取控件实例。

如任务一 project3_1 中，就用如下语句：

countButton = (Button)findViewById(R. id. ButtonOK)；

获取了实例名为 ountButton 的一个按钮 ButtonOK。

(2)为该控件注册 OnClickListener 监听。

(3)实现 onClick 方法。

代码实例：

```
// 以下是获取按钮单击事件
    // 以下是通过 findViewById 获取控件实例，获取了一个名为 Button 的按钮控件
实例 myButton
        Button myButton = (Button) findViewById(R. id. myButton);
          //以下为按钮控件注册 OnClickListener 监听
          myButton. setOnClickListener(new  OnClickListener() {
            public void onClick(View v) {
              //此处定义按钮事件的内容，实现 onClick 方法
              }
```

2)onLongClick：长按事件

实现步骤：

(1)通过控件 ID 获取控件实例，方法同上。

(2)为该控件注册 OnLongClickListener 监听。

(3)实现 onLongClick 方法。

代码实例：

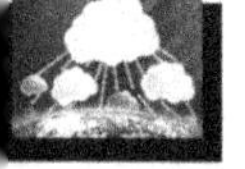

```
    // 以下是获取按钮长按事件
      // 以下是通过 findViewById 获取控件实例，获取了一个名为 Button 的按钮控
件实例 myButton
    Button myButton = (Button) findViewById(R. id. myButton);
    //以下为按钮控件注册 OnClickListener 监听
    myBotton. setOnClickListener(new OnLongClickListener() {
          public void onLongClick(View v) {
            //此处定义按钮长按事件，实现 onLongClick 方法
            }
```

3)onCreateContextMenu：上下文菜单事件

Android 的上下文菜单在概念上和其他面向对象的语言的右键弹出菜单(PopMemu)类似。当一个视图注册到一个上下文菜单时，如果在该对象上执行一个“长按”(按住不动差不多两秒钟)的动作，将出现一个提供相关功能的弹出菜单。例如在手机上最常见的长按文本时所弹出的“复制、粘贴、剪切”菜单。上下文菜单可以被注册到任何视图对象中。不过，最常见的是被注册于列表控件 ListView 的选项 item 中，在按中列表项时，一般会转换其背景色提示将呈现上下文菜单。比较常见的具体实例比如手机中的电话联系人列表。图 3. 16 所示就是一个一个电话联系人列表项上长按时所弹出的上下文菜单。

ContextMenu 上下文菜单创建方法：

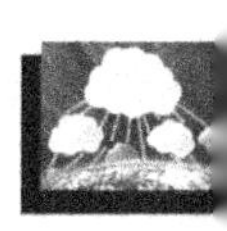

(1)重写 Activity 中的 onCreateContextMenu( )方法。

(2)调用 Menu 的 add 方法添加菜单项，通过扩充一个定义在 XML 中的菜单资源。

(3)重写 onContextItemSelected( )方法，响应菜单单击事件。

(4)调用 registerForContexMenu( )注册上下文菜单。

首先将上下文菜单中的 onCreateContextMenu( )方法声明中的参数做一个说明：

public void onCreateContextMenu (ContextMenu menu, View v, ContextMenu. ContextMenuInfo menuInfo)

参数说明：

menu：需要显示的快捷菜单，ContextMenu 类型。

v：v 是用户选择的界面元素，View 类型。

menuInfo：menuInfo 是所选择的视图元素的额外信息。有时视图元素需要向上下文菜单传递一些信息，比如该 View 对应 DB 记录的 ID 等。

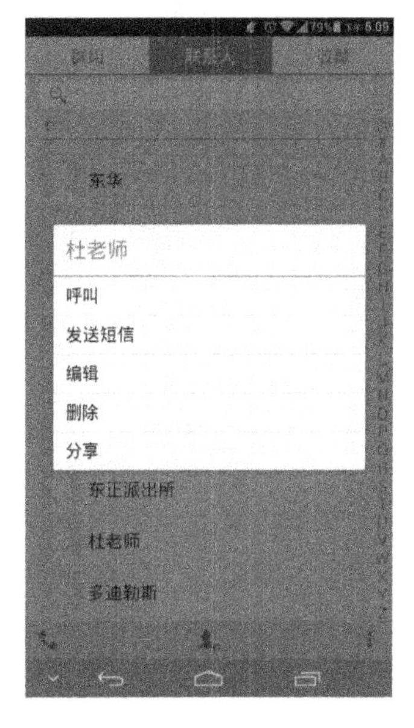

图 3.16　上下文菜单

说明

用 XML 文件来加载和响应菜单方法如下：在/res 目录下创建 menu 文件夹，在 menu 目录下使用与 menu 相关的元素定义 xml 文件，文件名随意，Android 会自动为其生成资源 ID。使用 xml 文件的资源 ID，将 xml 文件中定义的菜单项添加到 menu 对象中，响应菜单项时，使用每个菜单项对应的资源 ID。

下面代码重写 onCreateContextMenu( )方法，并用两种方法配置上下文菜单选项。

```
//重写 Activity 中的 onCreateContextMenu( )方法
public void onCreateContextMenu(ContextMenu menu, View v,
        ContextMenuInfo menuInfo) {
    //1. 通过手动添加来配置上下文菜单选项
    //menu.add(0, 1, 0, "复制");
    //menu.add(0, 2, 0, "剪切");
    //menu.add(0, 3, 0, "粘贴");
    //2. 通过 xml 文件来配置上下文菜单选项，用此方法需先创建一个菜单选项
文件
    MenuInflater mInflater = getMenuInflater();
    mInflater.inflate(R.menu.cmenu, menu);
    super.onCreateContextMenu(menu, v, menuInfo);
}
```

以下代码创建菜单选项文件：

```
<? xml version ="1.0" encoding ="utf-8"? >
    <menu
      xmlns: android ="http: //schemas.android.com/apk/res/android" >
       <item android: id ="@ +id/copy" android: title ="复制"/ >
       <item android: id ="@ +id/cut" android: title ="剪切"/ >
       <item android: id ="@ +id/pasta" android: title ="粘贴"/ >
    </menu >
```

当菜单的某个选项被选中时调用以下方法：

```
//重写 onContextItemSelected()方法，响应菜单单击事件
    public boolean onContextItemSelected(MenuItem item) {
          switch(item.getItemId()){
             case R.id.copy:
             Toast.makeText(this, "你选择复制", Toast.LENGTH_SHORT).show();
               break;
             case R.id.cut:
               Toast.makeText(this, "你选择剪切", Toast.LENGTH_SHORT).show();
               break;
             case R.id.pasta:
               Toast.makeText(this, "你选择粘贴", Toast.LENGTH_SHORT).show();
               break;
             }
             return super.onContextItemSelected(item);
       }
```

上述代码中的事件的结果是：当你把焦点移动到某控件上时，长按 2 秒左右会弹出上下文菜单列表。当选择上下文菜单的某个菜单项时，用 Toast 弹出相应的提示信息。

4) onFocusChange：焦点事件

当对象获得或失去焦点时被触发。有一点需要注意，焦点事件是只要该控件获得焦点，不需要点击就会触发相应的事件。

实现步骤：

(1)通过控件 ID 获取控件实例。

(2)为该控件注册 onFocusChange Listener 监听。

(3)实现 onFocusChange 方法。

首先将焦点事件中的 onFocusChange ()方法的参数做一个说明：

声明：protected void onFocusChanged (View v, boolean gainFocus, int direction, Rect previously FocusedRect)

参数 gainFocus 表示触发该事件的 View 是否获得了焦点，当该控件获得焦点时，

gainFocus 等于 true，否则等于 false。

参数 direction 表示焦点移动的方向，用数值表示，大家可以重写 View 中的该方法打印该参数进行观察。

参数 previouslyFocusedRect 表示在触发事件的 View 的坐标系中，前一个获得焦点的矩形区域，即表示焦点是从哪里来的。如果不可用则为 null。

组件与焦点事件相关的常用方法：

setFocusable 方法：设置对象是否可以拥有焦点。

isFocusable 方法：监测此对象是否可以拥有焦点。

setNextFocusDownId 方法：设置对象的焦点向下移动后获得焦点对象的 ID。

hasFocus 方法：返回对象的父控件是否获得了焦点。

requestFocus 方法：尝试让此对象获得焦点。

isFocusableTouchMode 方法：设置对象是否可以在触摸模式下获得焦点，在默认情况下是不可用获得的。

代码实例如下：

```
// 文本框的焦点事件
    //通过 ID 获得控件
        inputEditText = (EditText)findViewById(R.id.inputText);
    //为该控件注册 onFocusChange Listener 监听
        inputEditText.setOnFocusChangeListener(new OnFocusChangeListener() {
            //实现 onFocusChange 方法
            public void onFocusChange(View v, boolean hasFocus) {
Toast.makeText(getApplicationContext(), inputEditText.getText(),
Toast.LENGTH_LONG);
            }
        }
```

上述代码的作用是当文本编辑框控件获得焦点时，用 Toast 输出文本框中的相应内容。

5)onTouchEvent：触屏事件

触摸屏幕、按下屏幕、从屏幕上抬起或在屏幕上滑动，都会触发该事件。

实现步骤：

(1)通过控件 ID 获取控件实例。

(2)为该控件注册 onTouchEvent Listener 监听。

(3)实现 onTouchEvent 方法。

首先将触屏事件中的 onTouchEvent ()方法的参数做一个说明：

声明：public boolean onTouchEvent (MotionEvent event)

参数 event 为手机屏幕触屏事件封装类的对象，其中封装了该事件的所有信息，例如触摸的位置、触摸的类型以及触摸的时间等。该对象会在用户触摸手机屏幕时被创建。

返回值：该方法的返回值为一个 boolean 类型的变量，当返回 true 时，表示已经完整地处理了这个事件，并不希望其他的回调方法再次进行处理，而当返回 false 时，表示并

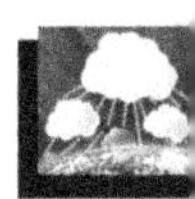

没有完全处理完该事件，更希望其他回调方法继续对其进行处理，例如 Activity 中的回调方法。

当屏幕被按下、从屏幕上抬起、在屏幕中滑动时，该事件都会被触发，可以调用 MotionEvent. getAction( )方法，判断动作值为 MotionEvent. ACTION_DOWN(按下屏幕)、MotionEvent. ACTION_UP(从屏幕上抬起)、MotionEvent. ACTION_MOVE(在屏幕上滑动)中的哪一个，来判断动作类型后再进行处理。如下例所示：

```
//重写的 onTouchEvent 方法
public boolean onTouchEvent(MotionEvent event) {
        switch(event. getAction( )) {
                case (MotionEvent. ACTION_DOWN):    // 当按下的时候
                    Display("ACTION_DOWN", event);
                    break;
                case (MotionEvent. ACTION_UP):      // 当抬起的时候
                    int historysize = ProcessHistory(event);
                    histroy. setText("历史数据" + historysize);
                    Display("ACTION_UP", event);
                    break;
               case (MotionEvent. ACTION_MOVE):     // 当滑动的时候
                    Display("ACTION_MOVE", event);
        }
        return super. onTouchEvent(event);
    }
```

6)onKeyUp、onKeyDown：键盘操作事件

onKeyUp 当按键弹起时触发，onKeyDown 在按键按下时触发。

声明：public boolean onKeyDown (int keyCode, KeyEvent event)

public boolean onKeyUp (int keyCode, KeyEvent event)

参数 keyCode 为被按下的键的键值，手机键盘中每个按钮都会有其单独的键值，应用程序都是通过键值来判断用户按下的是哪一个键。

参数 event 为按键事件封装类的对象，其中包含了触发事件的详细信息，例如事件的状态、事件的类型、事件发生的时间等。当用户按下按键时，系统会自动将事件封 装成 KeyEvent 对象供应用程序使用。

返回值：同上。

代码实例如下：

```
public boolean onKeyDown(int keyCode, KeyEvent event) {
      switch(keyCode) {
          case KeyEvent. KEYCODE_0:
             DisplayToast("你按下数字键0");
```

```
        break;
    case KeyEvent. KEYCODE_DPAD_CENTER:
        DisplayToast("你按下中间键");
        break;
    case KeyEvent. KEYCODE_DPAD_DOWN:
        DisplayToast("你按下下方向键");
        break;
    case KeyEvent. KEYCODE_ALT_LEFT:
        DisplayToast("你按下组合键 alt +←");
        break;
        }
  return super. onKeyDown(keyCode, event);
  }
```

任务实现

1. 新建 Android 项目

在 Eclipse 中，新建一个名为 project 3_2 的 Android 应用，并新建一个名为 TouchViewActivity. java 的 activity，其对应的布局文件为 Activity_touch_view. xml。具体方法可参考项目一。

此例的功能是测试 Android 中最常用的触屏事件。当用户手指操作触摸屏幕并划动时，记录手指触摸的位置坐标等。

2. 布局

项目的布局中用到了三个文本显示框来显示文本。同样，在利用布局文件进行布局之前，要先在/res/values/目录下的 string. xml 文件中定义好这三个显示文本框所显示的内容，如图 3. 17 及图 3. 18 所示。

图 3. 17　界面设计

图 3. 18　各控件命名

3. 代码

TouchViewActivity 的布局文件 Activity_touch_view. xml 如下所示，各个控件的具体设置见具体代码：

```
<LinearLayout xmlns: android = "http: //schemas. android. com/apk/res/android"
        android: orientation = "vertical"
        android: layout_width = "fill_parent"
        android: layout_height = "fill_parent" >
        <TextView
            android: id = "@ +id/touch_area"
            android: layout_width = "fill_parent"
            android: layout_height = "300dip"
            android: background = "#66FF66"
            android: textColor = "#000000"
            android: text = "@ string/TestArea"
            />
        <TextView
            android: id = "@ +id/history_label"
            android: layout_width = "fill_parent"
            android: layout_height = "wrap_content"
            android: text = "@ string/OldData"
            />
        <TextView
            android: id = "@ +id/event_label"
            android: layout_width = "fill_parent"
            android: layout_height = "wrap_content"
            android: text = "@ string/TouchEvt"
            />
</LinearLayout>
```

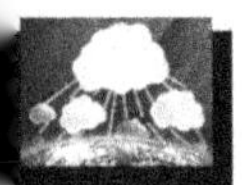

TouchViewActivity. java 文件如下所示：

```
package org. dglg. project3_2;
    import android. os. Bundle;
    import android. app. Activity;
    import android. view. Menu;
    import android. view. MotionEvent;
    import android. view. View;
    import android. widget. TextView;
    public class TouchViewActivity extends Activity {
        private TextView eventlable;
```

```
        private TextView histroy;
        private TextView TouchView;
        @ Override
        public void onCreate(Bundle savedInstanceState) {
            super. onCreate(savedInstanceState);
            setContentView(R. layout. activity_touch_view);
        TouchView = (TextView) findViewById(R. id. touch_area);
        histroy = (TextView) findViewById(R. id. history_label);
        eventlable = (TextView) findViewById(R. id. event_label);
    TouchView. setOnTouchListener(new View. OnTouchListener() {
                @ Override
                    public boolean onTouch(View v, MotionEvent event) {
                        int action = event. getAction();
                        switch (action) {              // 当按下的时候
                        case (MotionEvent. ACTION_DOWN):
                            Display("ACTION_DOWN", event);
                            break;                 // 当抬起的时候
                        case (MotionEvent. ACTION_UP):
                            int historysize = ProcessHistory(event);
                            histroy. setText("历史数据" + historysize);
                            Display("ACTION_UP", event);
                            break;
                        case (MotionEvent. ACTION_MOVE):   // 当滑动的时候
                            Display("ACTION_MOVE", event);
                        }
                        return true;
                    }
            });
        }
            public void Display(String eventType, MotionEvent event) {
            int x = (int) event. getX();                    // 触点相对坐标的信息
            int y = (int) event. getY();
            float pressure = event. getPressure();     // 表示触屏压力大小
            float size = event. getSize();                  // 表示触点尺寸
            int RawX = (int) event. getRawX();             // 获取绝对坐标信息
            int RawY = (int) event. getRawY();
            String msg = "";            // 定义显示信息 msg
            msg += "事件类型" + eventType + "\n";
            msg += "相对坐标" + String. valueOf(x) + "," + String. valueOf(y)
+ "\n";
```

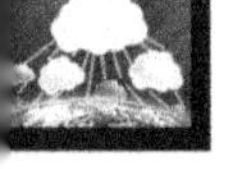

```
                msg += "绝对坐标" + String.valueOf(RawX) + "," + String.valueOf
(RawY) + "\n";
                msg += "触点压力" + String.valueOf(pressure) + ",";
                msg += "触点尺寸" + String.valueOf(size) + "\n";
                eventlable.setText(msg);
            }
            public int ProcessHistory(MotionEvent event) {
                int history = event.getHistorySize();
                for (int i = 0; i < history; i++) {
                    long time = event.getHistoricalEventTime(i);
                    float pressure = event.getHistoricalPressure(i);
                    float x = event.getHistoricalX(i);
                    float y = event.getHistoricalY(i);
                    float size = event.getHistoricalSize(i);
                }
                    return history;
                }
    @Override
    public boolean onCreateOptionsMenu(Menu menu) {
    // Inflate the menu; this adds items to the action bar if it is present.
    getMenuInflater().inflate(R.menu.touch_view, menu);
    return true;
    }
}
```

此外，还需在string.xml文件中添加如下常量定义：

```
<string name="TestArea">触屏事件测试区域</string>
<string name="OldData">历史数据</string>
<string name="TouchEvt">触屏事件</string>
```

运行后界面如图3.19所示。

图3.19　运行界面

项目四

Activity的生命周期

Activity的中文意思是“活动”，是Android应用的四大重要组成单元之一。Activity作为可视化用户交互的界面，一个Activity代表手机、平板电脑屏幕的一屏或一个窗口，为用户实现Android应用提供方便的界面的交互。本项目将走进Android的活动——Activity，帮助读者了解Android中Activity的基本知识及应用。

了解 Activity 的生命周期基本知识
掌握 Activity 的 4 个重要的回调方法，Log 日志信息输出，LogCat 日志查看器的使用，用 Bundle 在 Activity 间实现数据传递

任务一　用 LogCat 查看单个 Activity 的生命周期

任务描述

本任务主要是通过对 Activity 的 7 个回调方法重写，启动项目 Project4_1 的 Activity，在 LogCat(即日志查看器)中显示 Activity 生命周期的回调方法过程，从而了解 Activity 的生命周期及回调方法的次序。任务内容如下：

(1)在项目 Project4_1 的应用运行时，打开 LogCat 显示启动 Activity 的回调方法过程如图 4.1 所示。

(2)单击屏幕上的“后退键”按钮，这时会关闭 Activity，在 LogCat 中显示 Activity 的回调方法过程如图 4.2 所示。

(3)在启动 Activity 后，单击“主屏幕键”按钮，这时 Activity 会退到后台且不可见，相

Level	Time	PID	TID	Application	Tag	Text
I	09-23...	1..	1..	org.dglg.project4_1	System.out	Project4_1 Activity----onCreate
I	09-23...	1..	1..	org.dglg.project4_1	System.out	Project4_1 Activity----onStart
I	09-23...	1..	1..	org.dglg.project4_1	System.out	Project4_1 Activity----onResume

图 4.1 启动 Activity 的回调方法过程

Level	Time	PID	TID	Application	Tag	Text
I	09-23...	1..	1..	org.dglg.project4_1	System.out	Project4_1 Activity----onPause
I	09-23...	1..	1..	org.dglg.project4_1	System.out	Project4_1 Activity----onStop
I	09-23...	1..	1..	org.dglg.project4_1	System.out	Project4_1 Activity----onDestroy

图 4.2 关闭 Activity 的回调方法过程

应的 Activity 回调方法过程如图 4.3 所示。

Level	Time	PID	TID	Application	Tag	Text
I	09-25...	829	829	org.dglg.project4_1	System.out	Project4_1 Activity----onPause
I	09-25...	829	829	org.dglg.project4_1	System.out	Project4_1 Activity----onStop

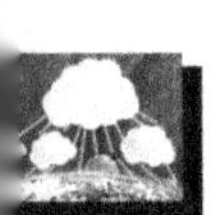

图 4.3 Activity 退到后台的回调方法过程

(4)当要重新启动 Activity 时，单击“最近应用键”按钮，在调出的任务管理器中选择 Activity，Activity 就会返回到前台，相应的 Activity 回调方法过程如图 4.4 所示。

Level	Time	PID	TID	Application	Tag	Text
I	09-25...	829	829	org.dglg.project4_1	System.out	Project4_1 Activity----onRestart
I	09-25...	829	829	org.dglg.project4_1	System.out	Project4_1 Activity----onStart
I	09-25...	829	829	org.dglg.project4_1	System.out	Project4_1 Activity----onResume

图 4.4 Activity 返回前台的回调方法过程

任务分析

本任务中的 Activity 是 MainActivity，通过在生命周期的各个回调方法添加 System. out. println()方法，将 MainActivity 的生命周期的回调方法过程，在 LogCat 日志查看器中打印出来。

(1)MainActivity 完成启动的过程，即是创建一个 Activity 完成的步骤，按次序调用了 onCreate()、onStart()、onResume()三个方法，其功能分别是创建、运行、获取 Activity。

(2)关闭 MainActivity 的过程，按次序调用了 onPause()、onStop()、onDestroy()三个

方法，其功能分别是暂停、停止、销毁 Activity。

(3) MainActivity 退到后台的过程，按次序调用了 onPause()、onStop()二个方法，其功能分别是暂停、停止 Activity。

(4) 重新启动该 MainActivity 返回到前台的过程，按次序调用了 onRestart()、onStart()、onResume()三个方法，其功能分别是重启、运行、获取 Activity。

MainActivity 的启动、关闭、退到后台和返回到前台的过程，正好是一个 Activity 的生命周期的过程，同时在 LogCat 中显示了 Activity 的回调方法及次序。

本任务的主要目的是对 Activity 的启动、关闭退到后台和返回到前台回调方法过程，通过 System. out. println()方法在 LogCat 日志查看器中打印显示出来，了解 Activity 的生命周期及回调方法的应用，重点和难点如下：

(1) 了解 Activity 的生命周期；
(2) 了解 Log 日志输出和 LogCat 日志查看器的基础知识；
(3) 掌握在 Activity 的生命周期中回调方法的应用。

任务准备

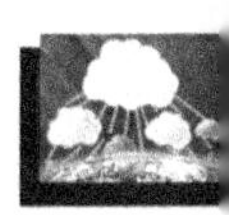

1. Activity 的生命周期

1) Activity 简介

Activity 的中文意思是"活动"，是 Android 应用的四大重要组成单元之一。Activity 作为可视化用户交互的界面，一个 Activity 代表手机、平板电脑屏幕的一屏或一个窗口；另外一个 Activity 中可加多个组件，为用户实现 Android 应用提供方便的界面的交互和功能。

实现 Android 应用时，用堆栈来实现多个 Activity 的活动。

活动的 Activity 置于栈顶，原来在栈顶的 Activity 就在其下面，和其他非栈顶的 Activity 处于非活动状态，等待被激活；再次被激活后，处于非活动状态的 Activity 又置于栈顶成为活动的 Activity，而最近一个活动的 Activity 就被压进栈位于当前活动的 Activity 的下面，成为非活动的 Activity；如此循环，直到该 Activity 被销毁。如图 4.5 所示。

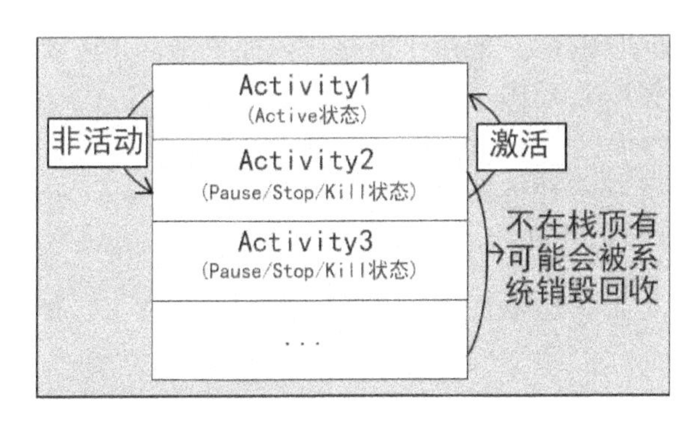

图 4.5　Activity 堆栈

2) Activity 的生命周期的重要状态及状态回调方法

在 Android 中，Activity 生命周期是指 Activity 从启动到销毁的过程。

(1) Activity 的 4 个重要状态。

每个 Activity 都有生命周期，有各种重要状态及其相应的回调方法。如表 4.1 所示就是 Activity 的 4 个重要状态。

表 4.1　Activity 的 4 个重要状态

状　态	描　　述
活动	Activity 位于 Activity 栈顶，可见，可被激活
暂停	非活动的 Activity，可见，内存低时不能被系统 killed(杀死)
停止	Activity 被其他 Activity 覆盖，不可见，虽仍保存所有的状态和信息，但内存低时将被系统 killed(杀死)
销毁	Activity 结束，或 Activity 所在的 Dalvik 进程被结束

(2) Activity 状态的保存和恢复事件的回调方法如表 4.2 所示。

表 4.2　Activity 状态保存和恢复事件的回调方法

方　法	描　　述
onSaveInstanceState()	这是保存事件的回调方法。当内存低时，系统在终止 Activity 前，调用该方法保存 Activity 的状态信息，以备 onRestoreInstanceState()或 onCreate()恢复
onRestoreInstanceState()	这是恢复事件的回调方法。用来恢复 onSaveInstanceState()方法保存的 Activity 状态信息，在 onStart()和 onResume ()之间被调用

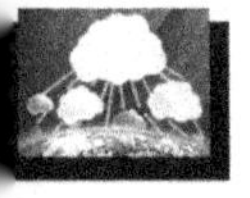

知识拓展

onPause()和 onSaveInstanceState()的区别在于这两个方法都可以用来保存界面的用户输入数据，onPause()一般用于保存持久性数据，并将数据保存在存储设备上的文件系统或数据库系统中。

onSaveInstanceState()主要用来保存动态的状态信息，信息一般保存在 Bundle 中。Bundle 是能够保存多种格式数据的对象，onSaveInstanceState()保存在 Bundle 中的数据，系统在调用 onRestoreInstanceState()和 onCreate()时，会同样利用 Bundle 将数据传递给函数。

(3) Activity 的生命周期及其回调方法如图 4.6 所示。

3) Activity 的生命周期的几个过程

由图 4.6 可知，Activity 的生命周期可概括为以下几个过程：

(1) 启动 Activity 时系统会按次序调用 onCreate()、onStart()、onResume()方法，之后

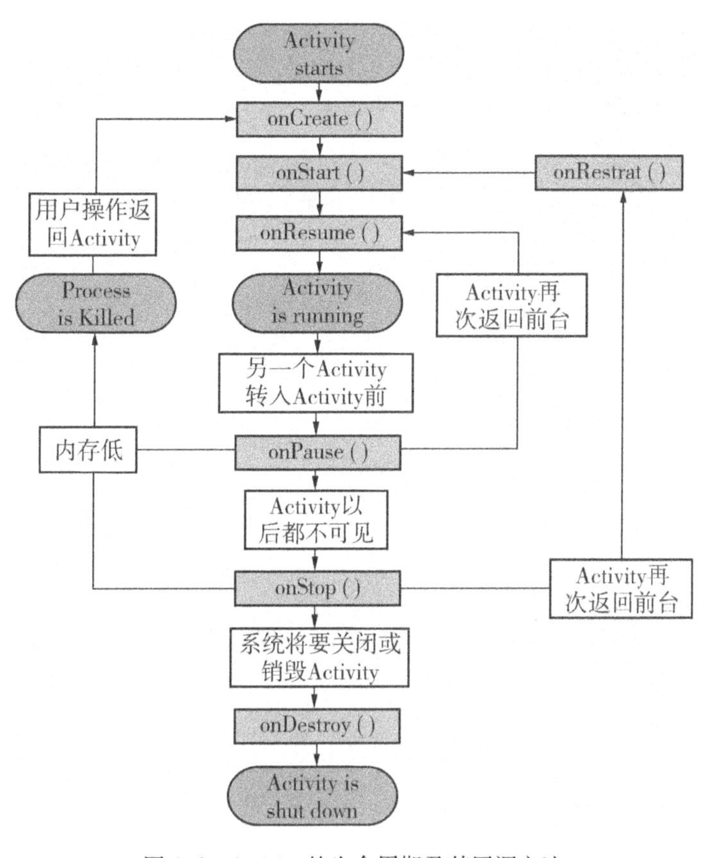

图 4.6 Activity 的生命周期及其回调方法

Activity 进入运行状态。

(2)活动的 Activity 被其他 Activity 覆盖其上或被锁屏时，系统会调用 onPause( )方法，暂停当前 Activity 的执行。

(3)活动的 Activity 由被覆盖状态回到前台或解锁屏时，系统会调用 onResume( )方法，再次进入运行状态。

(4)活动的 Activity 转到新的 Activity 或按 Home 键回到主屏，这时原来的 Activity 退到后台，系统会先后调用 onPause( )、onStop( )方法，进入等待重新激活状态。

(5)要重新激活 Activity 时，系统按次序调用 onRestart( )、onStart( )、onResume( )方法，再次进入运行状态。

(6)活动的 Activity 处于被覆盖状态或者后台不可见状态，系统内存低，杀死当前的 Activity 后，要重新激活当前 Activity 时，会再次调用 onCreate( )、onStart( )、onResume( )方法，进入运行状态。

(7)退出或关闭当前 Activity 时，系统按次序调用 onPause( )、onStop( )、onDestory( )方法，结束当前 Activity。

本任务在 LogCat 中显示的 Activity 的 4 个回调过程，正好是 Activity 生命周期的第 1、第 7、第 4 和第 5 个过程的回调方法及次序。

4）Activity 生命周期的三种生命周期

Activity 生命周期的回调方法是有顺序的，其顺序是按 1～9 的编号排序的，不同的回调方法又形成了三种生命周期，如图 4.7 所示。

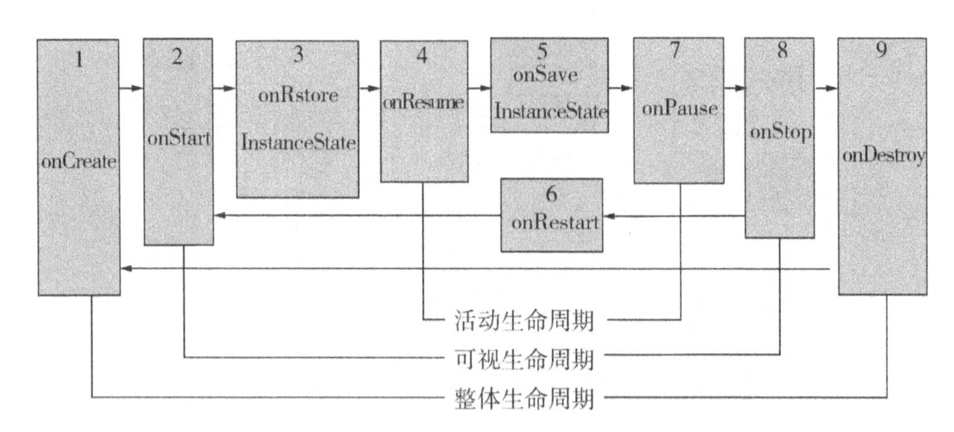

图 4.7　Activity 生命周期的分类及其回调方法顺序

Activity 三种生命周期分别是整体生命周期、可视生命周期和活动生命周期。

（1）整体生命周期。

整体生命周期是从 Activity 建立到销毁的全部过程，从 onCreate（）开始，到 onDestroy（）结束。

用户一般在 onCreate（）中初始化 Activity 所能使用的全局资源和状态，最后在 onDestroy（）中释放这些资源，但是在特殊的情况下，Android 系统会不调用 onDestroy（）函数，直接终止 Activity。

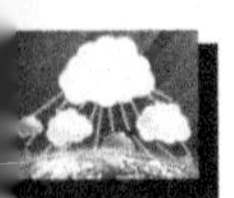

（2）可视生命周期。

可视生命周期是 Activity 在界面上从可见到不可见的过程，从 onStart（）开始，到 onStop（）结束。在此周期中的 onStart（）、onStop（）、onRestart（）方法都实现不同的功能。

onStart（）方法可用于初始化或启动与更新界面相关的资源。

onStop（）方法可用于暂停或停止一切与更新用户界面相关的线程、计时器和服务。

onRestart（）方法在 onStart（）方法前被调用时，在 Activity 从不可见变为可见的过程中，用来进行一些特定的处理过程。

onStart（）和 onStop（）除了会被多次调用，也经常被用来注册和注销 BroadcastReceiver。

（3）活动生命周期。

活动生命周期是 Activity 被置于栈顶（或屏幕的最上层）时，能够与用户交互的阶段，从 onResume（）开始，到 onPause（）结束。

在 Activity 的状态变换过程中 onResume（）和 onPause（）经常被调用，因此这两个方法中应使用更为简单、高效的代码。

onPause（）被调用后，onStop（）和 onDestroy（）随时能被 Android 系统终止，onPause（）常用来保存持久数据，如界面上的用户的输入信息等。

5）Activity 的回调方法

由于 Activity 实际上继承了 ApplicationContext 类，在使用到 Activity 的 onCreate（）、 onStart( )、onRestart( )、onResume( )、onPause( )、onStop( )、onDestroy( )这 7 个回调方法，可以对其进行重写，代码如下：

```
public class Activity extends ApplicationContext {
        protected void onCreate(Bundle savedInstanceState);
        protected void onStart();
        protected void onRestart();
        protected void onResume();
        protected void onPause();
        protected void onStop();
        protected void onDestroy();
    }
```

2. Activity 创建、配置、启动和关闭

1) Activity 的创建

Activity 的创建通常分 2 步实现。

(1)根据不同的应用，Activity 的创建会对不同的类或子类进行继承。通常是继承 android. app 包的 Activity 类；如要实现选项卡，可继承 TabActivity；如要实现列表，可继承 ListActivity。本任务中 Activity 的名称是 MainActivity，只要继承 android. app 包的 Activity 类即可，可通过以下代码实现：

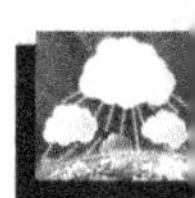

```
import android.app.Activity;
public class MainActivity extends Activity {
}
```

(2)重写回调方法且设置要显示的视图，本任务中要重写 onCreate 方法，并在 onCreate 方法中通过 setContentView 方法设置要显示的视图 activity_main，可通过以下代码实现：

```
@Override
  protected void onCreate(Bundle savedInstanceState) {
  super.onCreate(savedInstanceState);
  setContentView(R.layout.activity_main);
}
```

2) Activity 的配置

在 AndroidManifest. xml 文件中需对创建的 Activity 进行配置(或者称为注册)，否则启动所创建的 Activity 时会出现异常信息。配置方法是在 <application> </ application>标记中加入所创建的 Activity 信息的 <activity> </activity>标记，基本格式为：

```
<activity
android: icon = "@drawable/图标名"
android: name = "类名"
android: lable = "标签"
…
>
…
</activity>
```

本任务中对所创建的 Activity 配置的主要代码为:

```
<activity
  android: name = "org. dglg. project4_1. MainActivity"
  android: label = "@string/app_name"  >
</activity>
```

3) Activity 的启动

(1)对于只有一个 Activity 的项目，在 AndroidManifest. xml 文件配置(或者称为注册)了 Activity 作为程序入口后，项目运行时会自动启动 Activity。本任务中 Activity 的名称是 MainActivity，在创建 Activity 时就自动在 AndroidManifest. xml 文件配置好了。

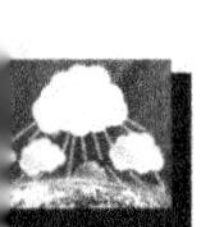

(2)如果项目不止一个 Activity，可用 startActivity( )方法来启动另一个 Activity，例如启动名称是 AnotherActivity 的 Activity，在 AndroidManifest. xml 文件配置好 Activity 后，MainActivity 中使用 startActivity( )方法的代码如下:

```
Intent intent = new Intent(MainActivity. this, AnotherActivity. class);
startActivity(intent);
```

以上代码中，关于 Intent 的内容简介可参考本书项目五的内容。

4) Activity 的关闭

可以通过自定义“关闭”按钮 button1 后，使用 Activity 的 finish( )方法来实现 Activity 的关闭，代码如下:

```
Button button1 = (Button)findViewByld(R. id. button1);
button1. setOnClickListener(new View. OnClickListener(){
  @Override
  public void onClick(View v){
    finish();
  }
});
```

本任务中，单击屏幕上的“返回”按钮就会关闭当前 Activity，这个操作和使用 Activity

的 finish( )方法关闭 Activity 的结果是一样的。

3. log 日志输出

由于在 Eclipse 的 Console 控制台中只能输出应用安装的信息，在 android 的应用中，System. out. println( )方法在 Console 控制台不会输出结果，而是在 LogCat 日志查看器中打印出结果。

在 android 中，对程序的信息进行输出，一般采用 android. util. Log 类的静态方法来实现，Log 类所输出的日志内容数量从少到多分别是 ERROR、WARN、INFO、DEBUG、VERBOSE，按首字母来分，对应有 Log. e( )、Log. w( )、Log. i( )、Log. d( )、Log. v( )五种静态方法，使用不同的方法输出信息的颜色各不相同。如表 4. 3 所示。

表 4. 3　Activity 状态保存和恢复的回调方法

Log 类静态方法	输出日志内容模式
Log. e( )	ERROR
Log. w( )	WARN
Log. i( )	INFO
Log. d( )	DEBUG
Log. v( )	VERBOSE

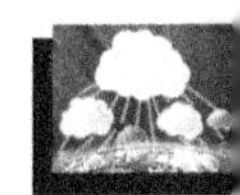

4. LogCat 日志查看器

通过 LogCat 日志查看器，可以查看 Activity 运行过程的日志信息。如果 LogCat 日志查看器未打开，那么可以通过以下的步骤打开。

STEP 1 单击【Window】菜单的【Show View】，再选择【Other】项，如图 4. 8 所示。

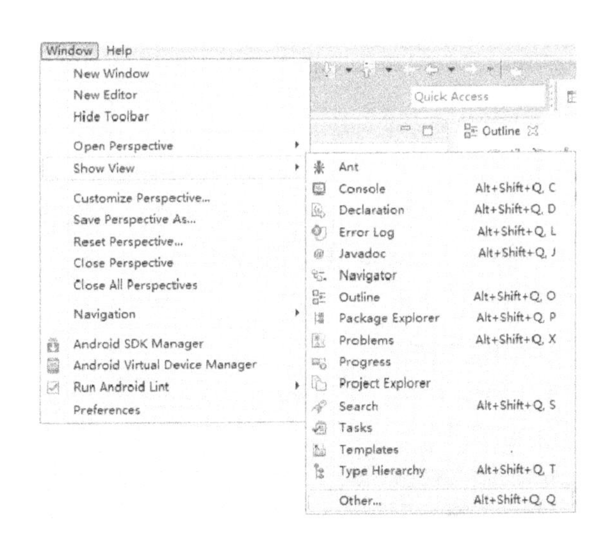

图 4. 8　打开 LogCat 日志查看器的步骤 1

STEP 2 在弹出的"Show View"的对话框，打开"Android"类型，选择"LogCat"项后单击【OK】按钮，如图4.9所示，这时LogCat日志查看器就被打开了，如图4.10所示。

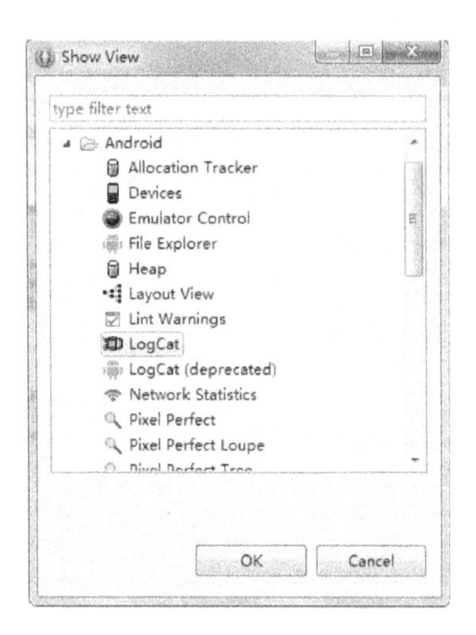

图4.9 打开LogCat日志查看器的步骤2

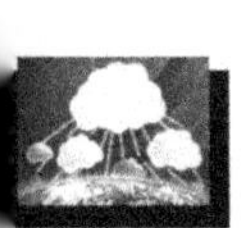

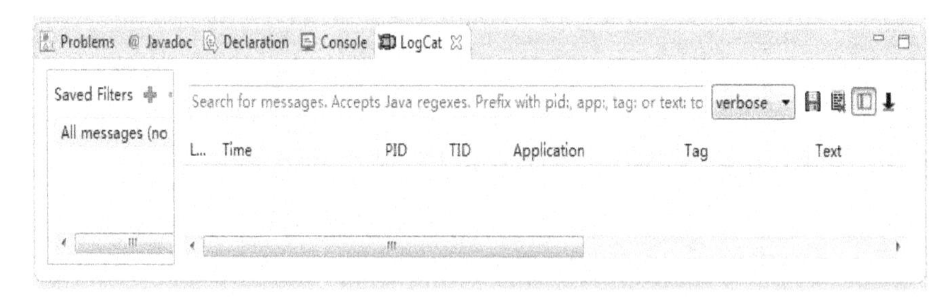

图4.10 打开LogCat日志查看器的步骤3

 任务实现

1. 新建Android项目

参考项目一创建项目的方法，在Eclipse中新建Android项目名为Project4_1和包为org.dglg.project4_1的Hello world程序，在新建Android项目时会自动创建一个Activity，本任务中Activity的名称是MainActivity。

2. 对MainActivity的代码进行编写

在本任务中，Activity_main.xml和strings.xml文件的内容是不需要修改的，因为只要对MainActivity.java文件作出修改。

(1)定义私有变量Ac赋值为"Project4_1 Activity - - - -"，代码如下：

```
private String Ac = "Project4_1 Activity - - - -";
```

(2)对onCreate、onStart、onPause、onRestart、onResume、onStop、onDestroy这7个回

调方法都进行重写后，在本任务中，为每个回调方法都添加 System. out. println( )方法输出，而不是使用 Log 日志输出，以便在执行某个回调方法时，在 LogCat(即命令列表)视窗显示该回调方法对应的输出内容。比如在 onCreate 回调方法中添加如下代码：

```
System. out. println( Ac  +  "onCreate" ) ;
```

那么，在执行 onCreate 回调方法时 LogCat(即命令列表) 视窗就会显示“Project4_1 Activity - - - - onCreate”。

(3) MainActivity. java 修改后的代码如下：

```
package org. dglg. project4_1;
import android. os. Bundle;
import android. app. Activity;
public class MainActivity extends Activity {
    private String Ac = " Project4_1 Activity - - - - "; //定义字符串变量 Ac 及
内容
//以下重写 7 个回调方法
@ Override
protected void onCreate( Bundle savedInstanceState) {
    System. out. println( Ac  +  "onCreate" ) ; //输出命令行的内容并在 LogCat 显示
    super. onCreate( savedInstanceState) ;
    setContentView( R. layout. activity_main) ;
    }
    @ Override
    protected void onStart( ) {
        System. out. println( Ac  +  "onStart" ) ;
        super. onStart( ) ;
    }
    @ Override
    protected void onResume( ) {
        System. out. println( Ac  +  "onResume" ) ;
        super. onResume( ) ;
    }
    @ Override
    protected void onPause( ) {
        System. out. println( Ac  +  "onPause" ) ;
        super. onPause( ) ;
    }
    @ Override
    protected void onRestart( ) {
```


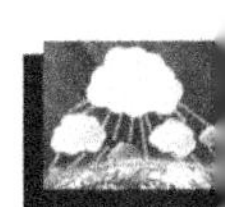

```
            System. out. println( Ac  + "onRestart" );
            super. onRestart( );
        }
    @ Override
        protected void onStop( ) {
            System. out. println( Ac  + "onStop" );
            super. onStop( );
        }
        @ Override
        protected void onDestroy( ) {
            System. out. println( Ac  + "onDestroy" );
            super. onDestroy( );
        }
    }
```

任务考核

参考本任务，创建一个新的项目名为ProjectTest4_1，在其自动创建的MainActivity中，对Activity的生命周期的回调方法进行重写，并使用System. out. println( )函数对MainActivity的生命周期的回调方法过程的Log日志打印输出，在LogCat日志查看器中显示MainActivity的生命周期的回调方法的过程。

任务二　用LogCat查看多个Activity的生命周期

任务描述

本任务主要是实现MainActivity与ShowActivity之间的跳转，通过Log日志静态方法Log. d( )输出日志信息，利用LogCat日志查看器查看MainActivity与ShowActivity这2个Activity的生命周期的回调方法过程，实现用LogCat查看多个Activity的生命周期的目的。

(1)在启动MainActivity后，即可在LogCat中看到MainActivity启动时生命周期的回调方法过程，如图4. 11所示。

Level	Time	PID	TID	Application	Tag	Text
D	10-...	7..	796	org.dglg.project4_2	project4_2 MainActivity	start onCreate()
D	10-...	7..	796	org.dglg.project4_2	project4_2 MainActivity	start onStart()
D	10-...	7..	796	org.dglg.project4_2	project4_2 MainActivity	start onResume()

图4. 11　启动MainActivity的回调方法过程

(2)在 MainActivity 显示的文本是“这是第一个 Activity”，单击“进入”按钮，屏幕跳转到 ShowActivity，其显示的文本是“这是第二个 Activity”，这时单击“返回”按钮，屏幕又跳转回到 MainActivity，如图 4.12、图 4.13 所示。此过程中，MainActivity 和 ShowActivity 是生命周期的回调方法过程，如图 4.14 所示。

图 4.12　MainActivity 界面

图 4.13　ShowActivity 界面

Level	Time	PID	TID	Application	Tag	Text
D	10-...	7..	780	org.dglg.project4_2	project4_2 MainActivity	start onPause()
D	10-...	7..	780	org.dglg.project4_2	project4_2 ShowActivity	start onCreate()
D	10-...	7..	780	org.dglg.project4_2	project4_2 ShowActivity	start onStart()
D	10-...	7..	780	org.dglg.project4_2	project4_2 ShowActivity	start onResume()
D	10-...	7..	780	org.dglg.project4_2	project4_2 MainActivity	start onStop()

图 4.14　MainActivity 转到 ShowActivity 的生命周期的回调方法过程

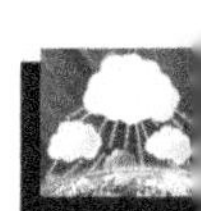

(3)单击“主屏幕键”按钮，这时 ShowActivity 会退到后台且不可见，相应的 Activity 回调方法过程如图 4.15 所示。

Level	Time	PID	TID	Application	Tag	Text
D	10-...	7..	796	org.dglg.project4_2	project4_2 ShowActivity	start onPause()
D	10-...	7..	796	org.dglg.project4_2	project4_2 ShowActivity	start onStop()

图 4.15　ShowActivity 退到后台的生命周期的回调方法过程

(4) 单击“最近应用键”按钮，在调出的任务管理器中选择“用户注册界面”应用，ShowActivity 就会返回到前台，相应的 Activity 回调方法过程如图 4.16 所示。

Level	Time	PID	TID	Application	Tag	Text
D	10-...	7..	796	org.dglg.project4_2	project4_2 ShowActivity	start onRestart()
D	10-...	7..	796	org.dglg.project4_2	project4_2 ShowActivity	start onStart()
D	10-...	7..	796	org.dglg.project4_2	project4_2 ShowActivity	start onResume()

图 4.16　ShowActivity 返回前台的生命周期的回调方法过程

(5) 单击屏幕上的“返回键”按钮或 ShowActivity 中的“返回”按钮，这时会关闭 ShowActivity，重新激活 MainActivity，此过程中，MainActivity 和 ShowActivity 的生命周期的

回调方法过程如图 4. 17 所示。

Level	Time	PID	TID	Application	Tag	Text
D	10-...	7..	780	org.dglg.project4_2	project4_2 ShowActivity	start onPause()
D	10-...	7..	780	org.dglg.project4_2	project4_2 MainActivity	start onRestart()
D	10-...	7..	780	org.dglg.project4_2	project4_2 MainActivity	start onStart()
D	10-...	7..	780	org.dglg.project4_2	project4_2 MainActivity	start onResume()
D	10-...	7..	780	org.dglg.project4_2	project4_2 ShowActivity	start onStop()
D	10-...	7..	780	org.dglg.project4_2	project4_2 ShowActivity	start onDestroy()

图 4. 17　关闭 ShowActivity，重新激活 MainActivity 生命周期的回调方法过程

(6)单击“主屏幕键”按钮，这时 MainActivity 会退到后台且不可见，相应的 Activity 回调方法过程如图 4. 18 所示。

Level	Time	PID	TID	Application	Tag	Text
D	10-...	7..	796	org.dglg.project4_2	project4_2 MainActivity	start onPause()
D	10-...	7..	796	org.dglg.project4_2	project4_2 MainActivity	start onStop()

图 4. 18　MainActivity 退到后台的生命周期的回调方法过程

(7)单击“最近应用键”按钮，在调出的任务管理器中选择“多个 Activity 跳转”应用，MainActivity 就会返回到前台，相应的 Activity 回调方法过程如图 4. 19 所示。

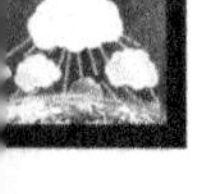

Level	Time	PID	TID	Application	Tag	Text
D	10-...	7..	760	org.dglg.project4_2	project4_2 MainActivity	start onRestart()
D	10-...	7..	760	org.dglg.project4_2	project4_2 MainActivity	start onStart()
D	10-...	7..	760	org.dglg.project4_2	project4_2 MainActivity	start onResume()

图 4. 19　MainActivity 返回前台的生命周期的回调方法过程

(8)单击屏幕上的“返回键”按钮，这时会关闭 MainActivity，此过程中 MainActivity 的生命周期的回调方法过程如图 4. 20 所示。

Level	Time	PID	TID	Application	Tag	Text
D	10-...	7..	760	org.dglg.project4_2	project4_2 MainActivity	start onStop()
D	10-...	7..	760	org.dglg.project4_2	project4_2 MainActivity	start onDestroy()

图 4. 20　关闭 MainActivity 的生命周期的回调方法过程

任务分析

本任务中的 Activity 是 MainActivity 和 ShowActivity，为 MainActivity 与 ShowActivity 重写 Activity 生命周期的各个回调方法，并添加 Log. d( ) 日志输出方法，通过 MainActivity 与 ShowActivity 之间的跳转，将 MainActivity 和 ShowActivity 的生命周期的回调方法的过程，在 LogCat 日志查看器中打印出来。

本任务主要是在掌握多个 Activity 的创建和布局设计，了解多个 Activity 的相互跳转的

实现，各个 Activity 生命周期中各回调方法的过程，从而加深对 Activity 生命周期的理解。重点和难点是：

(1)掌握多个 Activity 的创建和布局设计的知识；
(2)了解多个 Activity 的相互跳转的实现；
(3)了解 Log 日志输出和 LogCat 日志查看器的基础知识；
(4)掌握在多个 Activity 的生命周期回调方法的应用。

任务准备

1. 多个 Activity 的创建

在参考项目一创建项目的方法创建一个新项目后，会自动创建一个名称是 MainActivity 的 Activity。如果要创建多一个 Activity，例如：要创建名称为 SecondActivity 的 Activity，方法如下：

STEP 1 单击【文件】菜单的【新建】，在弹出的快捷菜单中选择【类】，如图 4.21 所示，Eclipse 将弹出如图 4.22 所示的“新建 java 类”的对话框。

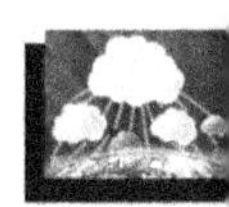

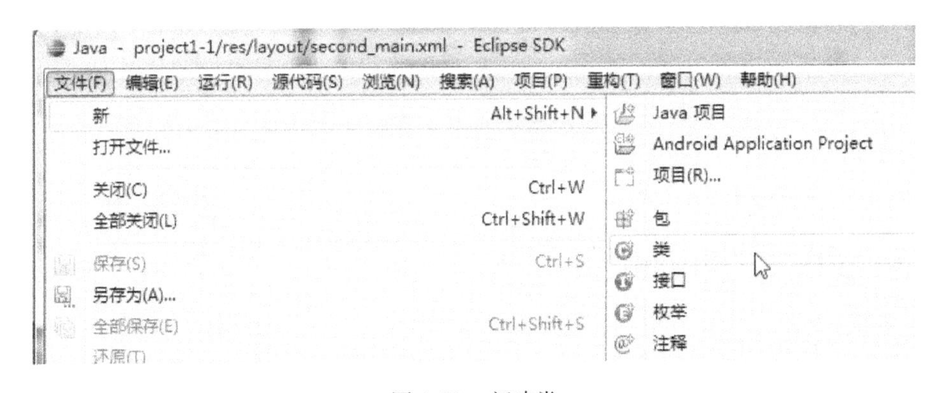

图 4.21　新建类

STEP 2 在弹出的“新建 java 类”的对话框中的单击【包】栏目右侧的【浏览】按钮，打开如图 4.23 所示“选择包”的对话框，双击选择要添加 Activity 的文件夹(本例为：org. dglg. project1_1)，Eclipse 将选择双击的文件夹并退出“选择包”对话框。

STEP 3 在图 4.22 中的【名称栏】中输入名称：“SecondActivity”。

知识拓展

在创建类的过程中，类的名称必须大写字母开头，而不能用小写字母开头！

在创建 XML 文件过程中，XML 文件名称必须小写字母开头，而不能用大写字母开头！

STEP 4 单击在【超类】右边的【浏览】按钮，打开如图 4.24 所示的【选择超类】对话

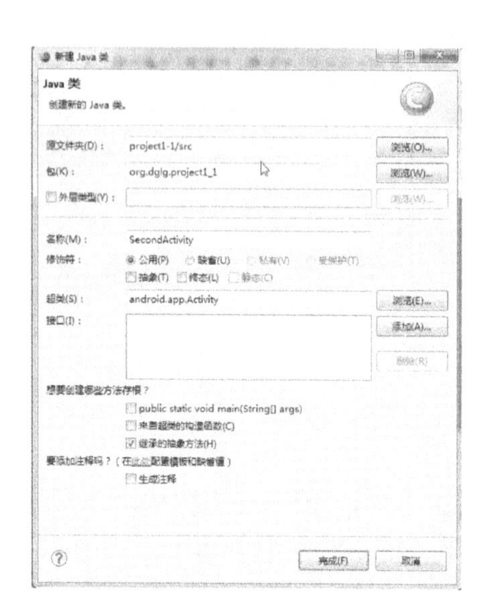

图 4.22　新建 java 类

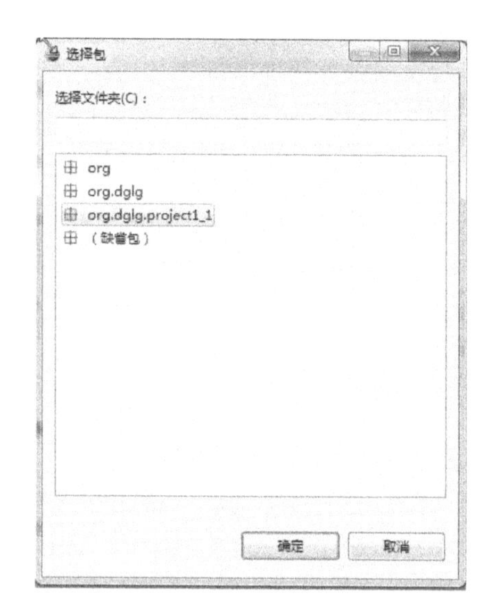

图 4.23　“选择包”对话框

框，在【选择类型】中输入“activity”，下方的【匹配项】中会自动列出所有的与在【选择类型】中输入的内容相匹配的类，本例中选择 Activity - Android. app，单击【确定】按钮确定选择。

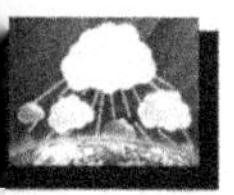

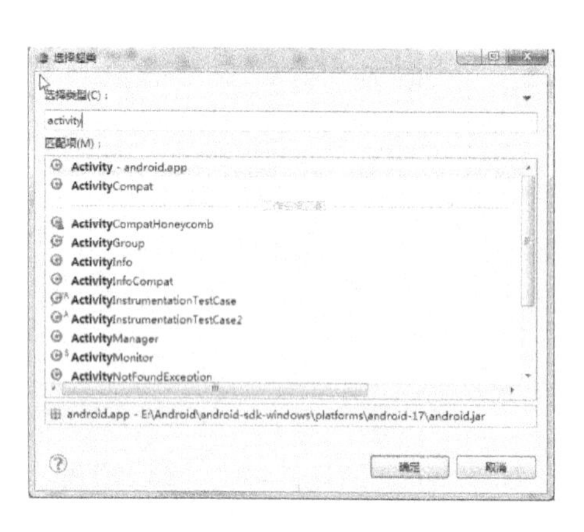

图 4.24　“选择超类”对话框

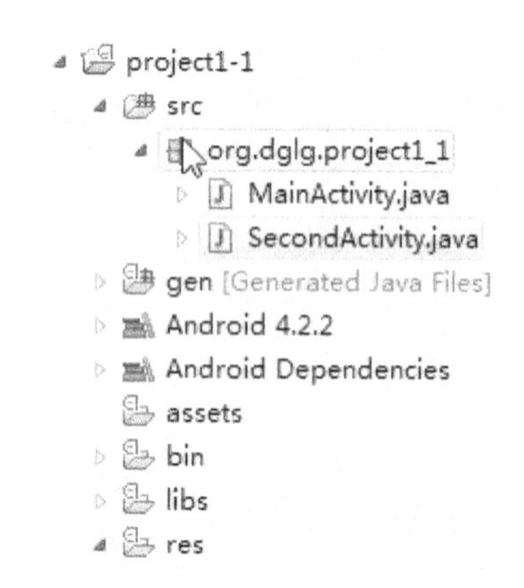

图 4.25　SecondActivity

STEP 5　单击【确定】按钮，完成 Activity 创建。此时在 MainActivity. java 下将新建一名为：SecondActivity. java 的类，如图 4. 25 所示。该类即为我们创建的 Activity。

2. 对创建的多个 Activity 进行布局设计

(1) 编写 strings. xml 定义字符串变量

对每个 Activity 布局文件中需要引用的字符串变量进行定义。

(2) 对第一个 Activity：MainActivity 进行布局

MainActivity 在创建时就自动新建名称为 activity_main. xml 的布局文件，可根据 MainActivity 的程序功能，对 activity_main. xml 进行布局。

(3)对第二个 Activity：SecondActivity 进行布局

在创建第二个 Activity 时，不会自动新建相应的布局文件。因此，要用户自行创建布局文件并与相应的 Activity 链接，这样所创建的布局文件才能成为相应的 Activity 布局文件，然后进行布局。例如：要为 SecondActivity 创建布局文件 second_main.xml，并与 SecondActivity 链接，然后进行布局，其步骤如下：

STEP 1 单击【文件】菜单的【新建】，在弹出的快捷菜单中选择【Android XML File】，将打开如图 4.26 所示的【New Android XML File】对话框。

在【New Android XML File】对话框中，对相关的各选项进行信息选择和输入：

【Resource Type】：选择资源类型，因为新建布局文件，所以选择“Layout”；

【Project】：选择项目，因为是在该项目中新建，所以选择默认即可；

【File】：输入 XML 文件名，此处输入 second_main.xml(首字母必须小写)；

【Root Element】：如果布局文件考虑采用线性布局，此处就选择“LinearLayout”。

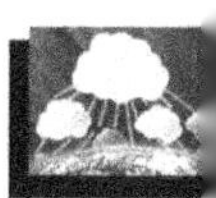

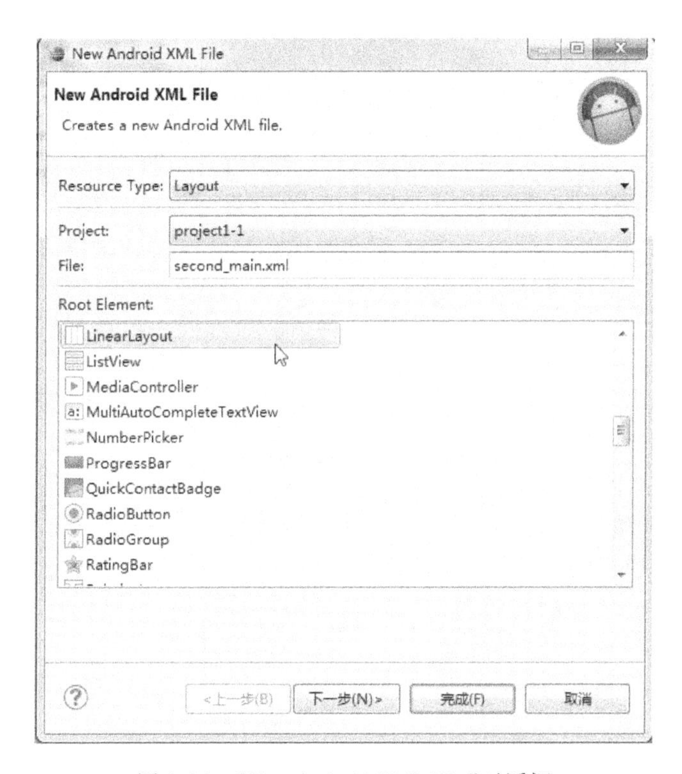

图 4.26 “New Android XML File”对话框

STEP 2 设定 Activity 与 XML 文件的引用链接，即 SecondActivity.java 与 second_main.xml 文件的链接，方法为在 SecondActivity.java 中 OnCreate()方法内添加如下代码：

```
setContentView(R.layout.second_main);
```

STEP 3 可根据 SecondActivity 的程序功能，对 second_main.xml 进行布局。

3. 多个 Activity 的相互跳转的实现

可以通过监听点击按钮事件，在自定义的点击按钮 onClick 函数内，添加 Intent 类实现从一个 Activity 跳转到另一个 Activity 的代码(详见项目 5 Intent 的内容)，再用 startActivity()启动相应的 Activity。例如：要实现 MainActivity 与 SecondActivity 的相互跳转，步骤如下：

STEP 1 实现从 MainActivity 跳转到 SecondActivity，代码片段如下：

```
Button MainActivitySubmit = (Button)findViewById(R.id.MainActivitySubmit); //获取"启动"按钮
MainActivitySubmit.setOnClickListener(new View.OnClickListener() //监听"启动"按钮
{@Override
public void onClick(View v)
{Intent MainIntent = new Intent(MainActivity.this, SecondActivity.class);
startActivity(MainIntent);    //启动 SecondActivity
}
});
```

STEP 2 实现从 SecondActivity 跳转到 MainActivity，代码片段如下：

```
Button SecondActivitySubmit = (Button)findViewById(R.id.SecondActivitySubmit); //获取"返回"按钮
SecondActivitySubmit.setOnClickListener(new View.OnClickListener() //监听"返回"按钮
{@Override
public void onClick(View v)
{Intent SecondIntent = new Intent(SecondActivity.this, MainActivity.class);
startActivity(SecondIntent);    //启动 MainActivity
}
});
```

4. Log 日志输出多个 Activity 的回调过程

可以使用 Log 类的 Log.d()方法，其在回调方法中重写的代码片段如下：

```
protected void onCreate(Bundle savedInstanceState) {
    Log.d(ActivityString,"onCreate"); //输出命令行的内容并在 LogCat 显示
    super.onCreate(savedInstanceState);
    setContentView(R.layout.activity_main);    }
    @Override
    protected void onStart() {
```

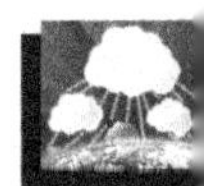

```
Log. d (ActivityString, "onStart");
super. onStart();   }
@ Override
protected void onResume() {
Log. d (ActivityString,"onResume");
super. onResume(); }
@ Override
protected void onPause() {
Log. d (ActivityString, "onPause");
super. onPause(); }
@ Override
protected void onRestart() {
Log. d (ActivityString , "onRestart");
super. onRestart(); }
@ Override
protected void onStop() {
Log. d (ActivityString ,"onStop");
super. onStop();   }
@ Override
protected void onDestroy() {
Log. d (ActivityString , "onDestroy");
super. onDestroy(); }
```

5. LogCat **日志查看器查看回调过程**

通过 LogCat 日志查看器查看 Activity 运行过程的日志信息。如果 LogCat 日志查看器未打开，那么可以参考本章任务一的步骤打开后再查看。

任务实现

1. **创建** Activity

本任务要创建 2 个 Activity，一个是 MainActivity，另一个是 ShowActivity。

参考本书项目 1 创建项目的方法，在 Eclipse 中新建 Android 项目名为 Project4_2 和包为 org. dglg. project4_2 的程序，在新建 Android 项目时会自动创建一个 Activity，本任务中 Activity 的名称是 MainActivity。

参考本章任务二"任务准备"部分第 1 项，创建新的 Activity 的方法，创建另一个名称是 ShowActivity 的 Activity。

2. **在** AndroidMinisfest. xml **文件中注册** ShowActivity

在 AndroidMinisfest. xml 文件，默认的 Activity 是 MainActivity，其他 Activity 要注册才能被激活。本任务中，对 ShowActivity 注册的代码如下：

```
    <activity
   android: label = "多个 Activity 跳转"
   android: icon = "@ drawable/ic_launcher"
   android: name = ". ShowActivity" >//"ShowActivity"前面的"."是必须的
    </activity>
```

3. 对 Activity 进行布局

1)对项目 Activity 的 strings. xml 进行布局

```
<? xml version = "1.0"  encoding = "utf-8"?  >
    <resources>
    <string name = "app_name" >多个 Activity 跳转</string>
    <string name = "TxtShowOne" >这是第一个 Activity</string>
    <string name = "TxtShowTwo" >这是第二个 Activity</string>
    <string name = "start" >进入</string>
    <string name = "back" >返回</string>
    </resources>
```

2)对 MainActivity 进行布局

本任务中，MainActivity 的布局文件是 activity_main. xml，用来实现界面的布局，代码如下：

```
<? xml version = "1.0"  encoding = "utf-8"?  >
    <LinearLayout xmlns: android = "http: //schemas. android. com/apk/res/android"
        android: layout_width = "match_parent"
        android: layout_height = "match_parent"
        android: orientation = "vertical"    >
    <TextView
        android: layout_width = "wrap_content"
        android: layout_height = "wrap_content"
        android: layout_marginTop = "30dp"
        android: layout_marginLeft = "60dp"
        android: text = "@ string/TxtShowOne"       />
    <Button
        android: id = "@ +id/GoBtn"
        android: layout_width = "wrap_content"
        android: layout_height = "wrap_content"
        android: layout_marginTop = "30dp"
        android: layout_marginLeft = "80dp"
```

```
    android: text = "@string/start"    />
</LinearLayout>
```

3)对ShowActivity进行布局

参考本章任务二"任务准备"部分第2项第3点，对第二个Activity创建布局文件的内容。为本任务的第二个Activity：ShowActivity创建布局文件show _main. xml，并设定ShowActivity. java与show _main. xml的引用链接，再进行布局，代码如下：

```
<? xml version = "1.0" encoding = "utf-8"? >
<LinearLayout xmlns: android = "http://schemas.android.com/apk/res/android"
    android: layout_width = "match_parent"
    android: layout_height = "match_parent"
    android: orientation = "vertical"    >
<TextView
    android: layout_width = "wrap_content"
    android: layout_height = "wrap_content"
      android: layout_marginTop = "30dp"
    android: layout_marginLeft = "60dp"
    android: text = "@string/TxtShowTwo"    />
<Button
    android: id = "@+id/BackBtn"
    android: layout_width = "wrap_content"
    android: layout_height = "wrap_content"
    android: layout_marginTop = "30dp"
    android: layout_marginLeft = "80dp"
    android: text = "@string/back"     />
</LinearLayout>
```

4. 实现Activity的程序功能

1)MainActivity的程序功能的实现

(1) MainActivity启动后，对onCreate()、onStart()、onPause()、onRestart()、onResume()、onStop()、onDestroy()这7个回调方法都进行重写后，为每个回调方法都添加log日志输出方法，并打印日志信息。

首先，定义私有变量MainActivityString赋值为"Project4_2 MainActivity ----"；代码如下：

```
private  String  MainActivityString = "project4_2 MainActivity ----";
```

其次，在MainActivity中对onCreate()进行重写时添加如下Log日志输出方法：

```
Log.d(MainActivityString, " start onCreate()");
```

这时，当回调 onCreate( )时，log 日志就会输出“Project4_2MainActivity - - - - start onCreate”。

其余的 6 个回调方法 onStart( )、onResume( )、onPause( )、onRestart( )、onStop( )、onDestroy( )都如 onCreate( )一样添加 Log 日志输出方法。

(2)通过获取并监听“进入”按钮，在自定义的点击按钮 onClick 函数内，添加 Intent 类实现从 MainActivity 跳转到 ShowActivity 的代码，再用 startActivity( )启动 ShowActivity。

2)MainActivity 的完整代码

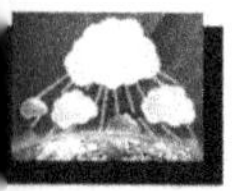

```
package org. dglg. project4_2;
//导入相关的类
import android. app. Activity;
import android. content. Intent;
import android. os. Bundle;
import android. view. View;
import android. widget. Button;
public class MainActivity extends Activity {
private   String   MainActivityString = "Project4_2MainActivity - - - - ";
    @ Override
    public void onCreate( Bundle savedInstanceState)
    {   super. onCreate( savedInstanceState) ;
      Log. d( MainActivityString, " start   onCreate( ) " ) ;
      setContentView( R. layout. activity_main) ; //载入 activity_main. xml 布局文件
      Button GoBtn  = ( Button) findViewById( R. id. GoBtn) ; //获取“进入”按钮
      GoBtn. setOnClickListener( new View. OnClickListener( )
      { @ Override
    public void onClick( View v)
    { Intent intent = new Intent( MainActivity. this, ShowActivity. class) ;
    startActivity( intent) ;    //启动 ShowActivity
    } } ) ; }
  @ Override
  protected void onStart( ) {
  Log. d ( MainActivityString, " start   onStart( ) " ) ;
  super. onStart( ) ;     }
  @ Override
  protected void onResume( ) {
  Log. d ( MainActivityString, " start   onResume( ) " ) ;
  super. onResume( ) ;          }
    @ Override
    protected void onPause( ) {
```

```
Log. d (MainActivityString ," start  onPause()");
super. onPause();    }
@ Override
protected void onRestart() {
Log. d (MainActivityString, " start  onRestart()");
super. onRestart();    }
@ Override
protected void onStop() {
Log. d (MainActivityString, " start  onStop()");
super. onStop();           }
@ Override
protected void onDestroy() {
Log. d (MainActivityString, " start  onDestroy()");
super. onDestroy();        }}
```

3)ShowActivity 的程序功能的实现

(1)参考本任务中 MainActivity 程序功能的实现，对 ShowActivity 的 onCreate()、onStart()、onPause()、onRestart()、onResume()、onStop()、onDestroy()这 7 个回调方法都进行重写后，为每个回调方法都添加 Log 日志输出方法。

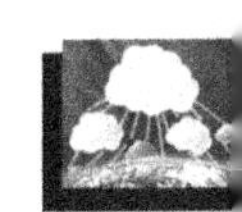

(2) 通过获取并监听“返回”按钮，在自定义的点击按钮 onClick 函数内，添加 Intent 类实现从 ShowActivity 跳转到 MainActivity 的代码，再用 startActivity()启动 MainActivity。

(3)ShowActivity 的程序功能完整代码如下。

```
package org. dglg. project4_2;
// 用 import 导入相关的类
import android. app. Activity;
import android. content. Intent;
import android. os. Bundle;
import android. widget. TextView;
public class ShowActivity extends Activity {
private String ShowActivityString = "Project4_2ShowActivity - - - -";
@ Override
protected void onCreate(Bundle savedInstanceState) {
super. onCreate(savedInstanceState);
Log. d(ShowActivityString, " start onCreate()");
setContentView(R. layout. show_main); ///载入 show_main. xml 布局文件
Button BackBtn = (Button)findViewById(R. id. BackBtn); //获取“返回”按钮
BackBtn. setOnClickListener(new View. OnClickListener()
{ @ Override
```

```
public void onClick(View v)
{Intent intent = new Intent(ShowActivity.this, MainActivity.class);
startActivity(intent); //启动 MainActivity
}});
  @Override
protected void onStart() {
Log.d (ShowActivityString ," start   onStart()");
super.onStart(); }
@Override
protected void onResume() {
Log.d (ShowActivityString , " start   onResume()");
super.onResume(); }
@Override
protected void onPause() {
Log.d (ShowActivityString ," start   onPause()");
super.onPause(); }
@Override
protected void onRestart() {
Log.d (ShowActivityString , " start   onRestart()");
super.onRestart(); }
@Override
protected void onStop() {
Log.d (ShowActivityString, " start   onStop()");
super.onStop(); }
@Override
protected void onDestroy() {
Log.d (ShowActivityString, " start   onDestroy()");
super.onDestroy(); }
}
```

任务考核

参考本任务，创建一个新的项目名为 ProjectTest4_2，在其自动创建 MainActivity 后，新建另一个名为 Test4_2Activity 的 Activity，然后利用按钮事件实现 MainActivity 与 Test4_2Activity 的相互跳转。再分别在 MainActivity 与 Test4_2Activity 中对各自的 Activity 生命周期的回调方法进行重写，并使用 Log.d() 方法对各自的 Activity 的生命周期的回调方法过程的 Log 日志打印输出，在 LogCat 日志查看器中显示各自的 Activity 的生命周期的回调方法的过程和次序。

项目五

走进Android的意图——Intent

Intent的中文意思是“意图”，在Android中Intent是在程序运行过程中连接两个或两个以上不同的组件。应用程序通过 Intent 向 Android 系统发出某种请求信息，Android 根据请求信息内容选择能够处理该请求的组件。本项目将走进Android的意图——Intent，帮助读者了解Android中Intent的基本知识及应用。

了解 Intent 的基本知识
掌握 Intent 的开发和应用
掌握利用 Intent 实现多个 Activity 间的跳转
掌握 Activity 之间参数传递机制
掌握并利用 Intent 实现拨号功能

任务一　Intent + Bundle 实现简易用户注册程序

任务描述

本任务主要是实现一个简单的用户注册程序，主要任务是对 Activity 中的 EditText 输入的数据进行有效性判断，并用 Intent 实现 Activity 跳转，用 Bundle 将数据从第一个 Activity 传递到第二个 Activity，如图 5. 1 ～图 5. 4 所示。

任务分析

在本任务中，共有 2 个 Activity。2 个 Activity 布局差不多，都是 1 个 TextView 和 1 个 Button，TextView 用来显示哪个 Activity，Button 用来跳转 Activity。

本任务的主要目的是通过 Intent 来实现 Activity 间的相互跳转，重点和难点如下：

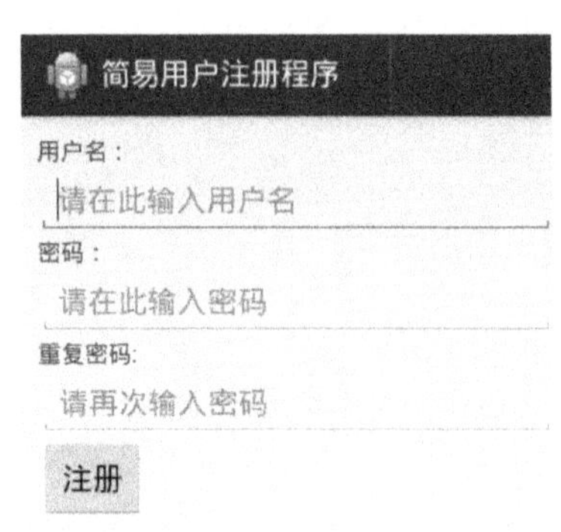

图 5.1　运行界面(一)

图 5.2　运行界面(二)

图 5.3　运行界面(三)

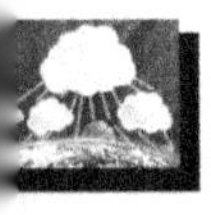

简易用户注册程序

您好，欢迎您的注册，请记住:
您的用户名是:admin
密码是:admin

图 5.4　运行界面(四)

(1)了解 Intent 基础知识；
(2)掌握多个 Activity 的创建和布局设计；
(3)掌握利用 Intent 来实现 Activity 间的相互跳转；
(4)掌握 Android 中数据有效性判断方法；
(5)掌握 Toast 信息提示；
(6)掌握利用 Bundle 实现多个 Activity 间的参数传递。

任务准备

1. Intent 介绍

Intent 的中文意思是“意图”，在 Android 中，Intent 是不同组件之间通信的一个“中介”，Android 开发者要做的就是：把你的意图告诉 Intent，其他的交给 Intent。

举个例子：有个房客想租房，于是他找了个中介，把他的要求告诉中介，然后中介按照他的要求去物色房子。在这里房客可以理解成一个源组件如 Activity，中介可以理解成 Intent，物色到的房子就是目标组件。通过这个例子可以进一步理解 Intent：我们只要把自己的意图(要求)告诉 Intent，其他的事情就不用管了，交给 Intent 来做。

Intent 是一种数据结构，在 Android 中被用来抽象描述一次将要被执行的操作，其作用是在程序运行过程中连接不同的 Android 组件。在 Android 系统中，应用程序通过 Intent 向 Android 系统发出某种请求信息，Android 根据请求的信息选择能够处理该请求的组件。

例如：拨打电话时，当按下拨号键就会向 Android 系统发送一个具有 CALL_BUTTON 行为的 Intent 对象。Android 系统根据该请求信息，找到能够处理该请求的电话号码拨号程序。当输入电话号码并再次按下拨号发送键时，拨号程序会将一个包含 ACTION_CALL 和电话号码数据的 Intent 请求发送给 Android 系统，Android 系统将查找合适的应用程序进行处理。

2. Intent 能做什么

Intent 可以用来启动一个 Activity，也可以启动一个 Service 服务，还可以发起一个 Broadcast 广播。具体方法如表 5.1 所示。

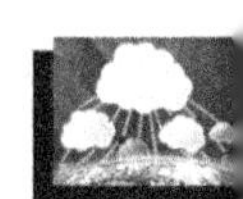

表 5.1 Intent 启动组件方法

启动组件名称	方法名称
Activity	startActivity( ) startActivityForResult( )
Service	startService( ) bindService( )
Broadcast	sendBroadcast( ) sendOrderedBroadcast( ) sendStickyBroadcast( )

3. Intent 对象组成

使用 Intent 能绑定应用程序代码，这样可以大大降低不同代码之间的耦合性，减少代码量，提高程序的集成性。Intent 封装了它要执行动作的属性，所以只要告诉 Intent 做什么就行了。

在 Android 中通常使用六个主要组成部分来抽象描述请求信息，它们分别是组件名称(ComponentName)、行为(Action)、数据(Data)、类别、附加信息和标示位。

1)组件名称(ComponentName)

Intent 对象使用组件名称(ComponentName)来描述传递消息的目标组件，这种明确指定目标组件名称的 Intent 称之为显式 Intent，系统会将显式 Intent 直接发送给目标组件。由于显式 Intent 需要指定目标组件，所以其更多的应用在内部组件间消息的传递。

组件名称通过 setComponent( )、setClass( )、setClassName( ) 设置，通过 getComponent( ) 获得。

如利用 Intent 实现 Activity 跳转的代码中，直接指定要跳转的目标组件的名称(下面代码为 OtherActivity. class)即可：

```
Intent intent = new Intent( );
intent. setClass( MainActivity. this, OtherActivity. class);
startActivity( intent);
```

2) 行为(Action)

Intent 行为(Action)是 Intent 要完成的动作，也即 Intent 要做的事情。Action 记录了 Intent 消息，是将要执行的行为字符串或者将要广播的行为字符串。

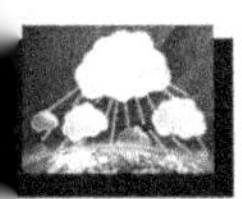

表 5.2 Intent 部分行为对应的目标组件和动作

行　为	目标组件	动　作
ACTION_CALL	activity	拨号动作
ACTION_DIAL	activity	显示拨号面板
ACTION_SEND	activity	发送短信
ACTION_EDIT	activity	向用户显示数据以供其编辑的动作
ACTION_MAIN -	activity	作为 task 中的初始 activity 启动
ACTION_BATTERY_LOW	broadcastReceiver	提醒手机电量过低
ACTION_SCREEN_ON -	broadcast	开启屏幕

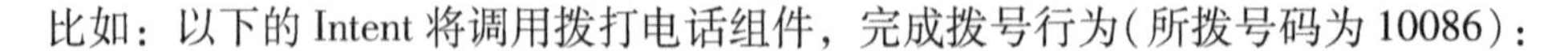

比如：以下的 Intent 将调用拨打电话组件，完成拨号行为(所拨号码为 10086)：

```
Uri uri = Uri. parse( "tel: 10086" ); // 参数分别为调用拨打电话组件的 Action 和获取 Data 数据的 Uri
Intent intent = new Intent( Intent. ACTION_DIAL, uri);
startActivity( intent);
```

知识拓展

Android 系统在 Intent 类中以静态常量的形式预先定义了一系列与系统有关的行为，而且同样支持使用自定义行为字符串来触发应用中其他的组件。

注意：自定义行为字符串需要使用应用的全包名形式命名。

3) 数据(Data)

数据(Data)属性是执行 Action 所需的信息，这些信息以 Uri 的形式存储在 Intent 中，不同的 Action 有不同的 Data 指定。表 5.3 列出了一些常见的 Intent 的 Action 和 Data 的使用。

表 5.3 常见的 Intent 的 Action 和 Data

Action 属性	Data 属性	说 明
ACTION_VIEW	content：// contacts/people/1	显示 id 为 1 的联系人信息
ACTION_CALL	tel：10086	拨号：10086
ACTION_VIEW	http：//www. dglg. net	打开网页
ACTION_VIEW	file：///sdcard/music. mp3	播放 sd 卡下的 mp3 文件

在上述的拨号行为中，ACTION_CALL(拨号行为)需要操作的电话号码(10086)就是以"tel：//number"的形式保存("tel：10086")。以下代码完成拨号：10086。

```
Intent intent = new Intent( );
intent. setAction(Intent. ACTION_CALL);
intent. setData(Uri. parse("tel：10086");
startActivity(intent);
```

名词解释

Uri 即通用资源标识符，Universal Resource Identifier，简称"URI"。

Uri 代表要操作的资源数据，Android 上各种资源：如图像、视频片段、音频资源等都可以用 Uri 来表示。

Uri 一般由三部分组成：

① 访问资源的命名机制。

② 资源自身的名称，由路径表示。

③ 存放资源的主机名。

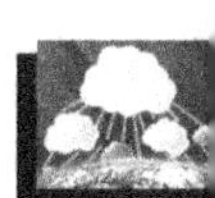

4)类别

类别属性主要描述 Intent 中被请求组件或执行行为动作的额外信息。Android 系统为类别定义了一系列的静态常量字符串来表示 Intent 不同类别。

如：android. intent. category. LAUNCHER 表示目标组件是应用程序中最优先被执行的组件。

5)附加信息

Android 为组件提供的扩展信息或额外的数据都保存在 Intent 对象的附加信息中。

附加信息采用键值对的结构，以 Bundle 对象的形式保存在 Intent 当中。附加信息其实是一个类型安全的容器，其实现就是将 HashMap 做了一层封装。与 HashMap 不同点在于，HashMap 可以将任何类型数据保存进去，值可以为任何的 java 对象。而附加信息虽采用键值对，但值只能保存基本数据对象类型，如 char、boolean、string 等。

6)标示位

标示位主要标示如何触发目标组件以及如何看待被触发的目标组件。例如用标示位标示被触发的组件应该属于哪一个任务或者触发的组件。标示位可以是多个标示符的组合。

4. Android 数据有效性判断

在本任务中，涉及到对 EditText 中输入的文本进行判断。在一般的程序语言中，我们判断数据是否相等一般用“= =”或者“! =”来判断，但在 Android 中不能简单地用“= =”或者“! =”来判断，因为 String 是引用类型的，不是基本数据类型，所以它们的比较是使用地址和值(相当于 C 中的指针)来比较的。因为它们是不同的对象，有不同的地址，所以 str1! =str2 永远都是 true，而 str1 = = str2 永远是 false。如果使用“= =”或者“! =”来判断的话，它们为真的条件是：不仅要求是同一对象，而值也要求相等。

在 Android 中一般使用语句 str1. equals(str2)来进行是否相等的判断，str1 为第一个字符串形式，str2 为第二个字符串形式。因此判断 EditText 输入是否为“admin”(假定该 EditText 的 id 为 yhtext)的代码如下：

```
mEditText = (EditText)findViewById(R. id. yhtext); //实例化 EditText
mEditText. getText( ). toString( ). equals("admin"); //判断 EditText 输入的文本是否为“admin”
```

思考

如何判断 EditText 中是否输入内容?

5. Toast 信息提示

Toast 英文含义是吐司，它就像烘烤机里做好的吐司弹出来，并持续一小段时间后慢慢消失。

Toast 是一个 View 视图，快速的为用户显示少量的信息。Toast 主要用在应用程序上浮动显示信息给用户，它永远不会获得焦点，不影响用户的输入等操作，主要用于一些帮助和提示。

Toast 是最常用、最简单的 Android 控件之一，其自动关闭的功能大大简化了代码量。Toast 最常见的创建方式是使用 Toast. makeText 方法。使用 Toast 时可以用工厂方法，它会自动构建一个带边框和文字的 Toast，格式和代码如下：

```
/* 利用工厂方法构造一个简单的 Toast */
Toast. makeText(this, "这是一个 Toast 提示", Toast. LENGTH_LONG). show();
```

显示效果如图 5. 5 所示。

这是一个Toast提示

图 5. 5　Toast 提示

参数解析：

Toast. makeText 方法涉及到 3 个参数：

第一个参数：当前的上下文环境。可用 getApplicationContext()或 this。

第二个参数：要显示的字符串。

第三个参数：显示的时间长短。Toast 默认的有两个

LENGTH_LONG(长)和 LENGTH_SHORT(短)。

6. Bundle 介绍

Bundle 是一个字符串值到各种 Parcelable 类型的映射，可用于保存数据包。在 Android 中可将要保存的数据存放在 Bundle 对象中，Bundle 类用作携带数据的对象，提供了如 putString( )、getString( )、putInt( )、getInt( )的 putXxx( )、getXxx( )类型的方法。通常 putXxx( )用于向 Bundle 对象放入数据，getXxx( )方法用于从 Bundle 对象里获取数据。

Bundle 的内部实际上是使用了 HashMap < String, Object >类型的变量来存放 putXxx( )方法放入的值，可通过如下代码实现。

```
publicfinal class Bundle implements Parcelable, Cloneable {
Map < String, Object > mMap; // 定义 HashMap < String, Object >类型的变量 mMap
public Bundle( ) {
mMap = new HashMap < String, Object > ( );
}
public void putString( String key, String value) {
mMap. put( key, value);
}
publicString getString( String key) {
    Object obj = mMap. get( key);
    return (String) obj;
}
}
```

在调用 Bundle 对象的 getXxx( )方法时，该方法内部会从该变量中获取数据，然后对数据进行类型转换，转换成什么类型由方法的 Xxx 决定，getXxx( )方法会把转换后的值返回。

7. Bundle 实现 Activity 间的数据传输

当 Activity 间要传输一些数据时，通过 Intent 将数据保存起来，就可以实现 Activity 间数据的传输。而将数据添加到 Intent，可以有以下四种方法来实现。

(1)批量添加数据到 Intent 的方法，其代码片断如下。

```
Intentintent = new Intent( );
Bundle bundle = new Bundle( ); //实例化一个携带数据的 bundle
bundle. putString( "name" ,"Project5_1" ); //将字符串"Project5_1"以"name"的名字
封装并存放在 bundle 中
intent. putExtras( bundle); //为 Intent 追加额外的数据，原来已经携带的数据不会
丢失
```

(2)把数据一个个地添加进 Intent 的方法，只需要编写如下少量代码。

```
Intent intent = new Intent();
intent.putExtra("name"," Project5_1");
```

(3)Activity间的数据传递：通过以下代码片段，可以实现将字符串"Project5_1"以"name"的名字封装在Bundle中，从名称为Main的Activity传输到名称为Second的Activity。

```
Intent intent = new Intent(Main.this, Second.class);
Bundle bundle = new Bundle();
bundle.putString("name","Project5_1");
intent.putExtrats(bundle);
startActivity(intent);
```

(4)接收端数据接收：接收端数据接收比较简单，只要知道传递端的封装名就可以接收了，如在上面是用"name"封装并传输的，接收端的代码如下。

```
Bundle bundle = this.getIntent().getExtras();
String JsName = bundle.getString("name");
```

任务实现

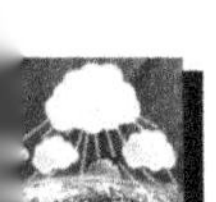

1. 新建 Android 项目

在Eclipse中，新建一Android应用程序。

2. 添加新 Activity

本任务涉及到两个Activity，所以需要在第一个MainActivity的基础上，添加第二个Activity，方法参考项目4。添加一名为ShowActivity的Activity，该Activity主要用来接收并显示从MainActivity传来的参数(用户名和密码)。

3. 对 ShowActivity 进行注册

在配置文件AndroidMainfest.xml里加上如下代码(注意代码添加位置，一般添加在</application>代码前)：

```
<activity android:name=".ShowActivity" // ShowActivity为新添加的Activity名，注意前面的"."
</activity>
```

4. 对 Activity 进行布局

1)编写strings.xml定义字符串变量

对布局文件中要引用的字符串变量进行定义，代码如下：

```
<? xml version = "1.0" encoding = "utf - 8"? >
<resources >
    <string name = "app_name" >简易用户注册程序 </string >
    <string name = "action_settings" > Settings </string >
    <string name = "TxUsername" >用户名： </string >
    <string name = "TxPassword" >密码： </string >
    <string name = "TxtLogin" >注册 </string >
    <string name = "InputUsername" >请在此输入用户名 </string >
    <string name = "InputPassword" >请在此输入密码 </string >
    <string name = "TxPasswordb" >重复密码： </string >
    <string name = "InputPasswordA" >请再次输入密码 </string >
<string name = "showdl" >注册信息 </string >
</resources >
```

2）对 MainActivity 布局：activity_main. xml

根据程序功能，MainActivity 布局文件为：activity_main. xml。可采用相对布局，代码如下：

```
<LinearLayout xmlns: android = "http: //schemas. android. com/apk/res/android"
    android: layout_width = "fill_parent"
    android: layout_height = "fill_parent"
    android: layout_marginLeft = "10dp"
    android: layout_marginTop = "10dp"
    android: orientation = "vertical" >
    <TextView
        android: layout_width = "wrap_content"
        android: layout_height = "wrap_content"
        android: text = "@ string/TxUsername" />
    <EditText
        android: id = "@ +id/UserName"
        android: layout_width = "match_parent"
          android: layout_height = "wrap_content"
          android: hint = "@ string/InputUsername"
        />
    <TextView
        android: layout_width = "wrap_content"
        android: layout_height = "wrap_content"
        android: text = "@ string/TxPassword" />
```


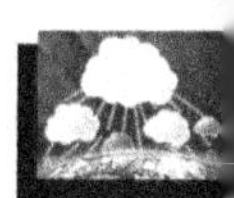

```
        <EditText
            android:id="@+id/FirstPwd"
            android:layout_width="match_parent"
            android:layout_height="wrap_content"
            android:inputType="textWebPassword"
            android:hint="@string/InputPassword"
          />
        <TextView
            android:layout_width="wrap_content"
            android:layout_height="wrap_content"
            android:text="@string/TxPasswordb" />
        <EditText
            android:id="@+id/SecondPwd"
            android:layout_width="match_parent"
            android:layout_height="wrap_content"
            android:inputType="textWebPassword"
            android:hint="@string/InputPasswordA"
          />
        <Button
            android:id="@+id/BtnZhuce"
            android:layout_width="wrap_content"
            android:layout_height="wrap_content"
            android:text="@string/TxtLogin" />
    </LinearLayout>
```

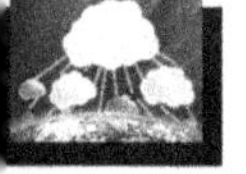

3）对 ShowActivity 布局：show_main. xml

show_main. xml 布局文件比较简单，只涉及到一个 TextView 用来显示接收到的数据，代码如下：

```
<RelativeLayout xmlns:android="http://schemas.android.com/apk/res/android"
xmlns:tools="http://schemas.android.com/tools"
android:layout_width="fill_parent"
android:layout_height="fill_parent"
android:orientation="vertical"
android:padding="10dp" >
<TextView
android:id="@+id/ShowDetails"
android:layout_width="wrap_content"
```

```
android: layout_height = "wrap_content"
android: textIsSelectable = "true"
android: text = "@ string/showdl"
/>
</RelativeLayout>
```

布局效果如图 5. 6 与图 5. 7 所示。

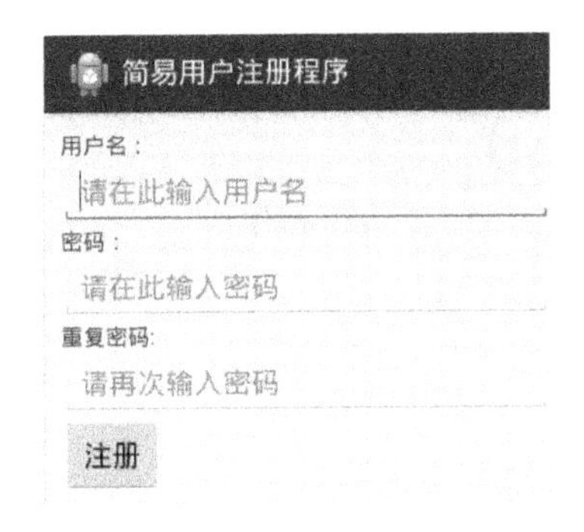

图 5. 6　MainActivity 布局效果

图 5. 7　ShowActivity 布局效果

5. 主程序功能实现

根据程序实现的功能以及 UI 界面的设计，先引用相关的类，然后声明并取得 UI 界面中的控件变量：EditText 和 Button，并设定 Button 的 onClickListener 动作是对 EditText 中的数据进行有效性判断，并用 Bundle 传值到 ShowActivity 中。

MainActivity 完整代码如下：

```
package org. dglg. project5_1;
import android. os. Bundle;
import android. app. Activity;
import android. content. Intent;
import android. view. View;
import android. view. View. OnClickListener;
import android. widget. Button;
import android. widget. EditText;
import android. widget. Toast;
public class MainActivity extends Activity {
    /* 声明对象 */
    EditText UserName;
    EditText FirstPwd;
    EditText SecondPwd;
    Button BtnZhuce;
```

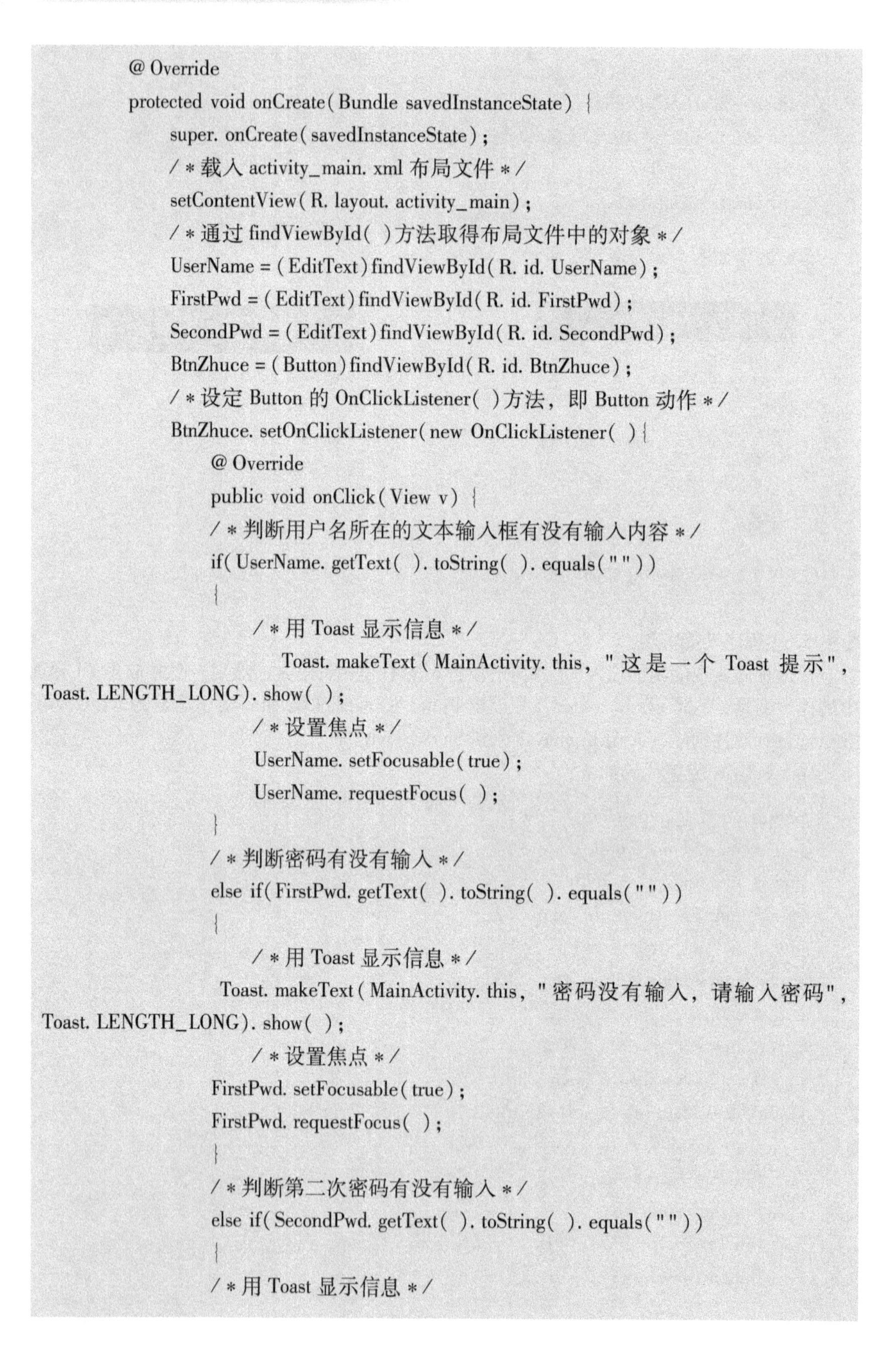

```
        @ Override
        protected void onCreate(Bundle savedInstanceState) {
            super. onCreate(savedInstanceState);
            /*载入 activity_main. xml 布局文件*/
            setContentView(R. layout. activity_main);
            /*通过 findViewById( )方法取得布局文件中的对象*/
            UserName = (EditText)findViewById(R. id. UserName);
            FirstPwd = (EditText)findViewById(R. id. FirstPwd);
            SecondPwd = (EditText)findViewById(R. id. SecondPwd);
            BtnZhuce = (Button)findViewById(R. id. BtnZhuce);
            /*设定 Button 的 OnClickListener( )方法，即 Button 动作*/
            BtnZhuce. setOnClickListener(new OnClickListener( ){
                @ Override
                public void onClick(View v) {
                /*判断用户名所在的文本输入框有没有输入内容*/
                if(UserName. getText( ). toString( ). equals(""))
                {
                    /*用 Toast 显示信息*/
                        Toast. makeText(MainActivity. this, "这是一个 Toast 提示",
Toast. LENGTH_LONG). show( );
                    /*设置焦点*/
                    UserName. setFocusable(true);
                    UserName. requestFocus( );
                }
                /*判断密码有没有输入*/
                else if(FirstPwd. getText( ). toString( ). equals(""))
                {
                    /*用 Toast 显示信息*/
                 Toast. makeText(MainActivity. this, "密码没有输入，请输入密码",
Toast. LENGTH_LONG). show( );
                    /*设置焦点*/
                FirstPwd. setFocusable(true);
                FirstPwd. requestFocus( );
                }
                /*判断第二次密码有没有输入*/
                else if(SecondPwd. getText( ). toString( ). equals(""))
                {
                /*用 Toast 显示信息*/
```

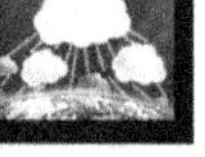

```
                    Toast. makeText( MainActivity. this, "密码没有确认，请确认密码",
Toast. LENGTH_LONG). show( );
                    /*设置焦点*/
                    SecondPwd. setFocusable(true);
                    SecondPwd. requestFocus( );
                    }
                    /*判断两次密码输入是否一致*/
                        else if (! FirstPwd. getText ( ) . toString ( ) . trim ( ) . equals
(SecondPwd. getText( ). toString( ). trim()))
                    {
                    /*用 Toast 显示信息*/
                    Toast. makeText(MainActivity. this, "两次密码输入不一样，请重新输
入", Toast. LENGTH_LONG). show( );
                    /*设置焦点*/
                    FirstPwd. setText("");
                    SecondPwd. setText("");
                    FirstPwd. setFocusable(true);
                    FirstPwd. requestFocus( );
                    }
                    else
                    {
                    /*定义字符串变量，并将文本输入框中的内容赋值给该变量*/
                        String StrUserName = UserName. getText( ). toString( );
                        String StrPassWord = FirstPwd. getText( ). toString( );
                        /*利用 Bundle 传值*/
                        Intent intent = new Intent( );
                        Bundle bundle = new Bundle( );
                        bundle. putString( "UserName" , StrUserName);
                        bundle. putString( "PassWord" , StrPassWord);
                            intent. putExtras(bundle);
                            intent. setClass(MainActivity. this, ShowActivity. class);
                            startActivity(intent);
                            /*关闭 Activity*/
                            MainActivity. this. finish( );
                        }}});}}
```

接收端 ShowActivity 代码如下：

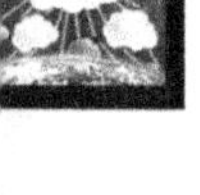

```
package org. dglg. project5_1;
/* 导入相关类 */
import android. os. Bundle;
import android. widget. TextView;
import android. app. Activity;
public class ShowActivity extends Activity {
    @Override
    protected void onCreate(Bundle savedInstanceState) {
        super. onCreate(savedInstanceState);
        /* 载入 activity_main. xml 布局文件 */
        setContentView(R. layout. show_main);
        /* 通过 findViewById( )方法取得布局文件中对象 */
        TextView ShowDetails = (TextView)findViewById(R. id. ShowDetails);
        /* 利用 Intent 接收 MainActivity 传来的参数 */
        Bundle bundle = this. getIntent(). getExtras();
        String JsUserName = bundle. getString("UserName");
        String JsPassWord = bundle. getString("PassWord");
        /* 利用 TextView 的 setText( )方法对 TextView 设置文本 */
        ShowDetails. setText("您好，欢迎您的注册，请记住:\n 您的用户名是:"
+ JsUserName + "\n 密码是:" + JsPassWord);
        }}
```

完成效果如图 5.8 与图 5.9 所示。

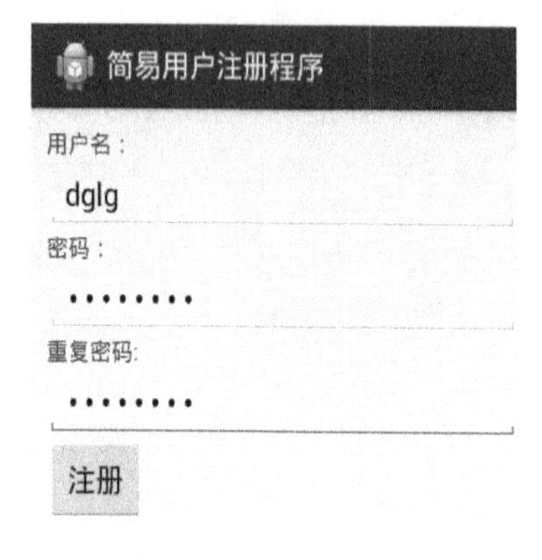

图 5.8　运行界面(一)

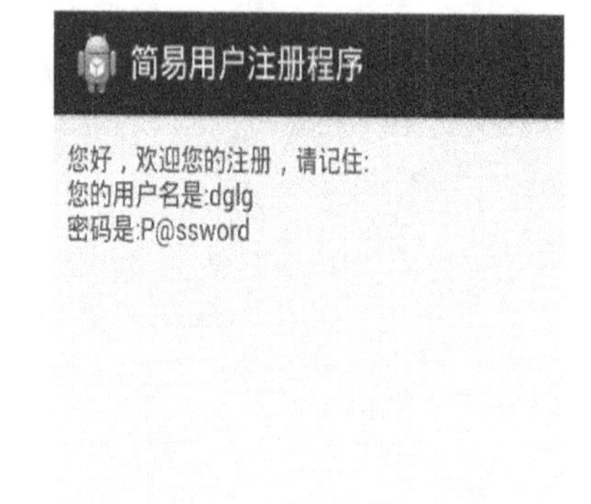

图 5.9　运行界面(二)

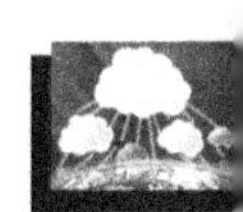

任务二　Intent 实现简易用户登录程序

任务描述

本任务主要制作一个简易用户登录程序。运行界面和效果如图 5. 10 ～图 5. 13 所示。

图 5. 10　MainActivity 运行界面(一)

图 5. 11　MainActivity 运行界面(二)

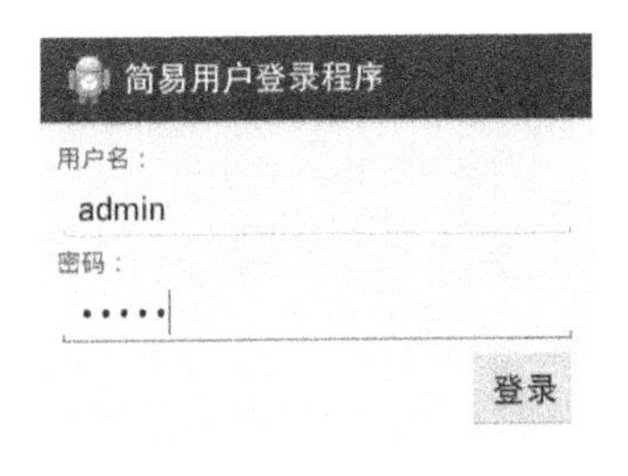

图 5. 12　MainActivity 运行界面(三)

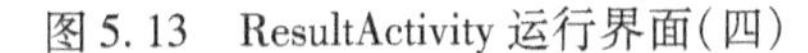

图 5. 13　ResultActivity 运行界面(四)

任务分析

在本任务中，主要是利用 Intent 实现 Activity 的传值(用户名和密码)。在 MainActivity 中，主要有如图 5. 10 ～图 5. 13 所示的登录界面，当用户在登录界面没有输入用户名或密码时，用 Toast 显示提示相应的信息，直到用户输入用户名和密码后，单击【登录】，跳转到 SecondActivity，SecondActivity 显示用户输入的用户名和密码(利用 Intent 实现传值)以及欢迎信息。

本任务的重点和难点如下：

(1)进一步了解 Intent;
(2)复习 Android 中数据有效性判断方法(本任务为判断用户名和密码是否输入);
(3)掌握利用 Intent 来实现 Activity 间参数传递。

任务准备

1. Intent 的声明

在 Intent 应用中，可以以两种形式来声明使用 Intent:

(1)显式 Intent：这种形式的 Intent 应用，指定了 component 属性 ，调用 setComponent (ComponentName)或者 setClass(Context，Class)来指定。通过指定具体的组件类，通知应用启动对应的组件。

(2)隐式 Intent：这种形式的 Intent 应用，没有指定 component 属性。所以这些 Intent 需要包含足够的信息，这样系统才能根据这些信息，在所有的可用组件中，选择确定满足 Intent 的组件。

2. Intent 的解析

根据 Intent 的声明方式不同，Android 解析方式也不大相同。

对于显式 Intent，因为目标组件已经很明确，Android 不需要去做解析。

Android 需要解析的是那些隐式 Intent，通过解析隐式 Intent 中的信息将 Intent 映射给可以处理此 Intent 的 Activity、Service 或 Broadcast Receiver。所以在隐式 Intent 声明中，需要包含足够的信息，这样系统才能根据这些信息在所有的组件中过滤 action、data 或者 category 来解析以匹配目标组件。

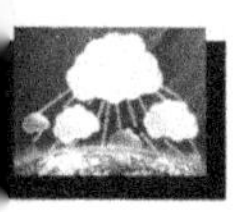

① 如果隐式 Intent 声明中指定了 action，则目标组件的 IntentFilter 的 action 列表中就必须包含有这个 action，否则不能解析。

② 如果隐式 Intent 没有提供 type，系统将从 data 中得到数据类型。和 action 一样，目标组件的数据类型列表中必须包含 Intent 的数据类型，否则不能解析。

③ 如果隐式 Intent 中的数据不是 content 类型的 URI，而且 Intent 也没有明确指定它的 type，将根据 Intent 中数据的 scheme (比如 http：或者 mailto：) 进行匹配。同上，Intent 的 scheme 必须出现在目标组件的 scheme 列表中。

④ 如果 Intent 指定了一个或多个 category，这些类别必须全部出现在组建的类别列表中。

3. Intent 实现 Activity 间通信

在任务一中，是直接 Activity 跳转而不需要通信。如果涉及到 Activity 间的通信即传值的话，要在发送端用到 Intent 的 putExtra( )方法传递参数，而在接收端用 getStringExtra( ) 方法来接收。

发送端(所传值为 fsnum，取自输入框 UserText)代码如下：

```
Sting fsnum = UserText. getText( ). toString( );
Intent intent = new Intent( );
Intent. putExtra( "one", fsnum); //将num1的变量值封装在“one”中
Intent. setClass( MainActivity, SecondActivity. class);
startActivity( intent);
```

接收端代码如下：

```
Intent intent = getIntent( ); //注意intent的定义方法，这样得到用于激活它的意图
String jsnum = intent. getStringExtra( "one"); //接收“one”传来的值，存于jsnum中
```

注意区别：发送端定义intent的方式与接收端定义intent的方式不一样。

```
发送端：Intent intent = new Intent( );
接收端：Intent intent = getIntent( );
```

知识提高

Intent实现传值，接收端接到的数值类型都为String(字符型)，所以如果涉及其他类型的变量，比如整型，需要在接收端转换。

如在上述代码中，可以用int jsint = Integer. parseInt(jsnum)代码来进行转换。

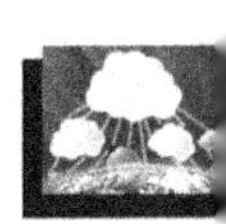

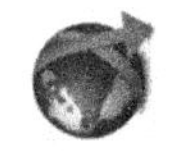

任务实现

本任务和任务一程序架构大致相仿，任务实现的步骤如下：

STEP 1 新建Android项目

在Eclipse中，新建一Android应用程序，方法可参考项目一。

STEP 2 添加新Activity：ResultActivity

参照任务一的方法，添加一新的Activity，命名为ResultActivity。

STEP 3 对ResultActivity进行注册

对ResultActivity进行注册，在配置文件AndroidMainfest. xml里加上如下代码：

```
<activity android: name = ". ResultActivity" </activity>
```

STEP 4 对Activity进行布局

1)编写strings. xml定义字符串变量

对布局文件中要引用的字符串变量进行定义，代码如下：

```
<? xml version = "1.0" encoding = "utf-8"? >
<resources>
    <string name = "app_name" >简易用户登录程序</string>
    <string name = "action_settings" >Settings</string>
    <string name = "TxUsername" >用户名: </string>
    <string name = "TxPassword" >密码: </string>
    <string name = "TxtLogin" >登录</string>
    <string name = "InputUsername" >请在此输入用户名</string>
    <string name = "InputPassword" >请在此输入用户名</string>
    <string name = "showdl" >登录信息</string>
</resources>
```

2)对 MainActivity 布局：activity_main. xml

根据程序功能，MainActivity 布局文件为：activity_main. xml。可考虑采用相对布局，代码如下：

```
<RelativeLayout xmlns: android = "http: //schemas. android. com/apk/res/android"
xmlns: tools = "http: //schemas. android. com/tools"
android: layout_width = "fill_parent"
android: layout_height = "fill_parent"
android: orientation = "vertical"
android: padding = "10dp" >
<TextView android: layout_width = "wrap_content"
android: layout_height = "wrap_content"
android: id = "@ +id/TxUsername"
android: text = "@string/TxUsername"/>
<EditText
android: id = "@ +id/UserName"
android: layout_width = "fill_parent"
android: layout_height = "wrap_content"
android: layout_below = "@id/TxUsername"
android: hint = "@string/InputUsername"
/>
<TextView
android: id = "@ +id/TxPassword"
android: layout_width = "wrap_content"
android: layout_height = "wrap_content"
android: layout_below = "@id/UserName"
android: text = "@string/TxPassword"/>
```

```
<EditText
android: id = "@ +id/PassWord"
android: layout_below = "@id/TxPassword"
android: layout_width = "fill_parent"
android: layout_height = "wrap_content"
android: inputType = "textWebPassword"
android: hint = "@string/InputPassword"/>
<Button
android: id = "@ +id/Login"
android: layout_width = "wrap_content"
android: layout_height = "wrap_content"
android: layout_gravity = "center_horizontal"
android: layout_below = "@id/PassWord"
android: layout_alignParentRight = "true"
android: text = "@string/TxtLogin" />
</RelativeLayout>
```

3)对 ResultActivity 布局：result_main. xml

(1)按任务一的方法对 ResultActivity 建立布局文件 result_main. xml；

(2)设定 Activity 与 XML 文件的引用链接，在 ResultActivity. java 中 OnCreate( )方法内添加如下代码：

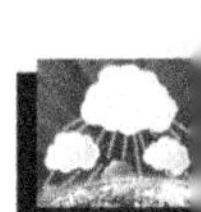

```
setContentView(R. layout. result_main);
```

(3)对 result_main. xml 进行布局，同 activity_main. xml 布局文件一样，result_main. xml 同样采用相对布局，代码如下：

```
<RelativeLayout xmlns: android = "http: //schemas. android. com/apk/res/android"
xmlns: tools = "http: //schemas. android. com/tools"
android: layout_width = "fill_parent"
android: layout_height = "fill_parent"
android: orientation = "vertical"
android: padding = "10dp" >
<TextView
android: id = "@ +id/ShowDetails"
android: layout_width = "wrap_content"
android: layout_height = "wrap_content"
android: textIsSelectable = "true"
android: text = "@string/showdl"
/>
</RelativeLayout>
```

(4) activity_main 与 result_main. xml 的布局效果如下：图 5. 14 为 activity_main 布局效果，图 5. 15 为 result_main. xml 布局效果。

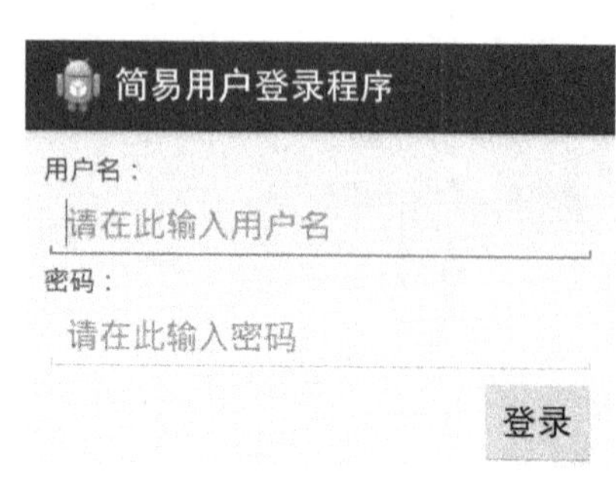

图 5. 14　activity_main 布局效果

简易用户登录程序

登录信息

图 5. 15　result_main. xml 布局效果

STEP 5 主程序功能实现

(1) MainActivity. java 程序文件编写：根据程序实现的功能以及 UI 界面的设计，引用相关的类，然后声明 UI 界面中的控件变量：EditText 和 Button，并设定 Button 的 onClickListener 动作实现功能：利用 Intent 实现 Activity 间传值通信。完整代码如下：

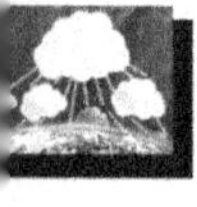

```
package org.dglg.project5_2;
/*导入相关类*/
import android.os.Bundle;
import android.app.Activity;
import android.content.Intent;
import android.view.View;
import android.view.View.onClickListener;
import android.widget.Button;
import android.widget.EditText;
import android.widget.Toast;
public class MainActivity extends Activity {
    /*声明对象*/
EditText UserName;
EditText PassWord;
Button LoginBtn;
    @Override
    protected void onCreate(Bundle savedInstanceState) {
        super.onCreate(savedInstanceState);
        /*载入 activity_main.xml 布局文件*/
        setContentView(R.layout.activity_main);
        /*通过 findViewById()方法取得布局文件中的对象*/
        UserName = (EditText)findViewById(R.id.UserName);
```

```
        PassWord = (EditText)findViewById(R. id. PassWord);
        LoginBtn = (Button)findViewById(R. id. Login);
        /*设定 Button 的 OnClickListener( )方法，即 Button 动作*/
        LoginBtn. setOnClickListener(new OnClickListener( ){
            @Override
            public void onClick(View v) {
            /*判断用户名所在的文本输入框有没有输入内容*/
            if(UserName. getText( ). toString( ). equals(""))
            {
                /*用 Toast 显示信息*/
                Toast. makeText(MainActivity. this, "用户名没有输入，请输入用
户名", Toast. LENGTH_LONG). show( );
                /*设置焦点*/
                UserName. setFocusable(true);
                UserName. requestFocus( );
            }
            else if(PassWord. getText( ). toString( ). equals(""))
            {
              Toast. makeText(MainActivity. this, "密码没有输入，请输入密码",
Toast. LENGTH_LONG). show( );
                /*设置焦点*/
                PassWord. setFocusable(true);
                PassWord. requestFocus( );
            }
            else
            {
            /*定义字符串变量，并将文本输入框中的内容赋值给该变量*/
                String StrUserName = UserName. getText( ). toString( );
                String StrPassWord = PassWord. getText( ). toString( );
                /*利用 Intent 传值*/
                Intent intent = new Intent( );
                intent. putExtra("username", StrUserName);
                intent. putExtra("password", StrPassWord);
                intent. setClass(MainActivity. this, ResultActivity. class);
                startActivity(intent);
                /*关闭 Activity*/
                MainActivity. this. finish( );
            }
```

```
                }
            });
        }
    }
```

上述程序首先通过findViewById( )方法从布局文件中取得对象，然后设定Button的事件。在Button事件内，首先判断用户名和密码的输入框EditText(电话号码)中有没有输入内容，如果没有则用Toast提示，如果有输入，则用Intent实现跳转，并将参数传给另一个Activity：ResultActivity。

(2)ResultActivity.java程序文件的编写：ResultActivity程序文件实现的功能比较简单，主要是接收MainActivity传递来的参数，完整代码如下：

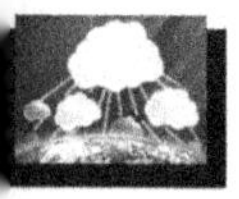

```
package org.dglg.project5_2;
/*导入相关类*/
import android.os.Bundle;
import android.widget.TextView;
import android.app.Activity;
import android.content.Intent;
public class ResultActivity extends Activity {
    @Override
    protected void onCreate(Bundle savedInstanceState) {
        super.onCreate(savedInstanceState);
        /*载入activity_main.xml布局文件*/
        setContentView(R.layout.result_main);
        /*通过findViewById( )方法取得布局文件中对象*/
        TextView ShowDetails = (TextView)findViewById(R.id.ShowDetails);
        /*利用Intent接收MainActivity传来的参数*/
        Intent intent = this.getIntent( );
        String JsUserName = intent.getStringExtra("username");
        String JsPassWord = intent.getStringExtra("password");
        /*利用TextView的setText( )方法对TextView设置文本*/
        ShowDetails.setText(JsUserName + "您好，欢迎您的登录，请记住您的密码
是:" + JsPassWord);
        }
    }
```

上述程序首先通过findViewById( )方法从布局文件中取得TextView对象，然后通过Intent接收从MainActivity传来的参数(用户名和密码)，并将相应的信息显示在TextView对象上。

任务考核

任务提高：在本任务中，对登录的用户名进行判断。要求如下：

1. 如果用户名为“admin”，则在ResultActivity中用Toast提示“管理员您好，欢迎您的登录”；

2. 如果输入其他用户名，则在ResultActivity中用Toast提示“您好，欢迎您的登录”。

任务三　简易拨号程序

任务描述

本任务主要是利用Intent实现拨号，单击拨号按钮后，将对EditText中输入的电话号码进行拨号(拨号前进行简易判断：EditText是否有内容，如果没有内容，则用Toast提示；如果有内容，则对内容进行拨号)。程序运行界面如图5.16、图5.17所示。

图5.16　程序界面

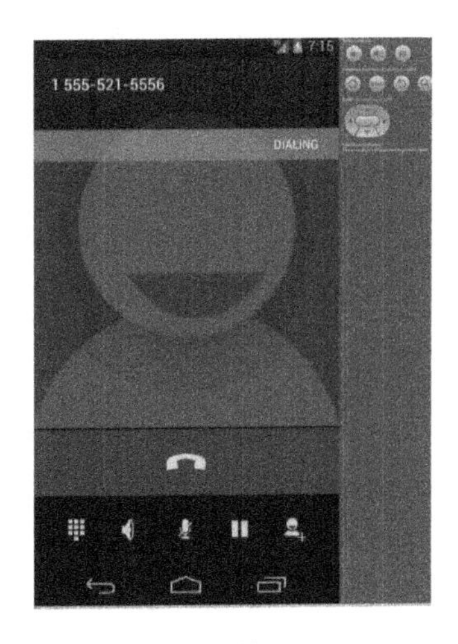

图5.17　拨号界面

任务分析

在本任务中，定义了1个TextView、1个EditText和1个Button。TextView用来说明(输入电话号码)，EditText和Button分别用于输入电话号码和实现拨号。

本任务的主要目的是通过Intent实现拨号功能，重点和难点如下：

(1)进一步了解Intent；
(2)熟悉Button事件；
(3)如何取得拨号权限；
(4)掌握利用Intent来实现拨号功能。

任务准备

1. 制作分析

本任务涉及的控件和功能比较简单。控件有EditText输入框和Button按钮，输入框中用来输入要拨号的电话号码，Button按钮用来拨号，在实现本任务功能方面，可以考虑用Intent来实现拨号功能。

2. 获得拨号权限

在Android开发中，如果要实现拨号功能，需要在androidManifest.xml注册，调用Android电话拨号的权限，才能拨号，否则程序将报错。

取得拨号权限，需要在androidManifest.xml中的<application>之前添加代码如下：

```
<uses-permission android:name="android.permission.CALL_PHONE"/>//取得拨号权限;
```

知识拓展

Android权限是一种安全机制。主要用于限制应用程序内部某些具有限制性特性的功能使用，以及应用程序之间的组件访问。

比如在Android开发中，如果程序需要联网，需要加上联网所需要的权限：

```
<uses-permission android:name="android.permission.INTERNET" />
```

在开发过程中，如果运行错误且提示：java.lang.SecurityException：Permission Denial... 根据此错误提示，一般情况下，在AndroidManifest.xml中通过uses-permission增加上相应权限即可。

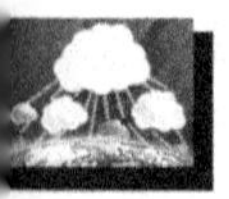

3. Intent实现拨号功能

在Android中，实现拨号功能最简单的方法是用Intent来实现，拨号功能代码如下(拨号：10086)：

```
Intent intent = new Intent( ); //定义新的intent
intent.setAction("android.intent.action.CALL"); //设置intent的Action为拨号
intent.setData(Uri.parse("tel:10086")); //设置拨号号码：10086
startActivity(intent); //开始拨号
```

知识提高

在 Intent 实现拨号功能时，可以考虑将号码存储在变量中，然后对拨号的 Uri 进行更改以实现任意号码拨号，可参考如下代码：

```
String PhoneNumber = "10086";
Intent intent = new Intent( );
intent. setAction("android. intent. action. CALL");
intent. setData(Uri. parse("tel: " + PhoneNumber));
startActivity(intent);
```

任务实现

本任务和任务一程序架构大致相仿，任务实现的步骤如下：

STEP 1 新建 Android 项目

在 Eclipse 中，新建一 Android 应用程序，方法可参考项目一。

STEP 2 对 MainActivity 进行布局

1）编写 strings. xml 定义字符串变量

对布局文件中要引用的字符串变量进行定义，代码如下：

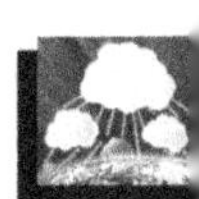

```
<? xml version = "1.0" encoding = "utf-8"? >
<resources>
    <string name = "app_name">简易拨号程序</string>
    <string name = "action_settings">Settings</string>
    <string name = "TxPhone">请输入电话号码: </string>
    <string name = "TxtCall">拨号</string>
</resources>
```

2）对 MainActivity 布局：activity_main. xml

根据程序功能，MainActivity 布局文件为：activity_main. xml。考虑采用线性布局，布局界面如图 3. 15 所示。代码如下：

```
<RelativeLayout xmlns: android = "http: //schemas. android. com/apk/res/android"
xmlns: tools = "http: //schemas. android. com/tools"
android: layout_width = "fill_parent"
android: layout_height = "fill_parent"
android: orientation = "vertical"
android: padding = "10dp" >
```

```
<TextView android: layout_width = "wrap_content"
android: layout_height = "wrap_content"
android: id = "@ +id/TxUsername"
android: text = "@string/TxPhone"/>
<EditText
android: id = "@ +id/InputPhoneNumber"
android: layout_width = "fill_parent"
android: layout_height = "wrap_content"
android: layout_below = "@id/TxUsername"
android: inputType = "phone"
android: hint = "@string/TxPhone"
/>
<Button
android: id = "@ +id/BtnCall"
android: layout_width = "wrap_content"
android: layout_height = "wrap_content"
android: layout_gravity = "center_horizontal"
android: layout_below = "@id/InputPhoneNumber"
android: layout_alignParentRight = "true"
android: text = "@string/TxtCall" />
</RelativeLayout>
```

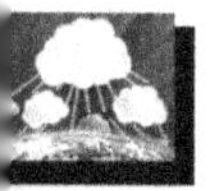

3)主程序功能实现

MainActivity. java 程序文件编写步骤：根据程序实现的功能以及 UI 界面的设计，引用相关的类，然后声明 UI 界面中的控件变量：EditText 和 Button，并设定 Button 的 onClickListener 动作实现功能，利用 Intent 实现拨号。完整代码如下：

```
package org.dglg.Project5_3;
/*导入相关类*/
import android.net.Uri;
import android.os.Bundle;
import android.app.Activity;
import android.content.Intent;
import android.view.View;
import android.view.View.OnClickListener;
import android.widget.Button;
import android.widget.EditText;
import android.widget.Toast;
public class MainActivity extends Activity {
```

```
    EditText InputPhoneNumber;
    Button BtnCall;
        @ Override
        protected void onCreate( Bundle savedInstanceState) {
            super. onCreate( savedInstanceState) ;
            /* 载入 activity_main. xml 布局文件 */
            setContentView( R. layout. activity_main) ;
            /* 通过 findViewById( )方法取得布局文件中的对象 */
            InputPhoneNumber = ( EditText) findViewById( R. id. InputPhoneNumber) ;
            BtnCall = ( Button) findViewById( R. id. BtnCall) ;
            /* 设定 Button 的 OnClickListener( )方法，即 Button 动作 */
            BtnCall. setOnClickListener( new OnClickListener( ) {
                @ Override
                public void onClick( View v) {
                // TODO Auto - generated method stub
                /* 取得文本框的值，并赋值给变量 */
                String PhoneNum = InputPhoneNumber. getText( ). toString( ) ;
                /* 首先判断电话号码有没有输入 */
                if( PhoneNum. equals( "" ) )
                {
                    /* Toast 提示相关信息 */
                        Toast. makeText ( MainActivity. this, "请输入电话号码",
Toast. LENGTH_LONG). show( ) ;
                    /* 设置焦点 */
                    InputPhoneNumber. setFocusable( true) ;
                    InputPhoneNumber. requestFocus( ) ;
                }
                else
                {
                    /* 拨号 */
                Intent intent = new Intent( ) ;
                intent. setAction( "android. intent. action. CALL" ) ;
                intent. setData( Uri. parse( "tel: " + PhoneNum) ) ;
                startActivity( intent) ;
                }
            }
        }) ;
    }
}
```

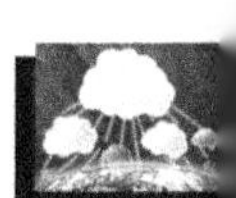

上述程序首先通过 findViewById( )方法从布局文件中取得对象，然后设定 Button 的事件。在 Button 事件内，首先判断输入框 EditText(电话号码)中有没有输入内容，如果没有，则用 Toast 提示用户输入，并设置焦点。如果有输入，则用 Intent 实现拨号。

STEP 4 程序运行

在模拟器中打开 2 个 AVD 窗口，其中一个运行本任务的拨号程序，所拨号码为另一个 AVD 窗口的号码(图例中的号码为：15555215554)，程序运行如图 5.18、5.19 所示。

图 5.18　拨号界面

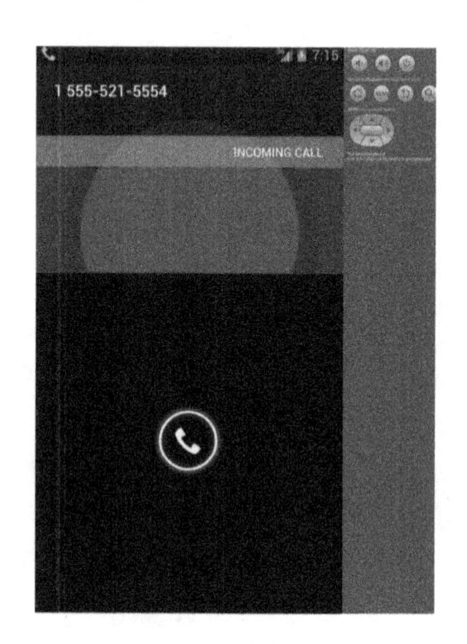

图 5.19　所拨号码界面

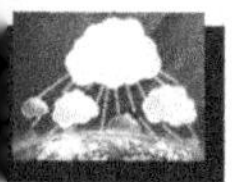

任务考核

任务提高：编写一简易短信发送程序，要求如下：

1. 2 个文本输入框，供输入短信发送号码和短信内容；
2. 1 个按钮，发送短信；
3. 界面简洁、美观。

项目六

走进Android的服务——Service

服务（Service）是Android的四大组件之一，是类似且与Activity同级的组件，它只能运用于后台，不能自己启动，也不能与用户交互。本项目将对Android服务进行多方位介绍，帮助读者了解Android中服务的基本知识以及应用。

熟悉 Service 组件的功能
熟悉 Service 组件的机制和用法
掌握用 Service 开发定时器功能程序方法
掌握 Service 组件开发流程和方法

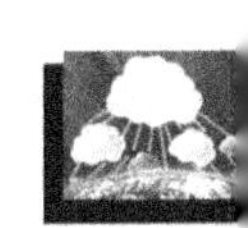

任务一　使用 Service 示例(一)

任务描述

本任务主要是使用 Service，包括创建 Service、利用 startService( )方法启动 Service、利用 stopService( )停止 Service，并结合模拟器的控制台的 LogCat 观察并熟悉 Service 的生命周期。运行界面和效果如图 6.1～图 6.3 所示。

图 6.1　使用 Service 示例运行界面

任务分析

在本任务中，完成的功能和效果主要是单击 Activity 中的“启动 Service”按钮，启动 Service，Service 没有实际功能，主要在控制台的 LogCat 上输出 Service 方法调用情况；单击 Activity 中的“结束 Service”按钮，结束 Service。

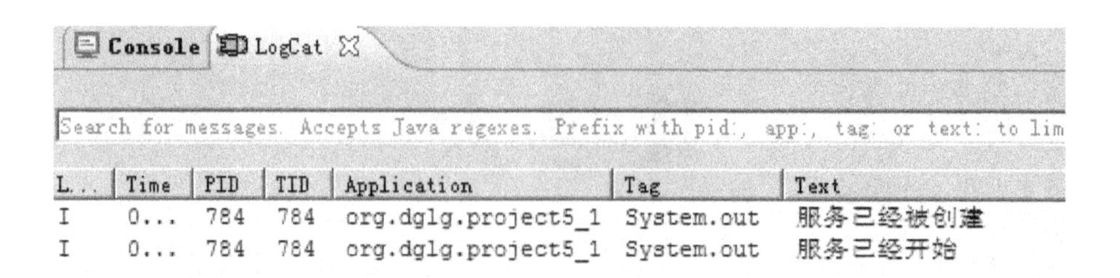

图 6.2　单击“开始 Service”按钮输出信息

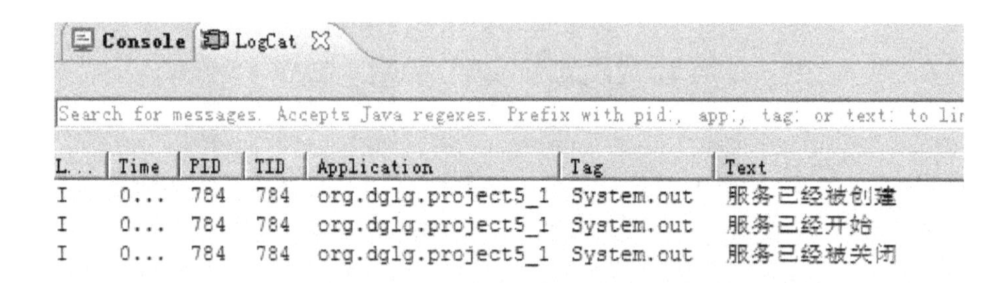

图 6.3　单击“停止 Service”按钮输出信息

本任务的主要目的是熟悉 Service，重点和难点如下：

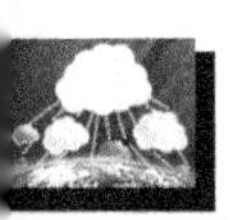

(1)熟悉 Service 开发和运行机制；

(2)学习 Service 的启动、运行、结束等方法；

(3)掌握调用和结束 Service 方法：startService( )和 stopService( )；

(4)熟悉 Activity 与 Service 相互开发程序的步骤和方法。

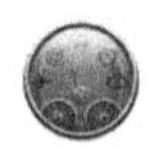

任务准备

1. Service 介绍

由于手机屏幕和资源的限制，通常情况下在同一时刻仅有一个应用程序处于激活状态，并能够显示在手机屏幕上，因此，应用程序需要一种机制，在没有用户界面的情况下，能够长时间在后台运行，实现应用程序的特定功能，并能够处理事件或更新数据。为此，Android 系统提供了(Service)服务组件，它不直接与用户进行交互，却能够长期在后台运行。

有很多情况需要使用 Service，典型的例子就是：MP3 播放器。

Service 是 Android 中四大组件之一，在 Android 开发中起到非常重要的作用。它一直在 Android 系统的后台运行，一般使用 Service 为应用程序提供一些服务或不需要界面的功能，例如：下载文件、控制音乐或视频播放等。

Android 中的 Service 和 Windows 中的 Service 是类似的东西，Service 一般没有用户操作界面，它运行于系统中不容易被用户发觉。

2. Service 特点

在 Android 中，Service(服务)是一个没有用户界面的在后台运行执行耗时操作的应用组件，其他应用组件能够启动 Service，并且当用户切换到另外的应用场景，Service 将持续在后台运行。另外，一个组件能够绑定到一个 Service 与之交互(IPC 机制)。

通过 Android 的 Service，可以做很多的操作，比如播放音乐，操作文件 I/O 或者与内容提供者(content provider)交互，而且这些活动都是在后台进行。

3. Service 类型

Android 中的 Service 分为本地服务(Local Service)和远程服务(Remote Service)2 种类型：

1)本地服务(Local Service)

本地服务用于应用程序内部，它的特点是可以启动并运行，直至有人停止或自己停止。

本地服务 Service 通过调用 startService( )启动，调用 stopService( )结束，也可以调用 Service. stopSelf( ) 或 Service. stopSelfResult( )来自己停止。不论调用了多少次 startService( )方法，只需要调用一次 stopService( )来停止服务。

本地服务主要用于实现应用程序的一些耗时任务，比如查询升级信息，并不占用应用程序；比如 Activity 所属线程，单开线程后台执行。这样用户体验比较好。

2)远程服务(Remote Service)

远程服务用于 Android 系统内部的应用程序之间，可以通过自己定义并暴露出来的接口进行程序操作。客户端建立一个到服务对象的连接，并通过那个连接来调用服务。

远程服务 Service 通过调用 bindService( )方法建立，调用 Context. unbindService( )关闭。多个客户端可以绑定至同一个服务。如果服务此时还没有加载，bindService( )会先加载它。

远程服务可被其他应用程序复用，比如天气预报服务，其他应用程序不需要再写这样的服务，调用已有的即可。

4. 调用 Service 的方法

调用 Service 的方法有两种方式：通过 Context 的 startService( )方法和 bindService( )方法。

startService( )方法：通过该方法启动 Service，访问者与 Service 之间没有关联，即使访问者退出，Service 仍然运行。同样，采用 startService( )方法启动 Service，也能调用 stopService( )方法结束服务。

bindService( )方法：使用该方法启动 Service，访问者与 Service 绑定在了一起，访问者一旦退出，Service 也就终止。

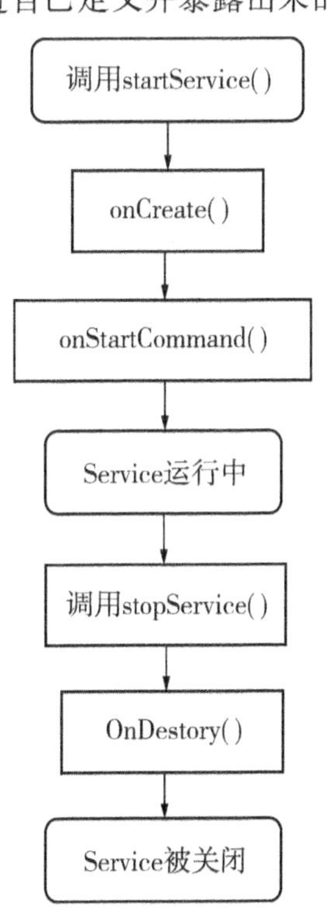

图 6.4 Service 生命周期

5. startService( )的生命周期

Service 和 Activity 一样，拥有自己的生命周期，分别是 onCreate( )、onStartCommand( )、onDestory( )。

Service 的整个生命周期始于 onCreate( )方法而止于 onDestroy( )方法。如果 Service 已经启动，当再次启动 Service 时，则不调用 onCreate( )方法而直接调用 onStart( )方法，示范代码如下：

```
Public void onCreate( );                    //创建服务
Public void onStartCommand(Intent intent, int flags, int startId);        //开始服务
Public void onDestroy( );                 //销毁服务
```

在本任务的运行效果中，可以从控制台的输出信息中清晰地了解到 Service 的生命周期，如图 6.4 所示。

6. Service 的开发步骤

Android 的 Service 服务的开发比较简单，步骤如下：

(1)创建一 Service 类继承 Service 类(下例为创建一名为 MainService 的 Service 类)：

```
public class MainService extends Service { }
```

(2)根据 Service 实现的功能编写 Service 类代码；

(3)在 AndroidManifest. xml 文件中的 <application> 节点里对服务进行配置(类名为 ExampleService)：

```
<service android: name = ". ExampleService " />
```

(4)在 Activity 中对 Service 进行调用。Service 不能自己运行，需要通过方法启动和调用，可以通过 startService( )或 bindService( )方法启动服务。

(5)运用 startService( )方法调用 Service(例如调用的 Service 名为：ExampleService. class)：

```
Intent intent = new Intent(MainActivity. this, ExampleService. class);
startService(intent);
```

(6)结束 Service：

```
Intent intent = new Intent(MainActivity. this, ExampleService. class);
stopService(intent);
```

任务实现

1. 新建 Android 项目

在 Eclipse 中，新建一 Android 应用程序，方法可参考项目一。

2. 添加新 Service

本任务除 Activity 外，还涉及 Service，所以需要新建 Service，新建方法如下：

STEP 1 单击【File】菜单的【New】，在弹出的快捷菜单中选择【Class】，如图 6.5 所示。

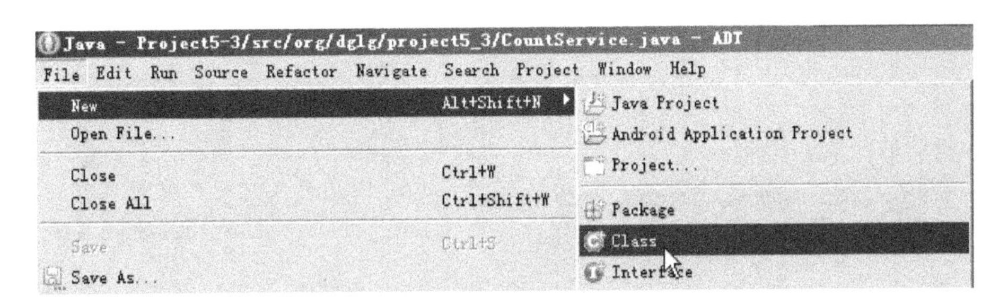

图 6.5 新建类

STEP 2 单击后，Android 将弹出如图 6.6 所示的“New Java Class”的对话框。单击【Package】栏目右侧的【Browse】按钮，打开如图 6.7 所示“Package Selection”的对话框，双击选择要添加 Service 的文件夹(本例为：org. dglg. project6_1)，Eclipse 将选择双击的文件夹并退出“Package Selection”对话框。

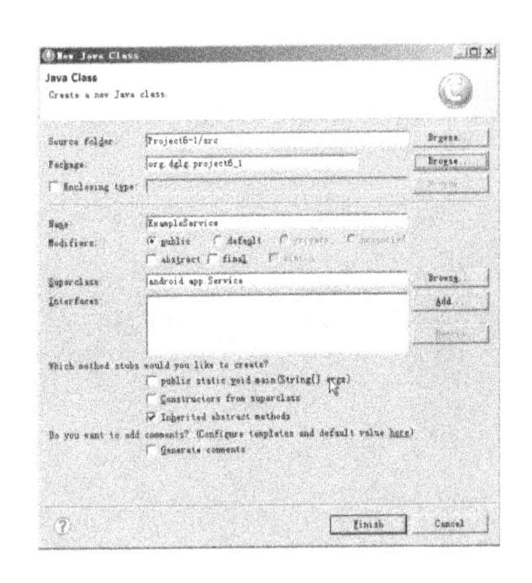

图 6.6 “New Java Class”对话框

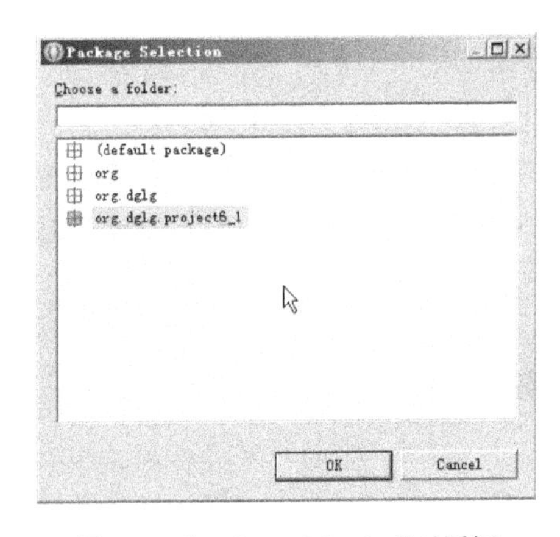

图 6.7 “Package Selection”对话框

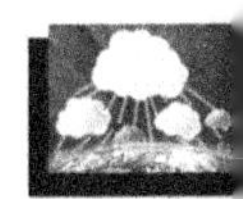

STEP 3 在图 6.6 中的【Name】中输入 Service 名称：“ExampleService”(注意首字母大写)。

STEP 4 单击在【Superclass】右边的【Browse】按钮，打开如图 6.8 所示的“Superclass selection”对话框，在【Choose a type】中输入“Service”，下方的【Matching items】中会自动列出所有与在【Choose a type】中输入的内容相匹配的类，本例中选择 Service - Android. app，单击【OK】按钮确定选择。

STEP 5 单击图 6.6 中的【Finish】按钮，完成 Service 创建。此时在 MainActivity. java 下将新建一名为 ExampleService. java 的类，如图 6.9 所示。该 Service 类即为创建的 Service。

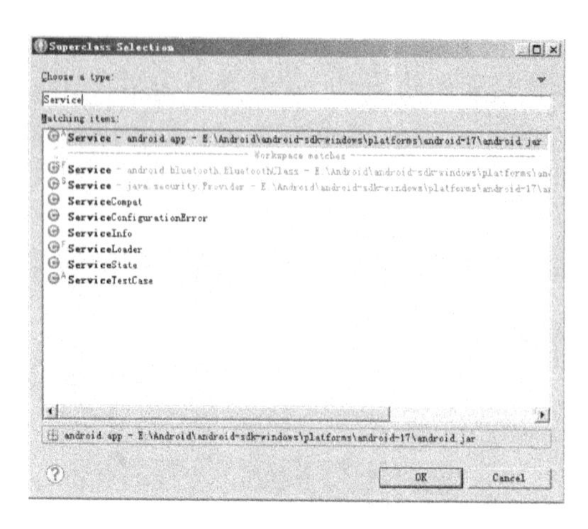

图6.8 “选择超类”对话框

图6.9 ExampleService

3. 在AndroidMainfest. XML中对Service进行注册

在Android中，对新建的Service都要进行注册，否则程序运行时将出错。注册的方法是在配置文件AndroidMainfest. xml中进行注册，需要在配置文件AndroidMainfest. xml里加上如下代码(注意代码添加位置，一般添加在</application>代码前)：

```
<activity android: name = ". ExampleService"  </activity>
```

4. 对Activity进行布局

1)编写strings. xml定义字符串变量

对布局文件中要引用的字符串变量进行定义，代码如下：

```
<? xml version = "1.0" encoding = "utf-8"? >
<resources>
    <string name = "app_name" >使用Service示例</string>
    <string name = "action_settings" >Settings</string>
    <string name = "start" >开始Service</string>
    <string name = "stop" >停止Service</string>
</resources>
```

2)对MainActivity布局：activity_main. xml

MainActivity布局文件为：activity_main. xml。据程序功能，可考虑采用线性布局，代码如下：

```
<? xml version = "1.0" encoding = "utf-8"? >
<LinearLayout xmlns: android = "http: //schemas. android. com/apk/res/android"
    android: layout_width = "match_parent"
    android: layout_height = "match_parent"
```

```
    android: orientation = "vertical" >
<Button
    android: id = "@ +id/StartBtn"
    android: layout_width = "wrap_content"
    android: layout_height = "wrap_content"
    android: layout_marginTop = "30dp"
    android: layout_marginLeft = "80dp"
    android: text = "@ string/start"
    />
<Button
    android: id = "@ +id/StopBtn"
    android: layout_width = "wrap_content"
    android: layout_height = "wrap_content"
    android: layout_marginTop = "30dp"
    android: layout_marginLeft = "80dp"
    android: text = "@ string/stop"
    />
</LinearLayout>
```

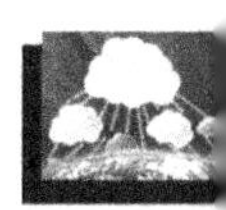

5. 主程序功能实现

(1)MainActivity. java 程序文件编写步骤：根据程序实现的功能以及 UI 界面的设计，先引用相关的类，然后声明并取得 UI 界面中的控件对象：Button，并设定 Button 的 onClickListener 动作。完整代码如下：

```
package org. dglg. project6_1;
/*导入相关类*/
import android. os. Bundle;
import android. app. Activity;
import android. content. Intent;
import android. view. View;
import android. view. View. OnClickListener;
import android. widget. Button;
public class MainActivity extends Activity {
    /*声明 intent*/
    Intent intent;
    @ Override
        protected void onCreate(Bundle savedInstanceState) {
        super. onCreate(savedInstanceState);
```

```
            /* 载入 activity_main. xml 布局文件 */
            setContentView(R. layout. activity_main);
            /* 通过 findViewById( )方法取得布局文件中 Button 对象 */
            Button StartBtn = (Button)findViewById(R. id. StartBtn);
            Button StopBtn = (Button)findViewById(R. id. StopBtn);
            /* 创建启动 ExampleService 的 Intent */
            intent = new Intent(MainActivity. this, ExampleService. class);
            /* 按钮事件，启动 Service */
            StartBtn. setOnClickListener(new OnClickListener( ){
                @ Override
                public void onClick(View v) {
              /* 启动指定的 Service */
                startService(intent);
                }
            });
            /* 按钮事件，启动 Service */
            StopBtn. setOnClickListener(new OnClickListener( ){
                @ Override
                public void onClick(View v) {
                    // TODO Auto - generated method stub
                /* 停止指定的 Service */
                stopService(intent);
                }
            });
        }
    }
```

程序说明如下：

① 上述程序主要采用 startService( )方法来启动 Service，实现程序功能的主要代码是：

```
    Intent = new Intent(MainActivity. this, ExampleService. class); //创建启动 Service 的
intent;
    startService(intent); //启动 Service;
```

② 停止上述 Service 的代码为：

```
    Intent = new Intent(MainActivity. this, ExampleService. class); //创建启动 Service 的
intent;
    stopService(intent); //停止 Service
```

(2) ExamleService. java 文件的编写：该 Service 的主要功能根据 Service 的启动和停止

在控制台的 LogCat 上显示相应信息来熟悉 Service 的生命周期，主要代码如下：

```
package org. dglg. project6_1;
import android. app. Service;
import android. content. Intent;
import android. os. IBinder;
public class ExampleService extends Service {
    @ Override
    public IBinder onBind(Intent intent) {
        // TODO Auto - generated method stub
        return null;
    }
    /* Service 被创建时回调该方法，并在控制台输出"服务已经被创建" */
    @ Override
    public void onCreate( ){
        super. onCreate( );
        System. out. println("服务已经被创建");
    }
    /* Service 被启动时回调该方法，并在控制台输出"服务已经开始" */
    @ Override
    public int onStartCommand(Intent intent, int flags, int startId)
    {
        System. out. println("服务已经开始");
        return START_STICKY;
    }
    /* Service 被关闭时回调该方法，并在控制台输出"服务已经被关闭" */
    @ Override
    public void onDestroy( ){
        super. onDestroy( );
        System. out. println("服务已经被关闭");
    }
}
```

程序中的 Service 功能是在控制台的 LogCat 上输出 Service 方法回调情况（因为 Service 在后台运行），以此来熟悉 startService( )的生命周期，说明如下：

①Public void onCreate( ){ }方法为 Service 被创建时回调的方法；

②System. out. println("服务已经被创建")语句在控制台上输出“服务已经被创建”；

③本任务运行后，可以从控制台观察到 startServic( )生命周期为 startService( ) → onCreate( ) →onStartCommand( )→Service running→onDestroy( )

④程序运行如图 6. 1 ~图 6. 3 所示。

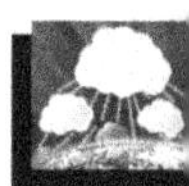

任务考核

参考本任务，创建一个新的项目名为 ProjectTest6_1，创建一 Service，并在 MainActivity 中对 Service 进行调用和结束(用 stratService 方法调用和用 endService 结束调用)，并在 LogCat 日志查看器中仔细查看 Service 的生命周期的回调方法的过程。

任务二　使用 Service 示例(二)

任务描述

本任务与任务一类似，主要是掌握利用 bindService( )方法绑定 Service、利用 unbindService( )解除绑定并结合模拟器控制台的 LogCat 观察、熟悉 bindService( )的生命周期。运行界面和效果如图 6.10～图 6.12 所示。

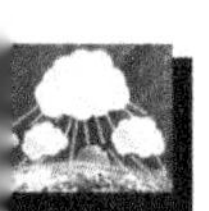

图 6.10　使用 Service 示例运行界面

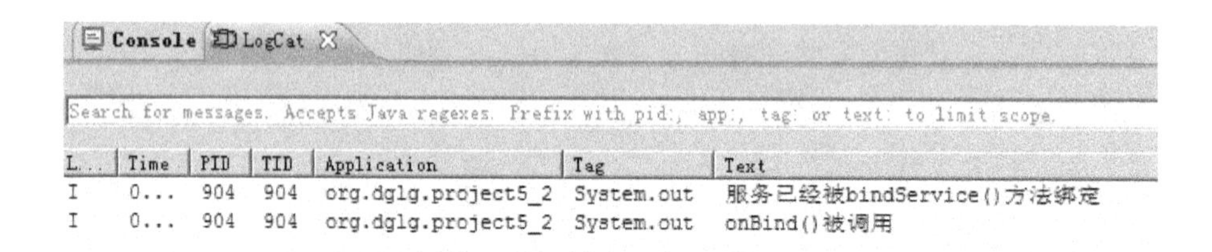

Console　LogCat

Search for messages. Accepts Java regexes. Prefix with pid:, app:, tag: or text: to limit scope.

L...	Time	PID	TID	Application	Tag	Text
I	0...	904	904	org.dglg.project5_2	System.out	服务已经被bindService()方法绑定
I	0...	904	904	org.dglg.project5_2	System.out	onBind()被调用

图 6.11　单击“开始 Service”按钮输出信息

Console　LogCat

Search for messages. Accepts Java regexes. Prefix with pid:, app:, tag: or text: to limit scope.

L...	Time	PID	TID	Application	Tag	Text
I	0...	904	904	org.dglg.project5_2	System.out	服务已经被bindService()方法绑定
I	0...	904	904	org.dglg.project5_2	System.out	onBind()被调用
I	0...	904	904	org.dglg.project5_2	System.out	服务已经被unbindService()解除绑定
I	0...	904	904	org.dglg.project5_2	System.out	服务已经被关闭

图 6.12　单击“停止 Service”按钮输出信息

任务分析

本任务与任务一功能类似，主要是利用 bindService( )方法绑定 Service、利用 unbindService( )解除绑定。在本任务中，主要完成的功能和效果是单击 Activity 中的“绑定 Service”按钮绑定 Service，单击 Activity 中的“解除绑定”按钮结束 Service。该 Service 没有实际功能，主要在控制台的 LogCat 上输出 Service 方法调用情况。

本任务的主要目的是熟悉 Service，重点和难点如下：

(1)继续熟悉 Service 开发和运行机制；
(2)继续学习 Service 的启动、运行、结束等方法；
(3)掌握绑定和解除绑定 Service 方法：bindService( )和 unbindService( )；
(4)熟悉 Activity 与 Service 相互开发程序的步骤和方法。

任务准备

1. bindService( )介绍

在 Android 中，如果用户端要建立一个与 Service 的连接，并使用此连接与 Service 进行通话，可以通过 bindService( )方法来绑定服务，通过 unbindService( )方法来解除绑定。

在 Android 中，多个用户端可以绑定同一个服务，如果 Service 还未被启动，bindService( )方法可以启动服务。

startService( )和 bindService( )两种模式是独立的。例如：一个播放音乐 Service 可以通过 startService(intend)对象来播放音乐。可以通过调用 bindServices( )方法与 Service 建立连接，bindServices( )可以执行一些其他的额外操作，比如获取歌曲的一些信息等。

2. bindService( )特点

相对来说，bindService( )比 startService( )复杂，bindService( )特点如下：

(1)如果访问者退出了，系统就会先调用 Service 的 onUnbind( )方法，接着调用 onDestroy( )方法。

(2)如果调用 bindService( )方法前 Service 已经被绑定，多次调用 bindService( )方法并不会导致多次创建 Service 及绑定。也就是说，onCreate( )和 onBind( )方法并不会被多次调用)。

(3)可以调用 unbindService( )方法与正在绑定的服务解除绑定，调用该方法也会导致系统调用服务的 onUnbind( )→onDestroy( )方法。

3. bindService( )的生命周期

bindService( )启动流程如下：

bindService( ) →onCreate( ) →onBind( ) →Service running→onUnbind( ) →onDestroy( )→Service stop

onBind( )将返回给客户端一个 IBind 接口实例，IBinder 允许客户端回调服务的方法，比如得到 Service 的实例、运行状态或其他操作。这个时候将把调用者(Context，例如 Activity)和 Service 绑定在一起，调用者(Context)退出了，Service 就会调用 onUnbind→

onDestroy 相应退出。

所以 bindService( )调用 Service 的生命周期为:

onCreate( )→ onBind( )(只可绑定一次) →onUnbind( )→ onDestory( ),如图 6.13 所示。

知识拓展

Android 系统中,Service 的 onBind( )方法返回的 IBinder 对象可以当成 Service 返回的代理对象,Service 允许通过 IBinder 对象来访问 Service 内部的数据,这样就实现了客户端与 Service 之间的通信。

4. bindService( )和 startService( )的区别

1)startService( )和 bindService( )区别一

使用 startService( )方法启用服务,访问者与服务之间没有关连,即使访问者退出了,服务仍然运行。

使用 bindService( )方法启用服务,访问者与服务绑定在了一起,访问者一旦退出,服务也就终止。

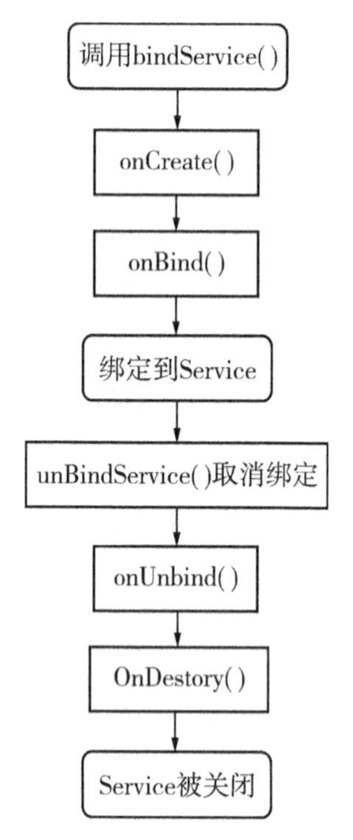

图 6.13 bindService 生命周期

startService( )采用 startService( )方法启动 Service,只能调用 stopService( )方法结束服务,服务结束时会调用 onDestroy( )方法。startService( )启动 Service 方式适用于服务和访问者之间没有交互的情况。如果服务和访问者之间需要方法调用或者传递参数,则需要使用 bindService( )和 unbindService( )方法启动关闭服务。

bindService( )方法启动服务,在服务未被创建时,系统会先调用服务的 onCreate( )方法,接着调用 onBind( )方法,这个时候访问者和服务绑定在一起。

2)startService( )和 bindService( )区别二

Service 的生命周期回调方法有所不同。

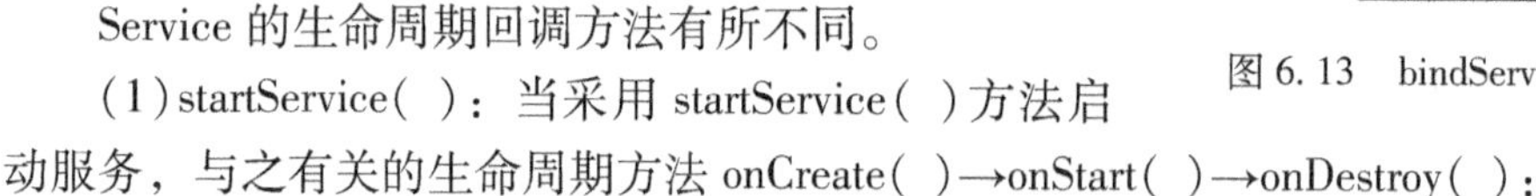

(1)startService( ):当采用 startService( )方法启动服务,与之有关的生命周期方法 onCreate( )→onStart( )→onDestroy( ):

- onCreate( ):该方法在服务被创建时调用,该方法只会被调用一次,无论调用多少次 startService( )或 bindService( )方法,服务也只被创建一次。
- onStart( ):只有采用 Context. startService( )方法启动服务时才会回调该方法。该方法在服务开始运行时被调用。多次调用 startService( )方法尽管不会多次创建服务,但 onStart( ) 方法会被多次调用。
- onDestroy( ):该方法在服务被终止时调用。

(2)bindService( ):当采用 Context. bindService( )方法启动服务,与之有关的生命周期方法为 onCreate( )→onBind( )→onUnbind( )→onDestroy( )。onBind( )只有采用 bindService( )方法启动服务时才会回调该方法。该方法在调用者与服务绑定时被调用,当

调用者与服务已经绑定，多次调用 bindService( )方法并不会导致该方法被多次调用。

onUnbind( )只有采用 bindService( )方法启动服务时才会回调该方法。该方法在调用者与服务解除绑定时被调用。

5. bindService( )方法参数

bindService( )方法设计到 3 个参数，其完整方法签名为：

bindService(Intent service，ServiceConnection conn，int flags)

该方法的三个参数分别如下：

① Intent Service：通过 Intent 制定要启动的 Service。

② ServiceConnection conn：该参数是一个 ServiceConnection 对象，用于监听 Service 与访问者的连接情况，该对象包含 2 个方法：onServiceConnected( )和 onServiceDisconnected( )。当访问者与 Service 连接成功时回调 ServiceConnection 对象的 OnServiceConnected( )方法；当 Service 由于异常终止，导致该 Service 与访问者之间断开连接，将回调 onServiceDisconnected( )方法。

③ Int flags：绑定 Service 时是否自动创建 Service。

知识拓展

ServiceConnection 对象的 onServiceDisconnected( )方法仅仅适用于 Service 的异常终止，当用 unBindService( )方法断开与 Service 连接时，onServiceDisconnected( )方法不会被调用。

6. bindService( )使用的一般步骤

Android 中，bindService( )的使用步骤相对 startService( )复杂一些。步骤如下：

STEP 1 创建一 Service 类继承 Service 类，并根据 Service 实现的功能编写 Service 类代码，并记得在 AndroidManifest. xml 文件中的 <application> 节点里对服务进行配置。

STEP 2 根据该 Service 类实现的功能编写相应代码；

STEP 3 在 Activity 中新建一个内部类(conn)，实现 ServiceConnection 接口，并添加方法 onServiceConnected 和 onServiceDisconnected，代码如下：

```
ServiceConnection conn = new ServiceConnection( ) {
public void onServiceConnected(ComponentName name, IBinder service) {
......
}
public void onServiceDisconnected(ComponentName name)
{
......
}
}
```

STEP 4 在onCreate( )中实现bindService( )以绑定Service，使用bindService( )代码如下（代码中调用的Service名为ExampleService. class，BIND_AUTO_CREATE表示自动创建）：

```
Intent intent = new Intent(MainActivity.this, ExampleService.class);
bindService(intent, conn, Service.BIND_AUTO_CREATE);
```

任务实现

1. 新建Android项目

在Eclipse中，新建一Android应用程序。

2. 添加新Service

按任务一的方法创建Service，命名为ExampleService. class。

3. 在AndroidMainfest. xml中对新建的Service进行注册

略。

4. 对Activity进行布局

1)编写strings. xml定义字符串变量

对布局文件中要引用的字符串变量进行定义，代码如下：

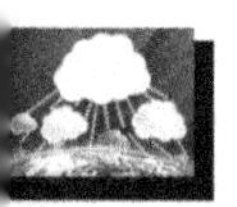

```
<?xml version="1.0" encoding="utf-8"?>
<resources>
    <string name="app_name">使用Service示例(二)</string>
    <string name="action_settings">Settings</string>
    <string name="start">绑定Service</string>
    <string name="stop">解除绑定</string>
</resources>
```

2)对MainActivity布局

MainActivity布局文件为activity_main. xml。根据程序功能，采用线性布局，代码如下：

```
<?xml version="1.0" encoding="utf-8"?>
<LinearLayout xmlns:android="http://schemas.android.com/apk/res/android"
    android:layout_width="match_parent"
    android:layout_height="match_parent"
    android:orientation="vertical" >

<Button
    android:id="@+id/StartBtn"
```

```
        android: layout_width = "wrap_content"
        android: layout_height = "wrap_content"
        android: layout_marginTop = "30dp"
        android: layout_marginLeft = "80dp"
        android: text = "@ string/start"
        />
    <Button
        android: id = "@ + id/StopBtn"
        android: layout_width = "wrap_content"
        android: layout_height = "wrap_content"
        android: layout_marginTop = "30dp"
        android: layout_marginLeft = "80dp"
        android: text = "@ string/stop"
        />
    </LinearLayout>
```

5. 主程序功能实现

(1)MainActivity. java 程序文件编写步骤：根据程序实现的功能以及 UI 界面的设计，先引用相关的类，接着新建一个内部类以实现 ServiceConnection 接口，然后声明并取得 UI 界面中的控件变量 Button，并设定 Button 的 onClickListener 动作。完整代码如下：

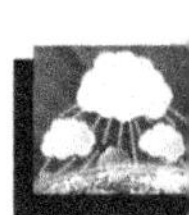

```
package org. dglg. project6_2;
/ * 导入相关类 * /
import android. os. Bundle;
import android. os. IBinder;
import android. app. Activity;
import android. app. Service;
import android. content. ComponentName;
import android. content. Intent;
import android. content. ServiceConnection;
import android. view. View;
import android. view. View. OnClickListener;
import android. widget. Button;
public class MainActivity extends Activity {
    / * 新建一个内部类，实现 ServiceConnection 接口 * /
        ServiceConnection conn = new ServiceConnection( ) {
        @ Override
        public void onServiceConnected( ComponentName name, IBinder service) {
```

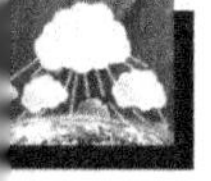

```
            // TODO Auto - generated method stub
        }
        @ Override
        public void onServiceDisconnected( ComponentName name) {
            // TODO Auto - generated method stub
        }
        };
    @ Override
    protected void onCreate( Bundle savedInstanceState) {
        super. onCreate( savedInstanceState) ;
        /* 载入 activity_main. xml 布局文件 */
        setContentView( R. layout. activity_main) ;
        /* 通过 findViewById( )方法取得布局文件中 Button 对象 */
        Button StartBtn = ( Button) findViewById( R. id. StartBtn) ;
        Button StopBtn = ( Button) findViewById( R. id. StopBtn) ;
        /* 创建启动 ExampleService 的 Intent */
        final Intent intent = new Intent( MainActivity. this, ExampleService. class) ;
        /* 按钮事件, bindService( )启动 Service */
        StartBtn. setOnClickListener( new OnClickListener( ) {
            @ Override
            public void onClick( View v) {
                // TODO Auto - generated method stub
                bindService( intent, conn, Service. BIND_AUTO_CREATE) ;
            }
        }) ;
        /* 按钮事件, unbindService( )停止 Service */
        StopBtn. setOnClickListener( new OnClickListener( ) {
            @ Override
            public void onClick( View v) {
                // TODO Auto - generated method stub
                unbindService( conn) ;
            }
        }) ;
    }
}
```

程序说明:

① 上述程序主要采用 bindService( )方法来启动 Service，实现程序功能的主要代码是:

```
Intent = new Intent(MainActivity.this, ExampleService.class); //创建启动 Service 的intent;
bindService(intent, conn, Service.BIND_AUTO_CREATE); //bindService( )方法的参数有3个，第一个为intent，第二个conn为实现ServiceConnection接口新建一个内部类，第三个Service.BIND_AUTO_CREATE比较固定，表示自动创建。
```

② 在新建的内部类conn中，定义了2个方法：

onServiceConnected和onServiceDisconnected，在任务三中将着重介绍。

③ 采用unbindService停止上述Service，unbindService( )方法只有1个参数：

```
Intent = new Intent(MainActivity.this, ExampleService.class); //创建启动 Service 的intent;
unbindService(intent); //停止 Service
```

(2) ExamleService.java文件的编写：该Service的主要功能是根据Service的启动和停止在控制台的LogCat上显示相应信息来熟悉Service的生命周期，主要代码如下：

```
package org.dglg.project6_2;
import android.app.Service;
import android.content.Intent;
import android.os.IBinder;
public class ExampleService extends Service{
/*bindService( )要实现的方法*/
@Override
public IBinder onBind(Intent intent) {
System.out.println("onBind()被调用");
return null;
}
/*Service被创建时回调该方法，并在控制台输出"服务已经被bindService()方法绑定"*/
@Override
public void onCreate(){
    super.onCreate();
    System.out.println("服务已经被bindService()方法绑定");
}
/*unbindService()停止Service时回调该方法，并在控制台输出"服务已经被unbindService()解除绑定"*/
@Override
public boolean onUnbind(Intent intent)
```

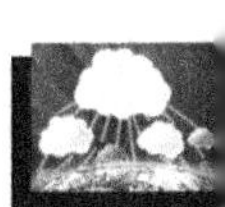

```
    {
        System.out.println("服务已经被unbindService()解除绑定");
        return true;
    }
    /*Service关闭时回调该方法，并在控制台输出"服务已经被关闭"*/
    public void onDestroy()
    {
        super.onDestroy();
        System.out.println("服务已经被关闭");
    }
}
```

该Service的功能是主要在控制台的LogCat上输出Service方法回调情况，说明如下：

① Public void onCreate( ){ }方法为Service被创建时回调的方法；

② bindService( )调用Service将触发onBind( )方法，onBind( )将返回给客户端一个IBind接口实例；

③ 程序运行如图6.10～图6.12所示。

任务考核

参考本任务，创建一个新的项目，并在创建一Service，并在MainActivity中对该Service进行调用和结束(用bindService方法绑定和用unbindService结束绑定)，并在LogCat日志查看器中仔细查看Service的生命周期的回调方法的过程。

任务三　Service应用：简易计时器

任务描述

本任务主要是运用Service服务制作一简易计时器，以实现计时(按每秒计时)，运行界面和效果如图6.14～图6.16所示。

图6.14　运行初始界面　　图6.15　开始计时　　图6.16　停止计时

任务分析

在本任务中，完成的功能是简易计时，涉及 1 个 TextView 和 2 个 Button，其中的 TextView 主要用来显示计时状态和所计时的时间，2 个 Button 分别为“开始计时”和“停止计时”，其中“停止计时”的状态初始化为不可用，初始界面如图 6. 14 所示。

效果主要是：

(1)单击 Activity 中的“开始计时”按钮绑定 Service(该 Service 主要完成每秒计时功能)，开始计时，同时将“停止按钮”状态转为可用状态，如图 6. 15 所示。

(2)单击 Activity 中的“停止计时”按钮对 Service 解除绑定，结束 Service。并将自己设为不可用状态(以备下次计时)，同时在 TextView 上显示计时信息，如图 6. 16 所示。

本任务的主要目的是通过制作简易计时器，来熟悉 Service 与 Activity 进行数据交互(在 Activity，取得 Service 中所计时间数)，重点和难点如下：

(1)Button 类的 setEnable( )方法；
(2)熟悉 bindService( )绑定和 unbindService( )解除绑定 Service；
(3)Service 新线程的计时方法；
(4)Activity 与 Service 进行数据交互方法。

任务准备

1. Android 系统中的进程与线程

在 Android 运行机制中，当一个应用程序开始运行它的第一个组件时，Android 会为它启动一个 Linux 进程，并在其中执行一个单一的线程。

1)Android 进程

在 Android 中，组件运行所在的进程由 Androidmanifest. xml 文件所指定。组件元素(如：Activity，Service 等)都由一个 process 属性来指定组件应当运行于哪个进程之内。这些属性可以设置为每个组件运行于它自己的进程之内，或一些组件共享一个进程。也可以设置为不同应用程序的组件在一个进程中运行。

Android 中 Application 元素也有一个 process 属性，以设定所有组件的默认值。所有的组件实例都位于特定进程的主线程内，而对这些组件的系统调用也将由那个线程进行分发。一般不会为每个实例创建线程。因此，Android 中的一些方法总是运行在进程的主线程内。这样组件在被系统调用的时候，不能进行长时间的耗时、阻塞的操作(例如上网下载，循环计算等)，因为这将阻塞同样位于这个进程的其他组件的运行。这个时候，应该为这些长时间操作建立一个单独的线程进行处理。

Android 在可用内存不足或者有一个正在为用户进行服务的进程需要更多内存的时候，有时会关闭一些进程以腾出内存。Android 会根据它们对于用户的相对重要性来决定关闭哪个进程，在被关闭进程中运行着的应用程序也因此被销毁。当再次需要该进程时，Android 会为这些组件重新创建进程。

2) Android 线程

在 Windows 系统中的应用程序当需要进行一些复杂的数据操作，而又不需要界面 UI 的时候，会为应用程序专门写一个线程去执行这些复杂的数据操作。通过线程，可以执行一些比较耗时的操作，同时对用户的界面不会产生影响。例如处理数据、下载数据等。在 Android 应用程序开发中，同样会遇到这样的问题。例如在 Android 中，要进行网络访问、数据下载等比较耗时的数据操作时，一般会使用后台线程去执行这些操作。

线程与进程相似，是一段完成某个特定功能的代码，是程序中单个顺序的流控制，一个进程中可以包含多个线程。与进程不同的是，同类的多个线程共享一块内存空间和一组系统资源，因此系统在各个线程之间切换时，资源占用要比进程小得多，因此，线程也被称为轻量级进程。

在 Android 中，线程在代码中是 Thread 对象创建的。Android 提供了很多便于管理线程的类：

① Handler 用于处理消息；

② Looper 用于在一个线程中运行一个消息循环；

③ HandlerThread 用于使用一个消息循环启用一个线程。

2. 创建线程

在 Android 中，提供了两种创建线程的方法：一种是通过 Thread 类的构造方法创建线程对象，重写 run( )方法实现。另一种是通过实现 Runnable 接口实现。本任务中主要采用的是 Thread 类来实现计时功能。下面对 Thread 类创建线程对象进行介绍。

Thread 类创建线程对象时，通过创建一个 Runnable 类的对象并通过 run( )方法来实现要执行的操作，代码如下：

```
Thread thread = new Thread( new Runnable( ) {
//run 方法为要线程要执行的操作
@ Override
Public void run ( ) {
……
}
});
```

上述代码中，通过 Thread 类创建了一个名称为 thread 的线程，该线程中的方法 run( )为该线程要执行的操作，可用以下代码启动和停止该线程：

```
thread. start( ); //启动线程
thread. stop( ); //停止线程
```

3. ServiceConnection 类

在 Android 中，客户端通过调用 bindService( )方法能够绑定服务，然后 Android 系统会调用服务的 onBind( )回调方法，这个方法会返回一个跟服务端交互的 IBinder 对象。但是这个绑定是异步的，bindService( )方法立即返回，并且不给客户端返回 IBinder 对象。如果客户端要与 Service 交互的话，客户端必须创建一个 ServiceConnection 类的实例，并且把这个实例传递给 bindService( )方法，以接收 IBinder 对象，从而实现客户端与 Service 的交互。

ServiceConnection 类中包含了一个系统调用的传递 IBinder 对象的回调方法。因此，客户端绑定一个 Service 后，并要与之交互，客户端必须实现：

1) 实现 ServiceConnection 类

ServiceConnection 类包括以下两个回调方法：

(1) onServiceConnected( )

系统能通过调用这个方法来发送由服务的 onBind( )方法返回的 IBinder 对象。

(2) onServiceDisconnected( )

Android 系统会在连接的服务突然丢失的时候调用这个方法，如服务崩溃时或被杀死时。在客户端解除服务绑定时不会调用这个方法。

2) 调用 bindService( )方法来传递 ServiceConnection 类的实现

当调用 onServiceConnected( )方法回调时，就可以开始使用接口中定义的方法来调用 Service 了。

3) 调用 unbindService( )方法断开与服务的连接

4. 客户端与 Service 交互方法及步骤

(1)定义一个继承 Service 的类(Service 名为 MyService)

```
public class MyService extends Service {
……
}
```

(2)在上面定义的 Service 类里定义一个实现 Binder 的类，用于程序绑定 Service，使得触发 onServiceConnected 事件，如：

```
public class MyBinder extends Binder{
    MyServiece getService( )
    {
        return MyService. this;
    }
}
```

(3)在 Activity 声明一个实现 ServiceConnection 接口的类：

```
ServiceConnection serviceConnection = new ServiceConnection( ) {
    @ Override
    public void onServiceConnected( ComponentName name, IBinder service) {
        // TODO Auto - generated method stub
        ……
    }
```


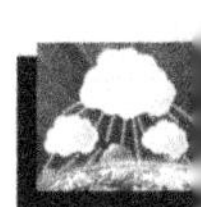

```
        @Override
        public void onServiceDisconnected(ComponentName name) {
            // TODO Auto-generated method stub
            ……
        }
    };
```

(4)采用绑定语句，绑定Service：

```
Intent intent = new Intent(ServiceTest.this, MyServiece.class);
bindService(intent, serviceConnection, BIND_AUTO_CREATE);
```

(5)用下面语句对Service进行解除绑定：
unbindService(serviceConnection);

任务实现

1. 新建Android项目

在Eclipse中，新建一Android应用程序。

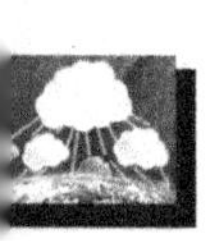

2. 添加新Service

创建Service，命名为CountService.class。

3. 在AndroidMainfest.xml中对Service进行注册

略。

4. 对Activity进行布局

1)编写strings.xml定义字符串变量

对布局文件中要引用的字符串变量进行定义，代码如下：

```
<?xml version="1.0" encoding="utf-8"?>
<resources>
    <string name="app_name">Service应用：简易定时器</string>
    <string name="action_settings">Settings</string>
    <string name="StartCount">开始计时</string>
    <string name="EndCount">停止计时</string>
    <string name="ShowTime">计时……</string>
</resources>
```

2)对MainActivity布局：activity_main.xml

根据程序功能，涉及的对象为1个TextView和2个Button，考虑使用线性布局，代码如下：

```
<?xml version="1.0" encoding="utf-8"?>
<LinearLayout xmlns:android="http://schemas.android.com/apk/res/android"
    android:layout_width="match_parent"
    android:layout_height="match_parent"
    android:orientation="vertical" >
<TextView
    android:id="@+id/tv"
    android:layout_width="wrap_content"
    android:layout_height="wrap_content"
    android:layout_marginTop="10dp"
    android:layout_marginLeft="80dp"
    android:text="@string/ShowTime"
    />
<Button
    android:id="@+id/BtnCount"
    android:layout_width="wrap_content"
    android:layout_height="wrap_content"
    android:layout_marginTop="10dp"
    android:layout_marginLeft="80dp"
    android:text="@string/StartCount"
      />
<Button
    android:id="@+id/BtnCountend"
    android:layout_width="wrap_content"
    android:layout_height="wrap_content"
    android:layout_marginTop="10dp"
    android:layout_marginLeft="80dp"
    android:enabled="false"
    android:text="@string/EndCount"
    />
</LinearLayout>
```

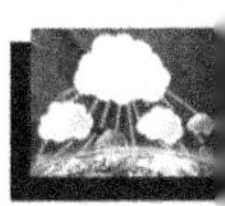

5. 主程序功能实现

1）MainActivity.java 程序文件编写步骤

根据程序实现的功能以及 UI 界面的设计，先引用相关的类，接着定义一个 ServiceConnection 对象 conn，然后声明并取得 UI 界面中的控件对象 TextView 和 Button，并设定 Button 的动作。完整代码如下：

```
package org. dglg. project6_3;
import android. os. Bundle;
import android. os. IBinder;
import android. app. Activity;
import android. app. Service;
import android. content. ComponentName;
import android. content. Intent;
import android. content. ServiceConnection;
import android. view. View;
import android. view. View. OnClickListener;
import android. widget. Button;
import android. widget. TextView;
public class MainActivity extends Activity {
    //保持所绑定的 Service 的 IBinder 对象
    CountService. MyBinder binder;
    /*定义变量及类*/
    int CountTime;
    TextView tv;
    Button BtnCount;
    Button BtnCountEnd;
    Intent intent;
    //定义一个 Serviceconnection 对象
    ServiceConnection conn = new ServiceConnection() {
        //Activity 与 Service 连接成功时回调该方法
        @ Override
        public void onServiceConnected(ComponentName name, IBinder service) {
            // TODO Auto - generated method stub
        //获取 Service 的 onBinder 方法所定义的 MyBinder 对象
            binder = (CountService. MyBinder) service;
        }
        //Activity 与 Service 断开连接时回调该方法
        @ Override
            public void onServiceDisconnected(ComponentName name) {
            // TODO Auto - generated method stub
        }
    };
    @ Override
    protected void onCreate(Bundle savedInstanceState) {
        super. onCreate(savedInstanceState);
```

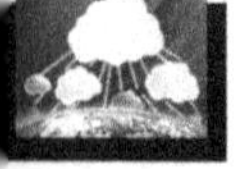

```
        /* 载入 activity_main. xml 布局文件 */
        setContentView(R. layout. activity_main);
        /* 获取布局对象中的 Button 和 TextView */
        BtnCount = (Button)findViewById(R. id. BtnCount);
        BtnCountEnd = (Button)findViewById(R. id. BtnCountend);
        tv = (TextView)findViewById(R. id. tv);
        /* 创建启动 Service 的 Intent */
        intent = new Intent(MainActivity. this, CountService. class);
        /* 开始计时按钮事件 */
          BtnCount. setOnClickListener ( new  OnClickListener ( ) {
@ Override
                public void onClick(View v) {
                // TODO Auto - generated method stub
                /* 用 bindServicce()绑定 Service */
                bindService(intent, conn, Service. BIND_AUTO_CREATE);
                /* 设定 TextView 文本 */
                tv. setText("正在计时……");
                /* 设定按钮状态为不可用 */
                BtnCount. setEnabled(false);
                /* 设定按钮状态为可用 */
                BtnCountEnd. setEnabled(true);
            }
        });
        /* 停止计时按钮事件 */
        BtnCountEnd. setOnClickListener(new OnClickListener() {
            @ Override
                public void onClick(View v) {
                // TODO Auto - generated method stub
                /* 利用 Service 中定义的 binder 类的 getCount 方法取得 Service 所
计时间 */
                String CountTime = String. valueOf(binder. getCount());
                /* 设定 TextView 文本:本次计时时间 */
                tv. setText("本次计时共" + CountTime. toString() + "秒");
                /* 解除 Service 绑定 */
                unbindService(conn);
                /* 设定按钮状态为可用 */
                BtnCount. setEnabled(true);
                /* 设定按钮状态为不可用 */
```

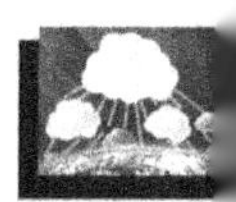

```
                    BtnCountEnd. setEnabled(false);
                }
            });
        }
    }
```

程序说明如下：

(1)在新建ServiceConnection类conn中，定义了2个方法：onServiceConnected和onServiceDisconnected，通过onServiceConnected方法来与Service进行通信。本任务中Activity与Service通信步骤如下(从Service中取得计时时间)：

① 在onServiceConnected中实现：获取CountService中的onBinder方法所定义的MyBinder对象，代码为：

```
CountService MyBinder binder = (CountService. MyBinder)service;
```

② 通过CountService中的getCount( )方法取得计时时间，代码为：

```
String CountTime = String. valueOf(binder. getCount());
```

(2)对按钮状态的设定：

可以在布局文件中直接设定按钮状态(按钮状态默认为true)，也可以在Activity中通过按钮的setEnabled( )方法进行设定，方法和代码如下：

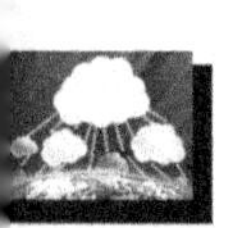

① 在布局文件中，对按钮状态的设定：

```
android: enabled = "false"//将按钮状态设定为不可用，true为可用
```

② 在Activity中对按钮状态的设定，用setEnabled( )方法设定：

```
Button. setEnabled(true); //将按钮状态设定为可用，true为可用
```

(3)采用unbindService停止上述Service，unbindService( )方法只有1个参数，如下代码：

```
Intent = new Intent(MainActivity. this, ExampleService. class); //创建启动Service的
intent;
unbindService(intent); //停止Service
```

2)CountService. java文件的编写步骤

该Service的主要功能是用线程实现计时，并定义方法getCount( )取得计时时间，以便Activity获取，主要代码如下：

```
package org. dglg. project6_3;
/ * 导入相关类 * /
import android. app. Service;
```

```
import android. content. Intent;
import android. os. Binder;
import android. os. IBinder;
public class CountService extends Service{
/ * 定义计时开始标志，false 为计时，true 为停止计时 * /
boolean threaddisable;
/ * 计时变量 * /
int CountTime;
Thread thread;
/ * 定义 onBinder 方法所返回的对象 * /
MyBinder binder = new MyBinder( );
/ * 通过继承 Binder 来实现 IBinder 类 * /
public class MyBinder extends Binder{
/ * 定义方法 getCount，取得所计时的时间(CountTime) * /
    public int getCount( )

  return CountTime;
}
}
@ Override
public IBinder onBind( Intent intent) {
/ * 返回 IBinder 对象 * /
return binder;
}
/ * Service 创建时回调 onCreate( )方法 * /
public void onCreate( ){
super. onCreate( );
/ * 启动一新线程，实现计时 * /
thread = new Thread( new Runnable( ){
@ Override
public void run( ) {
/ * 启动一条线程，计时每秒加一 * /
while( threaddisable = = false) {
try
{
{
Thread. sleep(1000);
} catch( InterruptedExceptione)
```

```
{
}
CountTime + + ;
}
}
});
thread. start( ) ;
}
/ * Service 断开连接时调用 * /
@ Override
public boolean onUnbind(Intent intent)
{
return true;
}
/ * Service 关闭前回调该方法 * /
public void onDestroy( )
{
super. onDestroy( ) ;
/ * 停止计时 * /
this. threaddisable = true;
}
}
```

程序中该Service的功能说明如下：

① 计时功能：在客户端(Activity)与Service绑定后的onCreate( )方法中，用线程的方法每秒计时，计时变量为CountTime。计时标志为boolean类型变量threaddisable，false为开始计时，true为停止计时。

② 与客户端交互：通过实现onBind( )方法，让该方法返回一个有效的IBinder对象，并定义一个类名为MyBinder的类，该类通过继承Binder来实现IBinder类，在里面定义一个方法getCount( )来返回计时时间值CountTime。

③ 程序运行如图6.14～图6.16所示。

任务考核

任务提高：参考本任务，将计时器功能改为定时器功能。要求如下：

1. 定时时间为10秒，10秒到后，用Toast提示定时时间到；
2. 1个按钮用来开始计时，1个文本框用来显示倒计时；
3. 定时过程中，定时按钮不可用；
4. 定时结束后，定时按钮可用，并可以再次定时。

项目七

Android数据存储

应用程序的开发过程离不开数据的输入、输出和保存。因此，Android应用程序的数据存储操作是必不可少的。本项目就如何在Android平台下进行数据存储而设计开发的，它由四个工作任务组成，分别是SharedPreferences类实现用户名和密码保存、SD卡文件的读取、SQLite数据表的操作及内容提供者ContentProvider数据共享和调用的设计。

掌握 SharedPreferences 轻量级存储数据的方法
掌握 SD 卡数据的存取方法
学会 SQLiteDatabase 操作数据库的方法
掌握使用 ContentProvider 实现数据共享的方法

任务一 SharedPreferences 轻量级数据存储——用户名和密码的存储

任务描述

本任务是利用 Android 简单的存储方式之 SharedPreferences 技术实现用户名和用户密码的存储，用户名和用户密码信息均存储项目的目录结构中。

任务分析

SharedPreferences 是 Android 平台上的轻量级的存储类，用于存储一些简单的 key/value 键值对。由于它的值只能是 int 型、boolean 型、string 型和 float 型，所以应用程序通常用它来保存一些常用的配置信息。本任务就是利用 SharedPreferences 来存储用户名和密码。任务分四步完成，第一步调用 context. getSharedPreferences(String name, int mode)方法获取 SharedPreferences 对象，第二步调用 SharedPreferences. Editor edit()方法获取 Editor 对

象，第三步通过 Editor 对象存储 key/value 键值对，第四步调用 commit( )方法提交需要保存的数据。

任务准备

1. Shared Preferences **是什么**

SharedPreferences 是一个接口，要获取 SharedPreferences 的实例对象，需要调用上下文的 getSharedPreferences(String name，int mode)方法。它的定义如下：

SharedPreferences spf = getSharedPreferences (String name，int mode)；

参数 name 表示如果该文件不存在，则会在调用 SharedPreferences. Editor commit( )后自动创建。新建的文件保存在"/data/data/ <包名> /shared_prefs"目录下。

参数 mode 分别有：

MODE_PRIVATE：指定该 SharedPreferences 里的数据只能被本应用程序读写。

MODE_WORLD_READABLE：指定该 SharedPreferences 里的数据可以被其他应用程序读，但不可以被其他应用程序写。

MODE_WORLD_WRITEABLE：指定该 SharedPreferences 里的数据可以被其他应用程序读写。

2. SharedPreferences **接口的常用方法**

Boolean contains(String key)：判断 SharedPreferences 是否包含特定 key 的数据。

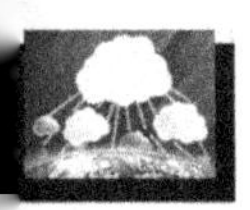

abstract Map <String? >getAll( )：获取 SharedPreferences 里的全部键值对。

boolean getBoolean(String key，boolean defValue)：获取 SharedPreferences 里指定 key 对应的 boolean 值。

int getInt(String key，boolean defValue)：获取 SharedPreferences 里指定的 key 对应的 int 值。

float getFloat(String key，float defValue)：获取 SharedPreferences 里指定的 key 对应的 float 值。

long getLong(String key，long defValue)：获取 SharedPreferences 里指定的 key 对应的 long 值。

string getString(String key，string defValue)：获取 SharedPreferences 里指定的 key 对应的 string 值。

3. SharedPreferences. Editor **对象**

SharedPreferences 对象本身只能获取数据，并不支持数据的存储和修改。数据的存储和修改是通过 SharedPreferences. Editor 对象来实现的。SharedPreferences. Editor 对象的相关方法如下。

SharedPreferences. Editor edit( )：创建一个 Editor 对象。

SharedPreferences. Editor clear( )：清空 SharedPreferences 里的所有数据。

SharedPreferences. Editor putString ( )：向 SharedPreferences 存入指定 key 对应的 String 值。

SharedPreferences. Editor putInt( )：向 SharedPreferences 存入指定 key 对应的 Int 值。

SharedPreferences. Editor putFloat()：向 SharedPreferences 存入指定 key 对应的 Float 值。

SharedPreferences. Editor putLong()：向 SharedPreferences 存入指定 key 对应的 Long 值。

SharedPreferences. Editor putBoolean()：向 SharedPreferences 存入指定 key 对应的 Boolean 值。

SharedPreferences. Editor remove(String key)：删除 SharedPreferences 指定 key 所对应的值。

Boolean commit()：调用该方法提交。

熟练掌握了上面的方法后，就可以开始任务设计了。

任务实现

新建项目 project7_1，在项目中添加处理用户名和密码读取的布局文件 activity_spfn。

1. 布局设计

activity_spfn. xml 很简单，两个 TextView 显示标题，两个 EditView 用于输入用户名和密码，一个存储数据按钮。代码如下：

```
<LinearLayout xmlns:android="http://schemas.android.com/apk/res/android"
    xmlns:tools="http://schemas.android.com/tools"
    android:id="@+id/LinearLayout1"
    android:layout_width="match_parent"
    android:layout_height="match_parent"
    android:orientation="vertical"
    tools:context=".SpfnActivity" >
    <TextView
        android:id="@+id/textView1"
        android:layout_width="wrap_content"
        android:layout_height="wrap_content"
        android:textSize="25sp"
        android:text="用户名:" />
    <EditText
        android:id="@+id/uName"
        android:layout_width="match_parent"
        android:layout_height="80dp"
        android:hint="输入用户名"
        android:background="#ffeefe"
        android:textSize="30sp"
        android:ems="10" >
        <requestFocus />
```

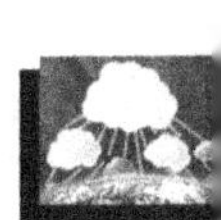

```
        </EditText>
        <TextView
            android: id = "@ + id/textView2"
            android: layout_width = "wrap_content"
            android: layout_height = "wrap_content"
            android: textSize = "25sp"
            android: text = "密码:" />
        <EditText
            android: id = "@ + id/uPwd"
            android: layout_width = "match_parent"
            android: layout_height = "80dp"
            android: ems = "10"
            android: textSize = "30sp"
            android: background = "#ffeefe"
            android: inputType = "textPassword" />
        <Button
            android: id = "@ + id/userName"
            android: layout_width = "wrap_content"
            android: layout_height = "70dp"
            android: textSize = "30sp"
            android: text = "保存用户信息" />
    </LinearLayout>
```

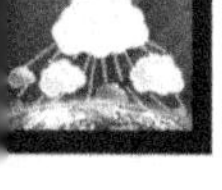

可以看出来，该文件利用线性布局定位的。

2. 存储用户名和密码信息的代码实现

对 SharedPerferences 编程需完成以下四步。

(1)根据 Context 获取 SharedPreferences 对象。

(2)利用 edit()方法获取 Editor 对象。

(3)通过 Editor 对象存储 key/value 键值对。

(4)通过 commit()方法提交数据。

因此，在 SpfnActivity. java 中添加如下代码：

```
uName = (EditText) findViewById(R. id. uName);
upwd = (EditText) findViewById(R. id. uPwd);
UserName = (Button)findViewById(R. id. userName);
//获取用户名和密码及按钮信息
userName = uName. toString();
userPwd = uPwd. toString();
//强制转化，把 Name 和 uPwd 转化为 String 型
```

```
    Name.setOnClickListener(new OnClickListener() {//监听事件编写
        public void onClick(View arg0) {
SharedPreferences spfn = getSharedPreferences("userInfo", MODE_WORLD_WRITEABLE);
    //获取 SharedPreferences 对象
        SharedPreferences.Editor edit = spfn.edit();
        //获取 Editor 对象
        edit.putString("userName", userName);
        edit.putString("pwd", userPwd);
        //通过 Editor 对象存储 key/value 键值对
        edit.commit();
        //commit()方法提交数据
    }
});
```

3. 实现效果及输出结果

实现效果如图 7.1。运行程序后，可以查看 userInfo.xml 文件的存储情况。在 Eclipse 环境中，点击界面右上角的【Open Perspective】按钮，打开 DDMS 视图，在【File Explorer】面板中找到 userInfo.xml 文件，导出并查看其数据情况。如图 7.2 所示。

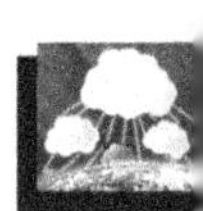

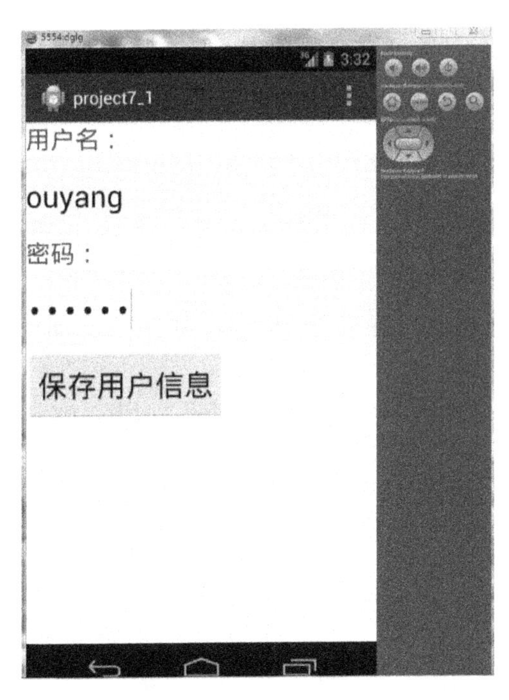

图 7.1　保存用户信息界面

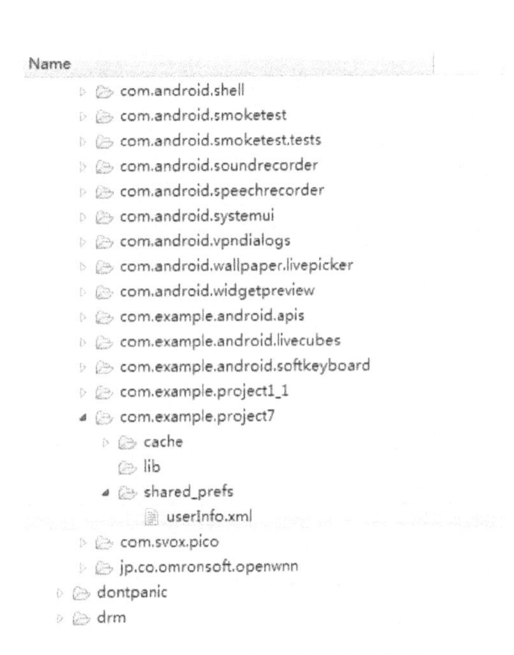

图 7.2　userInfo.xml 保存的位置

点击"File Explorer"视图中的图标，可以导出 userInfo.xml 文件。在 IE 浏览器中打开如图 7.3 所示。

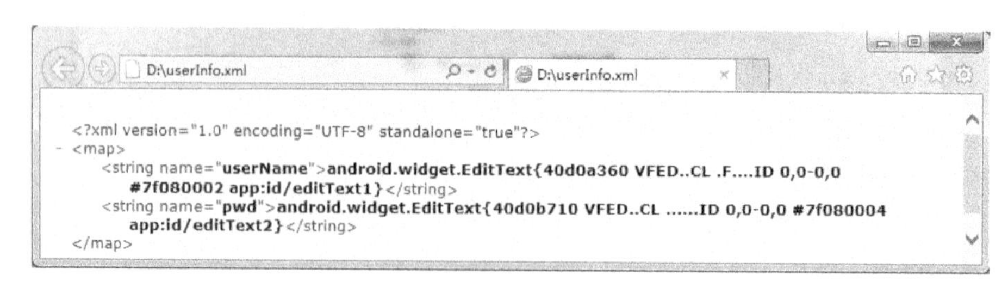

图 7.3　浏览器中打开的 userInfo. xml

任务二　读写 SD 卡中的数据

任务描述

本任务是完成 SD 卡上文件的读写。读者点击“存储 SD 卡数据”可以保存 EditText 编辑框的内容，点击“读取 SD 卡数据”就能读取保存在 SD 卡上的内容，读取和保存的信息均通过 Toast 对象显示。

任务分析

Android 系统支持使用文件存储数据，当手机本身的内存空间较小，不能满足用户的需求时，为了更好地存取应用程序的大文件、大数据，Android 系统也支持在 SD 卡的数据存储。本任务就是让读者掌握如何读取手机中 SD 卡文件的步骤和方法。

任务准备

1. 读写 SD 卡数据的步骤

(1)调用 Enviroment. getExternalStorageState()方法，判断手机中是否插入 SD 卡。如果程序具有读写 SD 卡的权限，那么

Enviroment. getExternalStorageState ( ) . equals (Enviroment. MEDIA _ MOUNTED) 返回 TRUE。

(2)调用 Enviroment. getExternalStorageDirectory()方法获取 Android 外部存储器的目录。

(3) 调用 FileReader、FlieWriter、FileInputStream、FileOutputStream 读写 SD 卡中的文件。

2. 读写 SD 卡要注意的事项

(1)手机中插入 SD 卡，用模拟器调试，要在创建模拟器时设置 SD 卡的容量。

(2)在 AndroidManifest. xml 文件中添加用户权限。

任务实现

1. 创建项目

创建项目工程 project7_2，在该布局文件里添加一个 EditText 用于输入存储文件信息；添加两个按钮，一个写入数据，一个读取数据。由于布局文件较简单，此处代码省略。

2. 添加用户权限

在系统的 AndroidManifest. xml 添加下列代码，旨在获取对 SD 卡读写的权限。

```
    < uses - permission
android: name = "android. permission. MOUNT_UNMOUNT_FILESYSTEMS"/ >
        //可在 SD 卡中创建或者删除文件
    < uses - permission
android: name = "android. permission. WRITE_EXTERNAL_STORAGE"/ >
        //可以在 SD 卡中写入文件内容
```

3. 业务的编写

在 SDActivity. java 代码中添加业务逻辑，为按钮绑定监听，重写 onClick( )方法，保存数据功能的代码为：

```
final EditText fileInfo = (EditText) findViewById(R. id. SDtext);
    Button btWriteBtn = (Button) findViewById(R. id. WriteSDK);
    Button btReaderBtn = (Button) findViewById(R. id. ReadSDK);
    btWriteBtn. setOnClickListener(new OnClickListener( ) {
    public void onClick(View arg0) {
        String state = Environment. getExternalStorageState( );
            if(state. equals(Environment. MEDIA_MOUNTED)) {
    File sdFile = Environment. getExternalStorageDirectory( );
                String filePath = sdFile. getAbsolutePath( );
                String msg = fileInfo. getText( ). toString( );
            try {
FileWriterfilewriter = new fileWriter(filePath + "/SDfile. txt", true);
        BufferedWriter bWriter = new BufferedWriter(filewriter);
                    bWriter. write(msg);
                    bWriter. flush( );
Toast. makeText(SDActivity. this, "写入成功", Toast. LENGTH_SHORT). show( );
                } catch (Exception e) {
                    e. printStackTrace( );
```

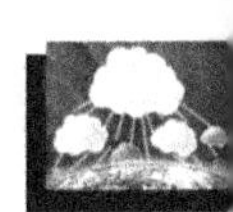

```
                    }
                }
            }
        });
```

为按钮绑定监听，重写 onClick( )方法，读取数据功能的代码为：

```
btReaderBtn. setOnClickListener( new OnClickListener( ) {
            public void onClick( View arg( ) ) {
                String state = Environment. getExternalStorageState( ) ;
                if( state. equals( Environment. MEDIA_MOUNTED) ) {
                    File sdFile = Environment. getExternalStorageDirectory( ) ;
                    String filePath = sdFile. getAbsolutePath( ) ;
                    try {
    FileReader fileReader = new FileReader( filePath + "/SDfile. txt" ) ;
    BufferedReader bReader = new BufferedReader( fileReader) ;
                String line;
                while( ( line = bReader. readLine( ) ) !  = null) {
                Toast. makeText( SDActivity. this, "内容是:"
     + line, Toast. LENGTH_SHORT ). show( ) ;
                        }

                    } catch (Exception e) {
                        e. printStackTrace( ) ;
                    }
                }
            }
        });
```

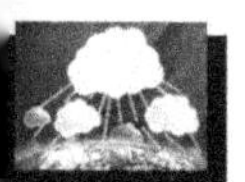

运行效果如图 7.4 和图 7.5 所示。

4. 项目运行导出结果

打开 DDMS 视图的“File Explorer”面板，找到 SDfile. txt 文件，点击 将其导出，如图 7.6 所示。在记事本打开 SDfile. txt，得到如图 7.7 所示的结果。

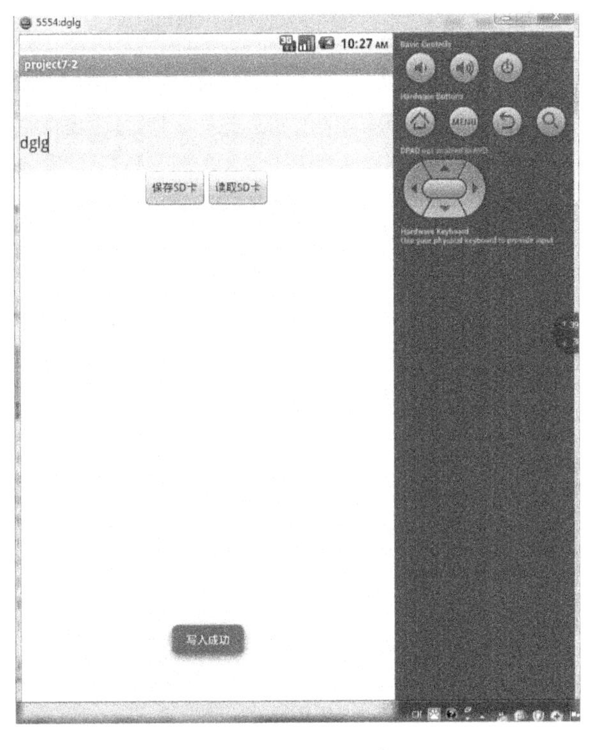

图 7.4　写入 SD 卡

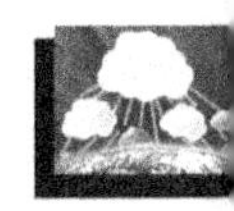

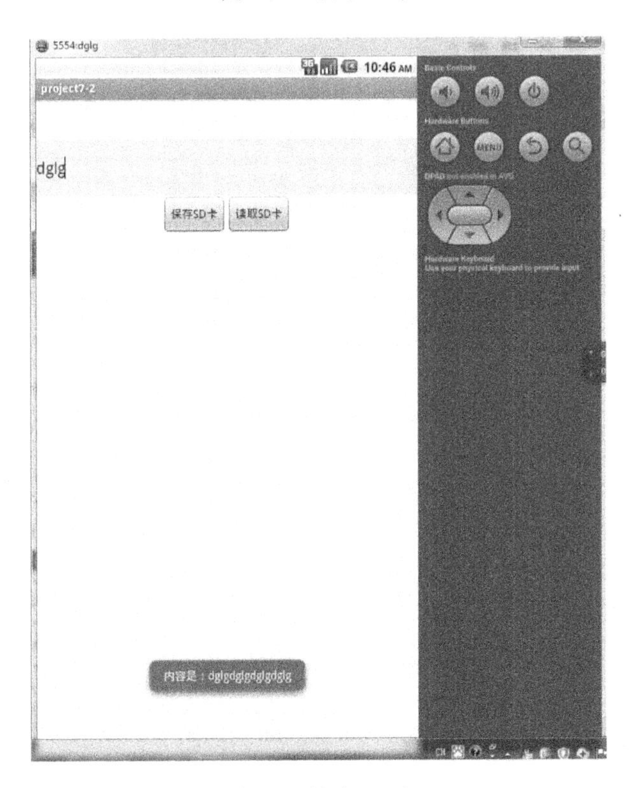

图 7.5　读取 SD 卡

Threads | Heap | File Explorer | Emulator Control

Name	Size	Date	Time	Permissions
proc		1969-12-31	19:00	dr-xr-xr-x
property_contexts	2161	1969-12-31	19:00	-rw-r--r--
root		2013-07-09	20:46	drwx------
sbin		1969-12-31	19:00	drwxr-x---
sdcard		2014-07-24	01:43	lrwxrwxrwx
seapp_contexts	656	1969-12-31	19:00	-rw-r--r--
sepolicy	74708	1969-12-31	19:00	-rw-r--r--
storage		2014-07-24	01:43	drwxr-x--x
sdcard		2014-07-24	02:19	drwxrwx--x
Alarms		2014-07-18	21:23	drwxrwx---
DCIM		2014-07-18	21:23	drwxrwx---
Download		2014-07-18	21:23	drwxrwx---
LOST.DIR		2014-07-19	01:19	drwxrwx---
Movies		2014-07-18	21:23	drwxrwx---
Music		2014-07-18	21:23	drwxrwx---
Notifications		2014-07-18	21:23	drwxrwx---
Pictures		2014-07-18	21:23	drwxrwx---
Podcasts		2014-07-18	21:23	drwxrwx---
Ringtones		2014-07-18	21:23	drwxrwx---
SDfile.txt	12	2014-07-24	02:19	-rwxrwx---
sys		2014-07-24	01:43	dr-xr-xr-x
system		1969-12-31	19:00	drwxr-xr-x

图 7.6　SDfile. txt 在 SD 卡存放的位置

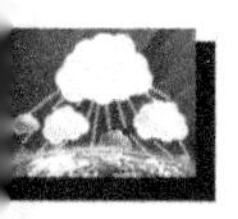

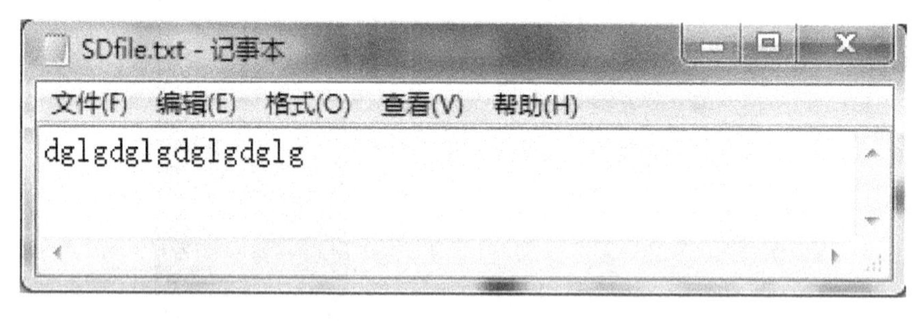

图 7.7　在记事本打开 SDfile. txt

任务三　数据表增删改查——学生信息管理

任务描述

本任务是完成在学生信息表的存储，项目有五个按钮，分别是创建学生信息表，插入一条学生信息，更改学生信息，删除学生信息，查询学生信息，Android 提供的 SQLiteDatabase 类的方法可以实现这些功能。

任务分析

Android 系统集成了一个轻量级数据库 SQLite。它是一个嵌入式的数据库引擎，专门适用于资源有限的设备上适量数据存储，但是它支持 SQL 语言，而且性能很高、很便捷。

我们要完成本任务，需要掌握 SQLiteDatabase 类创建数据库的方法和操作数据库的方法。

任务准备

1. SQLiteDatabase 类

在开发 SQLite 数据库的 APP 时，通常会用到 SQLiteDatabase 类。Android 系统提供了 SQLiteDatabase 类，该类提供大量的 API 来操控 SQLite 数据库，每一个 SQLiteDatabase 实例就代表一个数据库，对应底层的一个数据库文件，一旦应用程序获得了 SQLiteDatabase 对象，就可以通过该对象操作数据库了。

2. SQLiteDatabase 类创建数据库的方法

在 Android 平台上，可以通过 SQLiteDatabase 类来 Create 或 Open 数据库。

第一种方法

```
SQLiteDatabase data =                 //声明一个 SQLiteDatabase 对象
    openDatabase(                     //打开数据库
    String Path,                      //数据库存储的路径
    SQLiteDatabase. CursorFactory factory, //用于生产一个游标对象
    int flags)       //控制数据的可访问模式，0 表示可读可写，1 表示可读不可写。
```

第二种方法

```
SQLiteDatabase data =                 //声明一个 SQLiteDatabase 对象
    openorCreateDatabase(             //打开创建数据库
    String Path,                      //数据库存储的路径
    SQLiteDatabase. CursorFactory factory, //用于生产一个游标对象
    int flags)       //控制数据的可访问模式，0 表示可读可写，1 表示可读不可写。
```

第三种方法

```
SQLiteDatabase data =                 //声明一个 SQLiteDatabase 对象
    openorCreateDatabase(             //打开创建数据库
    File file,                        //File 数据库文件
    SQLiteDatabase. CursorFactory factory)//用于生产一个游标对象
```

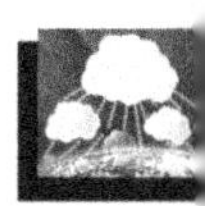

3. SQLiteDatabase 类操作数据库的方法

数据库创建之后，可以对数据库进行操作。SQLiteDatabase 类提供了下列方法来完成数据库的增、删、改、查。

1) execSQL(String sql)执行 SQL 语句

execSQL(String sql, Object[ ] bindArgs)执行带占位符的 SQL 语句。

Insert(String table, String nullColumnHack, ContentValues values) 向表中插入一条记录。

Update(String table, ContentValues values, String wherClause, String[ ] whereArgs) 更新表中的记录。

Delete(String table, String wherClause, String[ ] whereArgs) 删除表中的指定记录。

Query(String table, String[ ] columns, String selection, String[ ] selectArgs, String groupBy, String having, String orderBy) 查询表中的记录。

rawQuery(String sql, String[ ] selectionArgs) 查询带占位符的记录。

2)Cursor 对象的方法

对 Query(String table, String[ ] columns, String selection, String[ ] selectArgs, String groupBy, String having, String orderBy) 查询方法，其返回值是一个 Cursor 对象。它常用的方法有：

Move(int offset)从当前位置将游标向上或向下移动的行数。

moveToFirst()将游标移动到第一行，成功则返回 true。

moveToLast()将游标移动到最后一行，成功则返回 true。

moveToNext()将游标移动到下一行，成功则返回 true。

moveToPosition(int positon)将游标移动到指定行，成功则返回 true。

moveToPrevious()将游标移动到前一行，成功则返回 true。

任务实现

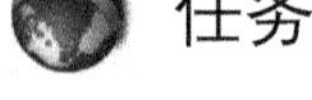

1. 新建项目 project7_3

在项目中添加 activity_sql 布局文件，在该布局文件里添加五个 Button，分别用来创建数据表、插入数据、更改数据、删除数据、查询数据。添加一个 ListView 用来显示查询结果，呈现学生信息。

2. 编写学生信息增删改查的逻辑代码

1)创建数据库

在 SQLActivity. java 中添加

SQLiteDatabase database = SQLiteDatabase. openOrCreateDatabase ("/data/data/com. example. project7/Stud_data. db" , null)

在/data/data/com. example. project7/目录下创建数据库 Stud_data. db。

2)创建学生信息表

在“创建学生信息表”按钮的监听事件中添加下列代码。创建 Stud_data 信息表，表中含有_id、name 和 age 三个字段。

```
String creatStr = "create table Stud_data(" +
                "_id int," +
                "name char(20)," +
                "age int)";
            database. execSQL(creatStr);
```

3)插入记录

在“插入一条学生信息”按钮的监听事件中添加下列代码。

```
public void onClick(View arg0) {
                String insertStr1 = "insert into
Stud_data(_id, name, age)" +" values(?,?,?)";
                Object[] valuesObjects = {1,"张学友", 22};
                database.execSQL(insertStr1, valuesObjects);
                String insertStr2 = "insert into
Stud_data(_id, name, age)" +"values(2,"洪金宝", 24)";
                database.execSQL(insertStr2);
        }
    });
```

4)更新记录

在“更新学生信息”按钮的监听事件中添加下列代码。

```
public void onClick(View arg0) {
                ContentValues values = new ContentValues();
                values.put("name", "刘德华");
                database.update("Stud_data", values, "_id = ?", new
String[]{"1"});
        }
    });
```

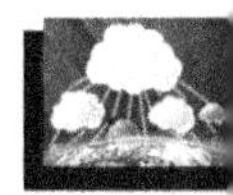

5)删除记录

在“删除学生信息”按钮的监听事件中添加下列代码。

```
public void onClick(View arg0) {
                database.delete("Stud_data", "_id = ?", new
String[]{"2"});
        }
    });
```

6)查询记录

在“查询学生信息”按钮的监听事件中添加用来查询表中的记录，并将结果以列表的形式显示出来。

```
public void onClick(View arg0) {
            Cursor cursor = database.query("Stud_data", new
String[]{"_id","name","age"}, null, null, null, null, "age desc");
            SimpleCursorAdapter sCursorAdapter = new SimpleCursorAdapter(
                SQLActivity.this,
                R.layout.users,
                cursor,
                new String[]{"_id","name","age"},
                new int[]{R.id.editText3, R.id.editText1, R.id.editText2},
        CursorAdapter.FLAG_REGISTER_CONTENT_OBSERVER);
            ListView listView = (ListView)findViewById(R.id.Stud_data);
            listView.setAdapter(sCursorAdapter);
        }
    });
```

3. 运行效果和数据库的查看

1)运行结果

运行结果如图 7.8 所示。

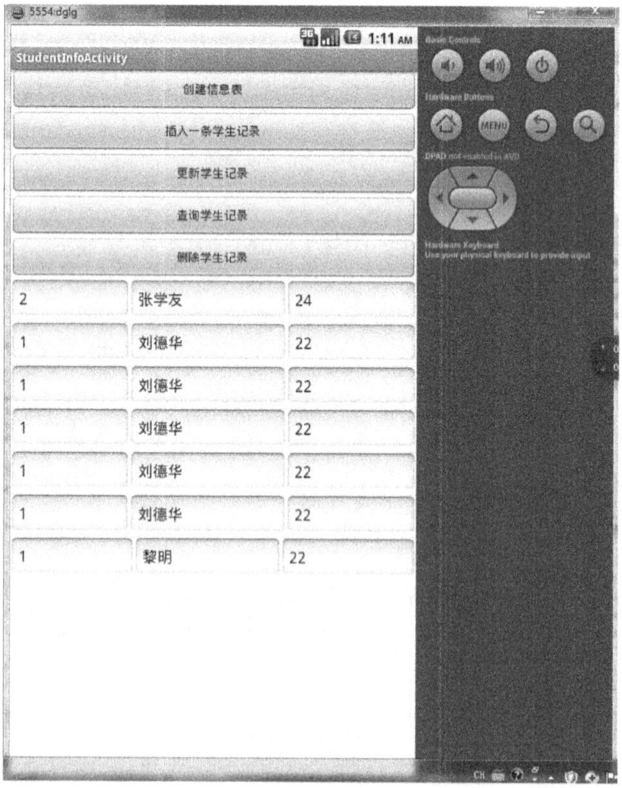

图 7.8　查询数据库

2）数据库查看

打开 DDMS 视图的“File Explorer”面板，找到 Stud_data 文件，点击将其导出，如图 7.9 所示。

Name	Size	Date	Time	Permissions	Info
com.android.packageinstaller		2014-09-24	10:24	drwxr-x--x	
com.android.phone		2014-09-24	10:25	drwxr-x--x	
com.android.protips		2014-09-24	10:24	drwxr-x--x	
com.android.providers.applications		2014-09-24	10:24	drwxr-x--x	
com.android.providers.contacts		2014-09-24	10:24	drwxr-x--x	
com.android.providers.downloads		2014-09-24	10:25	drwxr-x--x	
com.android.providers.drm		2014-09-24	10:24	drwxr-x--x	
com.android.providers.media		2014-09-24	10:25	drwxr-x--x	
com.android.providers.settings		2014-09-24	10:24	drwxr-x--x	
com.android.providers.subscribedfeeds		2014-09-24	10:24	drwxr-x--x	
com.android.providers.telephony		2014-09-24	10:25	drwxr-x--x	
com.android.providers.userdictionary		2014-09-24	10:24	drwxr-x--x	
com.android.quicksearchbox		2014-09-24	10:25	drwxr-x--x	
com.android.sdksetup		2014-09-24	10:24	drwxr-x--x	
com.android.server.vpn		2014-09-24	10:24	drwxr-x--x	
com.android.settings		2014-09-24	10:24	drwxr-x--x	
com.android.soundrecorder		2014-09-24	10:24	drwxr-x--x	
com.android.spare_parts		2014-09-24	10:24	drwxr-x--x	
com.android.speechrecorder		2014-09-24	10:24	drwxr-x--x	
com.android.term		2014-09-24	10:24	drwxr-x--x	
com.android.wallpaper.livepicker		2014-09-24	10:24	drwxr-x--x	
com.example.android.apis		2014-09-24	10:24	drwxr-x--x	
com.example.android.livecubes		2014-09-24	10:24	drwxr-x--x	
com.example.android.softkeyboard		2014-09-24	10:24	drwxr-x--x	
com.example.project7_2		2014-09-24	10:25	drwxr-x--x	
com.example.sdcardstore		2014-09-24	10:40	drwxr-x--x	
com.example.sqlitedatabase		2014-09-25	01:20	drwxr-x--x	
Stud_data.db	4096	2014-09-25	01:20	-rw-r--r--	
lib		2014-09-25	00:36	drwxr-xr-x	
uesrs.db	4096	2014-09-25	01:17	-rw-r--r--	
com.svox.pico		2014-09-24	10:24	drwxr-x--x	
jp.co.omronsoft.openwnn		2014-09-25	00:36	drwxr-x--x	

图 7.9　数据库存储位置

要查看数据库和表结构，需要下载安装 SQLite Expert Personal。将导出的 Stud_data. db 放在 SQLite Expert Personal 中，就可以看到如图 7.10 所示。

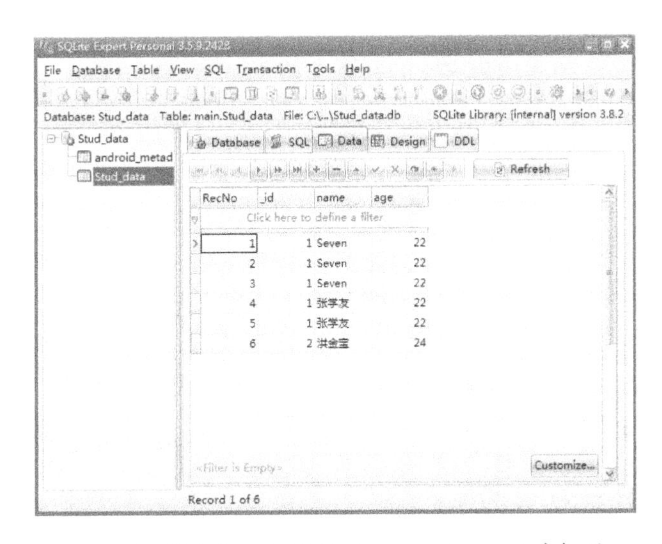

图 7.10　学生信息表在 SQLite Expert Personal 中打开

任务四　测试ContentProvider共享数据

 任务描述

本任务是完成ContentProvider子类FirstProvider的开发，重写ContentProvider的onCreate( )等方法，实现增删改查的方法，注册并暴露ContentProvider的数据。最后创建另外一个应用程序FirstResolver调用FirstProvider的增、删、改、查的方法来操作它的数据。当应用继承ContentProvider类，并重写该类用于提供数据和存储数据的方法，就可以向其他应用共享其数据。虽然使用其他方法也可以对外共享数据，但数据访问方式会因数据存储的方式而不同，例如采用文件方式对外共享数据，需要进行文件操作读写数据，还可能需要使用SharedPreferences API读写数据，而采用SharedPreferences共享数据就没有这些麻烦了，因为使用ContentProvider共享数据是使用了统一的数据访问方式。

 任务分析

内容提供者ContentProvider是Android平台的一大组件。它提供了标准的API接口，非常方便实现不同应用程序之间的数据共享。一个程序可以通过实现ContentProvider的抽象接口，将自己的数据以URI的形式完全暴露出来，然后通过其他程序使用ContentResolver根据URI访问操作指定的共享数据。因此，完成此任务需实现下列三步：第一是开发继承于ContentProvider的子类，第二是注册ContentProvider的子类，第三是开发另一个应用程序操作共享数据。

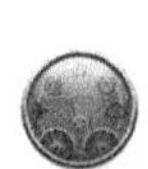 任务准备

1. ContentProvider简介

ContentProvider是不同应用程序之间数据交换的标准API，ContentProvider以某种URI的形式对外提供数据，允许其他应用访问或修改数据，其他应用程序使用ContentResolver根据URI去访问操作的指定数据。

如果把ContentProvider看作一个卖东西的商店，开商店要选址，如果找不着地址，那这个商店还能经营下去吗？如此看来，我们使用ContentProvider其实很简单。接下来我们就在Android世界里面开个店吧！步骤是这样的：

STEP 1 定义自己的ContentProvider类，该类需要继承Android系统提供的ContentProvider基类。

STEP 2 申请开这个“店”的许可证，就是在AndroidMainifest. xml文件中注册这个ContentProvider。注册的时候，ContentProvider需要为它绑定一个位置。

向Android系统中注册ContentProvider只要再<application. . . >元素下添加如下元素就行：

```
<provider android: name = ". DictProvider"
    android: authorities = "dg. dglg. project7. Dictprovider"
    android: exported = "true"/ >
```

上面的配置片段可以看出，配置 ContentProvider 时通常指定如下属性。

• name：指定 ContentProvider 的实现类的类名，基类是 Android 系统提供的。

• authorities：指定 ContentProvider 对应的 URI，相当于为 ContentProvider 分配一个域名。

• exported：指定 ContentProvider 是否允许其他应用调用。如果将该属性设为 false，那么 ContentProvider 将不允许其他应用调用。

STEP 3 在 Android 系统注册了 ContentProvider 之后，其他应用程序就可通过 URI 来访问 DictProvider 所出售的数据了。

STEP 4 那么 DictProvider 是怎么出售自己的数据呢？很简单，应用程序对数据的操作就是 CRUD 操作，因此 DictProvider 除了需要继承 ContentProvider 之外，还需要实现如下几个方法。

• Pubilc boolean onCreate( )：方法在 ContentProvider 创建后就被调用，用户第一次访问 ContentProvider 时，就会立即回调 onCreate( )方法。

• Public Uri insert(Uri uri, ContentValues values)：根据 URI 插入 values 对应的数据。

• Public int delete(Uri uri, String selection, String[ ] selectionArgs)：根据 URI 删除 select 条件所匹配的全部记录。

• Public int update (Uri uri, ContentValues values, String selection, String [ ] selectionArgs)：根据 URI 修改 select 条件所匹配的全部记录。

• Public Cursor query (Uri uri, String [ ] Projection, String selection, String [ ] selectionArgs, String sortOrder)：根据 URI 查询出 select 条件所匹配的全部记录，其中 Projection 就是一个列名列表，表明只选择出指定的数据列。

• Public String getType(Uri uri)：该方法用于返回当前 URI 代表的数据的 MIME 类型，如果 URI 对应数据可能包括多条记录，那么 MIME 类型字符串应该以 vnd. android. cursor. dir/开头；如果该 URI 对应数据只包含一条记录，那么返回的 MIME 类型的字符串应该以 vnd. android. cursor. item/开头。

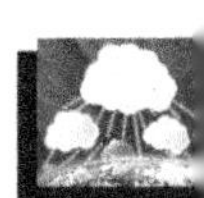

2. URI 简介

对于 ContentProvider 来讲，URI 是一个非常重要的概念，在介绍 Android 系统的 URI 之前，先看看一个最常用的互联网的 URL，我们经常访问百度的网址 http：//www. baidu. com/index. php。

百度的这个 URL，可分为以下三个部分：

• http：//：URL 的协议部分，只要通过 HTTP 协议来访问网站，均要用到。

• www. baidu. com：网站域名部分。只要访问指定网站，均要用到。

• index. php：网站资源部分。只要访问网站下特定内容，均要用的。

Android 的 URI 跟这个类似，例如以下 URI：

Content：//dg. dglg. project7. Dictprovider/words

同样可以分为三个部分：

- Content：//：这个部分是 Android 所规定的。
- dg. dglg. project7. Dictprovider/：这个部分就是 ContentProvider 的 authority。系统就是由这个部分来找到操作哪个 ContentProvider。只访问指定的 ContentProvider。
- words：资源部分（或者说数据部分），访问者访问不同资源时的标志符。如果有学习数据库的同学可以把这个看成一个数据库。

URL 是一个类，需要把字符串转换成 URI，contentProvider 才能够使用，Android 提供的 URL 工具类提供了 parse( ) 静态方法，把字符串转换成 URI 就可以写成：

URI uri = Uri. parse(“Content：//dg. dglg. project7. Dictprovider/words”)。

3. 使用 ContentResolver 操作数据

Android 系统的数据在 Android 当中是私有的，这些数据包括文件数据和数据库数据以及一些其他类型的数据。读者会问，难道两个程序之间就没有办法对于数据进行交换？Android 这么优秀的系统是不会让这种情况发生的。为了解决这个问题，一个 ContentProvider 类诞生了，它实现了一组标准的方法接口，从而能够让其他的应用保存或读取此 ContentProvider 的各种数据类型。就是说，一个程序可以通过实现一个 ContentProvider 的抽象接口将自己的数据暴露出去。外界根本看不到，也不用看到这个应用暴露的数据在应用当中是如何存储的，或者是用数据库存储还是用文件存储，还是通过网上获得，这些都不重要，重要的是外界可以通过这一套标准及统一的接口和程序里的数据打交道，可以读取程序的数据，也可以删除程序的数据，当然，中间也会涉及一些权限和权限设置的问题。

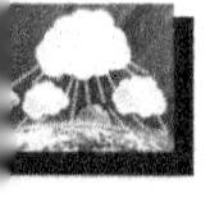

一个程序可以通过实现一个 ContentProvider 的抽象接口将自己的数据完全暴露出去，而且 ContentProviders 是以类似数据库中表的方式将数据暴露，也就是说 ContentProvider 就像一个“数据库”。外界获取其提供的数据，与从数据库中获取数据的操作基本一样，只不过是采用 URL 来表示外界需要访问的“数据库”。

ContentProvider 相当于一个网站的数据，它的作用是暴露可供操作的数据；其他应用程序则通过 ContentResolver 来操作 ContentProvider 所暴露的数据，ContentResolver 相当于 HttpClient。

Context 提供了 getContentResoler( ) 方法来获取 ContentResolver 对象。一旦在程序中获得 ContentResolver 对象后，接下来就可以调用如下方法来操作数据了。

Insert(Uri uri，ContentValues values)：向 Uri 对应的 ContentProvider 中插入 values 对应的数据。

Delete(Uri uri，String where，String[ ] selectionArgs)：删除 Uri 对应的 ContentProvider 中 where 提交匹配的数据。

Update(Uri uri，ContentValues values，String where，String[ ] selectionArgs)：更新 Uri 对应的 ContentProvider 中匹配的数据。

Query(Uri projection，String selecetion，String[ ] selectionArgs，String sortOrder)：查询 Uri 对应的 ContentProvider 中 where 提交匹配的数据。

此外，ContentProvider 还提供了一种多应用间数据共享的方式，比如：联系人信息可以被多个应用程序访问。例如

content：//media/internal/images 表示 URI 将返回设备上存储的所有图片。

content：//contacts/people/表示 URI 将返回设备上的所有联系人信息。

content：//contacts/people/45 表示 URI 返回单个结果(联系人信息中 ID 为 45 的联系人记录)。

任务实现

1. 项目 project7_4 添加 MyContentProvider

在 MyContentProvider. java 中创建 ContentProvider 子类，由于该文件主要是为了暴露数据，共享给其他应用程序使用，所以这里不用设计布局文件。

2. 编写 ContentProvider 子类 MyContentProvider

开发 ContentProvider 的子类，重写了增、删、改、查的方法。该类以独立的文件存储在项目 project7_4 中，文件名是 MyContentProviderProvider. java。

```
public class MyContentProvider extends ContentProvider {
private SQLiteDatabase      sqlDB;
private DatabaseHelper      dbHelper;
private static final String   DATABASE_NAME  =  "Users. db";
private static final int   DATABASE_VERSION = 1;
private static final String TABLE_NAME =  "User";
private static final String TAG  =  "MyContentProvider";
private static class DatabaseHelper extends SQLiteOpenHelper {
    DatabaseHelper(Context context) {
        super(context, DATABASE_NAME, null, DATABASE_VERSION);
    }
    @ Override
    public void onCreate(SQLiteDatabase db) {
        //创建用于存储数据的表
    db. execSQL("Create table "  + TABLE_NAME +  "( _id INTEGER PRIMARY
KEY AUTOINCREMENT, USER_NAME);");
    }
    @ Override
    public void onUpgrade(SQLiteDatabase db, int oldVersion, int newVersion) {
        db. execSQL("DROP TABLE IF EXISTS "  + TABLE_NAME);
        onCreate(db);
    }
}

@ Override
```

```
    public int delete(Uri uri, String s, String[] as) {
        return 0;
    }
    @Override
    public String getType(Uri uri) {
        return null;
    }
    @Override
    public Uri insert(Uri uri, ContentValues contentvalues) {
        sqlDB = dbHelper.getWritableDatabase();
        long rowId = sqlDB.insert(TABLE_NAME, "", contentvalues);
        if (rowId > 0) {
                Uri rowUri = ContentUris.appendId(MyUsers.User.CONTENT_URI.
buildUpon(), rowId).build();
            getContext().getContentResolver().notifyChange(rowUri, null);
            return rowUri;
        }
        throw new SQLException("Failed to insert row into" + uri);
    }
    @Override
    public boolean onCreate() {
        dbHelper = new DatabaseHelper(getContext());
        return (dbHelper == null) ? false : true;
    }
    @Override
    public Cursor query(Uri uri, String[] projection, String selection, String[]
selectionArgs, String sortOrder) {
        SQLiteQueryBuilder qb = new SQLiteQueryBuilder();
        SQLiteDatabase db = dbHelper.getReadableDatabase();
        qb.setTables(TABLE_NAME);
        Cursor c = qb.query(db, projection, selection, null, null, null, sortOrder);
        c.setNotificationUri(getContext().getContentResolver(), uri);
        return c;
    }
    @Override
    public int update(Uri uri, ContentValues contentvalues, String s, String[] as) {
        return 0;
    }
}
```

3. 注册和配置 ContentProvider

为了暴露 ContentProvider 的数据，使其他的应用程序可以共享 ContentProvider 的数据，需在 AndroidManifast. xml 注册 FirstProvider，即增加下列代码：

```
    <providerandroid: name = "MyContentProvider"  android: authorities = "com. example.
MyContentProvider" />
                android: exported = "true"/>
```

4. 编写 MyUsers 类

该类主要完成些全局常量的定义，包括访问 ContentProvider 的 URI 以及返回 ContentProvider 的数据类型、数据列表等信息。

```
    public class MyUsers {
    public static final String AUTHORITY    = "com. wissen. MyContentProvider";
        // BaseColumn 类中已经包含了 _id 字段
        public static final class User implements BaseColumns {
        public static final Uri CONTENT_URI    =
        Uri. parse("content: //com. wissen. MyContentProvider");
        // 表数据列
        public static final String  USER_NAME   = "USER_NAME";
    }
}
```

5. 开发 MyContectDemo 项目验证 MyContentProvider 暴露数据是否成功

在 MyContectDemo 项目中，通过 Toast 类显示结果，所以不用设计布局文件。MyContentDemo 业务代码为：

```
public class FirstResolver extends Activity
{
    ContentResolver contentResolver;
    Uri uri  =
  Uri. parse("content: //org. crazyit. providers. firstprovider/");
    public void onCreate(Bundle savedInstanceState)
    {
        super. onCreate(savedInstanceState);
        setContentView(R. layout. main);
        // 获取系统的 ContentResolver 对象
        contentResolver  =  getContentResolver();
    }
    public void query(View source)
    {
```

```
        // 调用 ContentResolver 的 query()方法。
        // 实际返回的是该 Uri 对应的 ContentProvider 的 query()的返回值
Cursor c = contentResolver.query(uri, null, "query_where", null, null);
Toast.makeText(this, "远程 ContentProvide 返回的 Cursor 为:" + c,
        Toast.LENGTH_LONG).show();
    }

    public void insert(View source)
    {
        ContentValues values = new ContentValues();
        values.put("name", "fkjava");
        // 调用 ContentResolver 的 insert()方法。
        // 实际返回的是 Uri 对应的 ContentProvider 的 insert()的返回值
        Uri newUri = contentResolver.insert(uri, values);
        Toast.makeText(this, "远程 ContentProvide 新插入记录的 Uri 为:"
            + newUri, Toast.LENGTH_LONG).show();
    }

    public void update(View source)
    {
        ContentValues values = new ContentValues();
        values.put("name", "fkjava");
        // 调用 ContentResolver 的 update()方法。
        // 实际返回的是 Uri 对应的 ContentProvider 的 update()的返回值
        int count = contentResolver.update(uri, values
            , "update_where", null);
        Toast.makeText(this, "远程 ContentProvide 更新记录数为:"
            + count, Toast.LENGTH_LONG).show();
    }

    public void delete(View source)
    {
        // 调用 ContentResolver 的 delete()方法。
        // 实际返回的是 Uri 对应的 ContentProvider 的 delete()的返回值
        int count = contentResolver.delete(uri
            , "delete_where", null);
        Toast.makeText(this, "远程 ContentProvide 删除记录数为:"
            + count, Toast.LENGTH_LONG).show();
    }
}
```

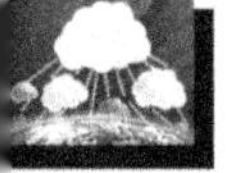

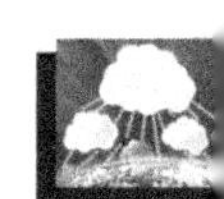

6. 运行结果

运行结果如图 7.11 所示。

图 7.11　自定义 ContentProvider 运行结果

本项目针对 Android 系统中数据存储的四种方式分别制作了四个任务，每个任务讲授一种存储方式和它们各自的适用场合。最后把四个任务制作整合到一个大项目中。四种存储方式中 SharedPreferences 轻量级存储数据和 SD 卡的读写比较简单，比较容易掌握，SQLite 数据库存储方式在 Android 开发中比较常用，而 ContentProvider 数据共享是难点，需要读者细心揣摩，努力掌握。

任务考核

(1)新建项目“SHP”，使用 SharedPreferences 实现数据存储。在“File Explorer”面板上查看生成的文件 shpFile. xml。

(2)新建项目“FileStore”，在编辑框中输入内容并保存，点击“读取”，将保存的信息显示在界面上。

项目八

Android多媒体开发

Android在多媒体与娱乐方面的应用是Android的一个重要组成部分，本项目将对Android开发多媒体方面进行多方位介绍，帮助读者了解Android中多媒体与娱乐开发的基本知识。

了解 Android 多媒体与娱乐开发的基本知识
熟悉及掌握 Android 的 MediaPlayer 类
熟悉及掌握 Android 的 VideoView 类
熟悉及掌握 Android 的 MediaRecorder 类
掌握 Android 多媒体基本开发流程

任务一　简易 mp3 音乐播放程序

任务描述

在 Android 手机应用中，播放音乐是主要功能之一。本任务通过一个具体的播放 mp3 音乐程序的实现过程，介绍在 Android 手机中播放 mp3 音乐文件的具体流程。

程序运行界面如图 8.1 所示(音乐文件为 AVD 虚拟的 SD 卡中的 mp3 文件)。

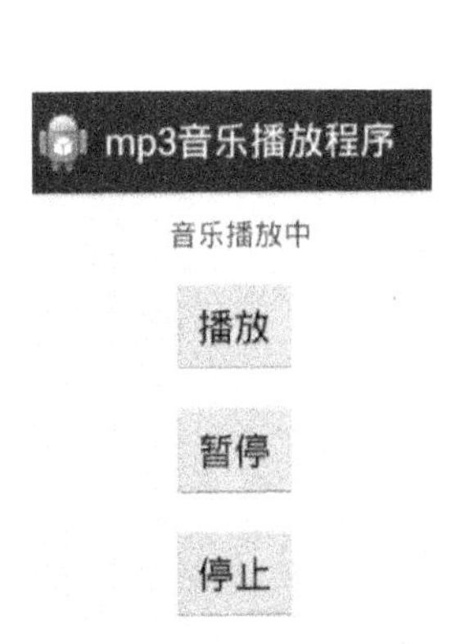

图 8.1　mp3 音乐播放程序

任务分析

在本任务中，定义了 3 个按钮，分别用于 mp3 音乐的播放、暂停、停止。

①"播放"按钮开始播放音乐文件。

②"暂停"按钮控制 mp3 音乐的暂停播放。

③"停止"按钮控制 mp3 音乐的停止播放。

本任务的主要目的通过制作简易 mp3 音乐播放器熟悉 MediaPlayer 类的开发，重点和难点如下：

(1)熟悉 MediaPlayer 类；
(2)学习和掌握 MediaPlayer 类的一些常用方法等；
(3)掌握 MediaPlayer 类开发的基本步骤和方法；
(4)掌握 Android 开发播放音乐类程序的基本方法。

任务准备

1. 任务解析

在 Android 平台下，要实现声音的播放可以考虑使用一个 MediaPlayer 类，并调用它的方法，从而达到对声音播放的控制。

在本任务中，可以考虑运用 MediaPlayer 类来对 mp3 资源文件进行操作。其中，定义 3 个按钮(Button)来对 MediaPlayer 类进行控制。用 MediaPlayer 类的 create( )方法用来创建播放器资源；MediaPlayer 类的 play( )方法用来播放播放器资源；MediaPlayer 类的 pause( )方法用来暂停播放资源；MediaPlayer 类的 stop( )方法用来停止播放资源。

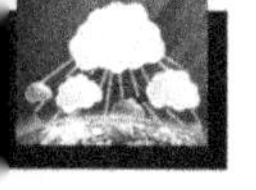

2. 虚拟 SD 卡

在 Android 开发中经常遇到与 SD 卡有关的调试，比如 mp3 文件、图片文件等，所以在创建 AVD 时，一般会创建一个虚拟 SD 卡。AVD 中的 SD 卡为虚拟 SD 卡，该虚拟 SD 卡实际是硬盘中的一块区域，起到模拟手机中的 SD 卡作用。

1)通过 Eclipse 的 DDMS 工具查看虚拟 SD 卡

STEP 1 打开 DDMS 工具：在 Eclipse 里，单击【Window】菜单的【Open Perspective】，在弹出的二级菜单中选择【DDMS】，如图 8.2 所示，点击打开 DDMS 工具。

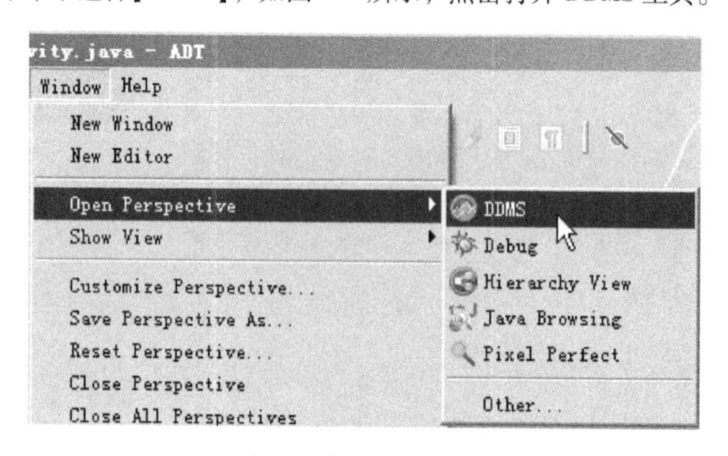

图 8.2　打开 DDMS 工具

STEP 2 DDMS 工具如图 8.3 所示，可以在“File Explor”标签页中查看；

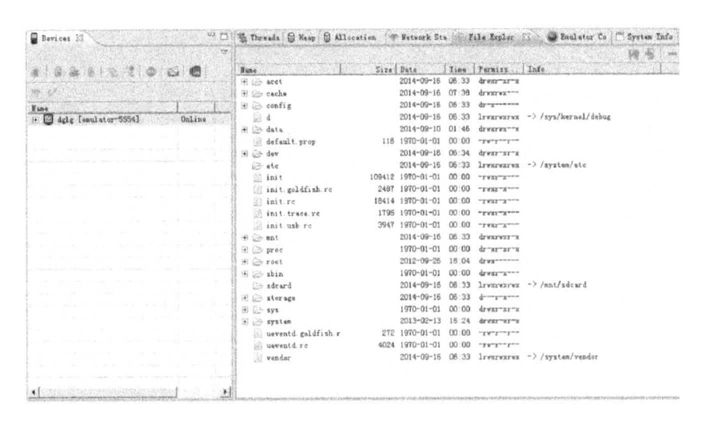

图 8.3 DDMS 工具

STEP 3 虚拟 SD 卡的位置：虚拟 SD 卡的位置位于“File Explor”标签页中的【mnt】中的【sdcard】项中，如图 8.4 所示，可以从中看到其目录结构。

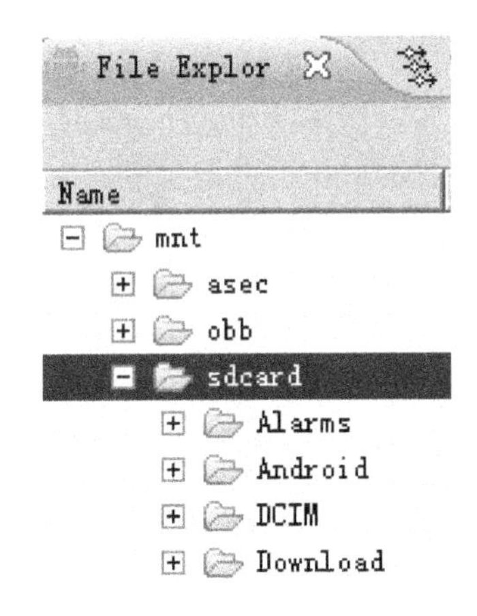

图 8.4 sdcard 目录结构

图 8.5 “Put File on Device”对话框

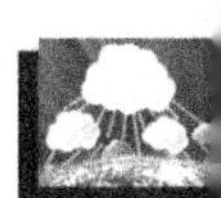

2) 虚拟 SD 卡中的文件导入

STEP 1 选定文件导入位置：在“File Explor”标签页中单击【mnt】中的【sdcard】项，如果【sdcard】项处于选中状态，文件将导入【sdcard】中。如果【sdcard】中的子目录项处于选中状态，文件将导入到该处。

STEP 2 导入文件：单击【push a file onto the devide】按钮（该按钮为“导入文件到设备”按钮），弹出如图 8.5 所示的“Put File on Device”对话框。

STEP 3 在“名称”栏中选择要导入的文件，单击【打开】按钮，Eclipse 会将文件导入到虚拟 SD 卡中。

3. MediaPlayer 类

MediaPlayer 类使用机制如下：

① 当一个 MediaPlayer 类被创建或调用 reset( )方法之后会处于空闲状态之下，调用 release( )方法之后，才会处于结束状态。

② 当一个 MediaPlayer 类不再使用之后，最好使用 release( )方法来释放使之处于结束状态，以免造成不必要的错误。

③ 当 MediaPlayer 类处于结束状态中，便不能再使用了。

④ 当一个 MediaPlayer 类被创建后处于空闲状态中，如果通过 create( )方法创建便处于准备状态之中。

⑤ 任何 MediaPlayer 类都必须先处于准备状态之中，然后才开始播放。

⑥ 要开始播放 MediaPlayer 类都必须成功调用 start( )方法。可以通过 isPlaying( )方法来判断当前是否正在播放。

⑦ 当 MediaPlayer 类正在播放时，可以进行暂停和停止等操作，pause( )方法来暂停播放，stop( )方法来停止播放。

⑧ 处于暂停状态时可以通过 start( )方法来恢复播放，但是处于停止状态必须先调用 pause( )方法来使之处于准备状态，然后再通过 start( )方法来开始播放。

⑨ 可以通过 setLooping(boolean)方法来设置是否循环播放。

⑩ MediaPlayer 类可以用于音频的播放控制，同时它也支持对视频媒体的控制操作，MediaPlayer 类支持从以下 3 种不同方式的数据源加载资源：

a. Res/drawable 资源目录：这种方式是播放的文件(假如引用名为 file_name)存放于 Android 应用的 raw 资源目录中，然后通过 R. drawable. File_Name 方式进行加载，加载代码如下：

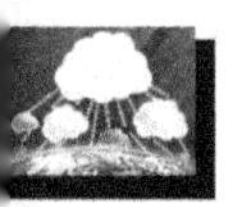

```
MediaPlayer MPlayer = new MediaPlayer( );
MPlayer. create(context, R. drawable. File_Name);
```

b. 本地文件系统：这种方式是播放的文件存放于文件系统中，如 SD 卡中。然后通过 MediaPlayer 的 setDataSource 方法制定相应路径进行加载，代码如下(假设文件为音频文件 File_Name. mp3，存放于/mnt/sdcard/mp3 目录中)：

```
MediaPlayer MPlayer  = new MediaPlayer( );
MPlayer. setDataSource("/mnt/sdcard/mp3/File_Name. mp3");
```

c. 网络资源文件：对于网络资源文件的播放，可以通过该网络资源文件的 URL(网络资源地址)方式加载，这种加载方式与在本地文件系统中的加载类似，区别是 MediaPlayer 的 setDataSource 方法对网络资源文件的加载传入的是网络资源文件的网络资源地址，即 URL，代码如下：

```
MediaPlayer MPlayer  = new MediaPlayer( );
MPlayer. setDataSource("http: //www. dglg. net/File_Name. mp3");
```

(11) MediaPlayer 类的常用方法如表 8. 1 所示。

表 8.1 MediaPlayer 类方法及描述

方　　法	描述说明
create(Context context, Uri uri)	静态方法，通过 Uri 创建一个多媒体播放器
create(Context context, int resid)	静态方法，通过资源 ID 创建一个多媒体播放器
create(Context context, Uri uri, SurfaceHolder holder)	静态方法，通过 Uri 和指定 SurfaceHolder（抽象类）创建一个多媒体播放器
getCurrentPosition( )	返回 Int，得到当前播放位置
getDuration( )	返回 Int，得到文件的时间
getVideoHeight( )	返回 Int ，得到视频的高度
getVideoWidth( )	返回 Int，得到视频的宽度
isLooping( )	返回 boolean ，是否循环播放
isPlaying( )	返回 boolean，是否正在播放
pause( )	无返回值 ，暂停
prepare( )	无返回值，准备同步
prepareAsync( )	无返回值，准备异步
release( )	无返回值，释放 MediaPlayer 类
reset( )	无返回值，重置 MediaPlayer 类
seekTo(int msec)	无返回值，指定播放的位置(以毫秒为单位的时间)
setAudioStreamType(int streamtype)	无返回值，指定流媒体的类型
setDataSource(String path)	无返回值，设置多媒体数据来源(根据 路径)
setDataSource(FileDescriptor fd, long offset, long length)	无返回值，设置多媒体数据来源(根据 FileDescriptor)
setDataSource(FileDescriptor fd)	无返回值，设置多媒体数据来源(根据 FileDescriptor)
setDataSource(Context context, Uri uri)	无返回值，设置多媒体数据来源(根据 Uri)
setDisplay(SurfaceHolder sh)	无返回值，设置用 SurfaceHolder 来显示多媒体
setLooping(boolean looping)	无返回值，设置是否循环播放
setScreenOnWhilePlaying(boolean screenOn)	无返回值，设置是否使用 SurfaceHolder 显示
setVolume(float leftVolume, float rightVolume)	无返回值，设置音量
start( )	无返回值，开始播放
stop( )	无返回值，停止播放

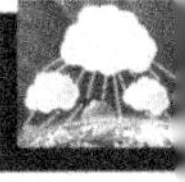

4. MediaPlayer 类使用步骤

通过 MediaPlayer 类播放媒体资源步骤如下。

(1)构建 MediaPlayer 类，代码如下：

```
MediaPlayer Mplayer = new MediaPlayer( )
```

(2)根据要播放媒体资源的类型，调用相应 MediaPlayer 类的相应方法加载要播放的媒体资源(以下的媒体资源文件为 File_Name. mp3)：

① Res/drawable 资源目录，用 create( )方法加载，代码如下：

```
MPlayer. create(context, R. drawable. File_Name);
```

② 本地文件系统：用 setDataSource( )方法加载，代码如下：

```
MPlayer. setDataSource("/mnt/sdcard/mp3/File_Name. mp3");
```

③ 网络资源文件：用 setDataSource( )方法加载，代码如下：

```
MPlayer. setDataSource("http://www. dglg. net/File_Name. mp3");
```

(3)调用 MediaPlayer 类的 prepare( )方法进行播放前准备(以 Res/drawable 资源方式加载时此步骤可省略)，代码如下：

```
Mplayer. prepare( )
```

(4)调用 MediaPlayer 类的 start( )方法开始播放，代码如下：

```
Mplayer. start( )
```

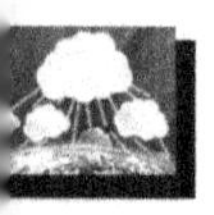

(5)调用 MediaPlayer 类的其他方法控制 MediaPlayer 类的播放，如 pause( )、stop( )等。

(6)程序结束前调用 MediaPlayer 类的 release( )方法释放所占用资源，代码如下：

```
Mplayer. release( )
```

任务实现

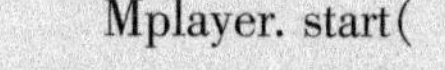

1. 新建 Android 项目

在 Eclipse 中，新建一 Android 应用程序。

2. 导入音乐文件入虚拟 SD 卡

准备 mp3 音乐文件素材，在 DDMS 工具中，导入 SD 卡。本任务中准备的是名为“ceshi. mp3”的 mp3 音乐文件，导入后，“ceshi. mp3”会添加在图 8. 4(sdcard 目录结构)中。本任务为导入到 sdcard 目录中，在 Eclipse 中引用的资源路径为“/mnt/sdcard/ceshi. mp3”。

3. 对 Activity 进行布局

1)编写 strings. xml 定义字符串变量

对布局文件中要引用的字符串变量进行定义，代码如下：

```
<?xml version="1.0" encoding="utf-8"?>
<resources>
    <string name="app_name">mp3 音乐播放程序</string>
    <string name="action_settings">Settings</string>
    <string name="showdetails">android 手机 mp3 播放程序</string>
    <string name="play">播放</string>
    <string name="pause">暂停</string>
    <string name="stop">停止</string>
</resources>
```

2）对 MainActivity 布局：activity_main. xml

界面布局文件：该程序界面比较简洁，涉及的类有 TextView（显示 mp3 音乐播放状态）和 3 个 Button（控制音乐的播放、暂停、停止），考虑用线性布局。布局文件 activity_main. xml 如下：

```
<?xml version="1.0" encoding="utf-8"?>
<LinearLayout xmlns:android="http://schemas.android.com/apk/res/android"
    android:layout_width="match_parent"
    android:layout_height="match_parent"
    android:orientation="vertical" >
    <TextView
        android:id="@+id/ShowTv"
        android:layout_width="wrap_content"
        android:layout_height="wrap_content"
        android:layout_marginTop="10dp"
        android:layout_marginLeft="80dp"
        android:text="@string/showdetails" />
    <Button
        android:id="@+id/BtnPlay"
        android:layout_width="wrap_content"
        android:layout_height="wrap_content"
        android:layout_marginTop="10dp"
        android:layout_marginLeft="80dp"
        android:text="@string/play" />
    <Button
        android:id="@+id/BtnPause"
        android:layout_width="wrap_content"
        android:layout_height="wrap_content"
        android:layout_marginTop="10dp"
```

```
            android: layout_marginLeft = "80dp"
            android: text = "@string/pause" />
        <Button
            android: id = "@+id/BtnStop"
            android: layout_width = "wrap_content"
            android: layout_height = "wrap_content"
            android: layout_marginTop = "10dp"
            android: layout_marginLeft = "80dp"
            android: text = "@string/stop" />
    </LinearLayout>
```

4. 主程序功能实现

MainActivity.java 程序文件编写：根据该程序实现的功能以及 UI 界面的设计，先引用相关的类，然后分别声明 UI 界面中的控件变量 Button、TextView、MediaPlayer 等，再声明一变量 IsStop 以标志音乐播放的状态(是否停止默认值为 False)，该变量为 True 代表处于停止状态。具体代码如下：

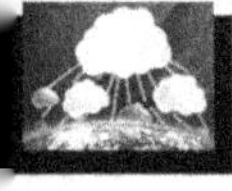

```
package org.dglg.project8_1;
/*导入相关类*/
import java.io.IOException;
import android.media.MediaPlayer;
import android.os.Bundle;
import android.annotation.SuppressLint;
import android.app.Activity;
import android.view.View;
import android.view.View.OnClickListener;
import android.widget.Button;
import android.widget.TextView;
@SuppressLint("SdCardPath")
public class MainActivity extends Activity {
/*导入变量及类*/
Button BtnPlay;
Button BtnPause;
Button BtnStop;
TextView ShowTv;
MediaPlayer MPlayer;
/*定义一状态变量，以标志播放状态*/
boolean IsStop;
```

```
        @ Override
        protected void onCreate( Bundle savedInstanceState) {
            super. onCreate( savedInstanceState) ;
            /* 载入 activity_main. xml 布局文件 */
            setContentView( R. layout. activity_main) ;
            /* 获取布局对象中的 Button 和 TextView */
            BtnPlay = ( Button) findViewById( R. id. BtnPlay) ;
            BtnPause = ( Button) findViewById( R. id. BtnPause) ;
            BtnStop = ( Button) findViewById( R. id. BtnStop) ;
            ShowTv = ( TextView) findViewById( R. id. ShowTv) ;
            /* 初始化 MPlayer 类 */
            MPlayer = new MediaPlayer( ) ;
            try {
        MPlayer. setDataSource( "/mnt/sdcard/ceshi. mp3" ) ;
        MPlayer. prepare( ) ;
                    } catch (IOException e) {
                // TODO Auto - generated catch block
                e. printStackTrace( ) ;
            }
            /* 初始化播放标志 */
            IsStop = false;
            /* 音乐播放按钮事件 */
            BtnPlay. setOnClickListener( new OnClickListener( ) {
            @ Override
            public void onClick( View v) {
            // TODO Auto - generated method stub
                /* 根据 IsStop 值判断 MPlayer 是否处于暂停还是停止状态, 如果暂
停则继续播放, 如果停止, 重新载入文件, 重新播放 */
                if( IsStop = = false) {
                /* MPlayer 处于暂停状态, 播放 */
                MPlayer. start( ) ;
                }
                else
                {
                /* MPlayer 处于停止状态, 先暂停, 再载入文件播放 */
                MPlayer. pause( ) ;
                try {
                MPlayer. setDataSource( "/mnt/sdcard/ceshi. mp3" ) ;
```

```
            MPlayer. prepare( ) ;
            } catch (IOException e) {
            e. printStackTrace( ) ;
            }
            MPlayer. start( ) ;
            IsStop = false;
            }
            /* 设置 TextView 文本值 */
            ShowTv. setText( "音乐播放中" ) ;
        } });
        /* 音乐暂停按钮事件 */
        BtnPause. setOnClickListener( new OnClickListener( ) {
            @ Override
            public void onClick( View v) {
                // TODO Auto - generated method stub
                /* 暂停播放 */
                if( MPlayer!  = null)
                {
                MPlayer. pause( ) ;
                ShowTv. setText( "音乐暂停播放中" ) ;
                }}});
        /* 音乐停止按钮事件 */
        BtnStop. setOnClickListener( new OnClickListener( ) {
            @ Override
            public void onClick( View v) {
                // TODO Auto - generated method stub
                /* 停止播放 */
                if( MPlayer!  = null)
                {
                MPlayer. stop( ) ;
                IsStop = true;
                ShowTv. setText( "音乐停止播放中" ) ;
                }} });}
    protected void onDestroy( )
    {
    /* 释放资源 */
    if( MPlayer!  = null)
    {
```

```
            MPlayer. release( ) ;
        }
        super. onDestroy( ) ;
    }
}
```

程序说明如下:

① 导入资源文件(音乐文件)进入虚拟 SD 卡:注意导入的目录及在 Eclipse 中引用路径。如导入到 sdcard 目录中,Eclipse 中引用路径为“/mnt/sdcard/ceshi. mp3”;如果是导入到 sdcard 目录的子目录中,则在路径中添加相应的目录。

② MediaPlayer 类运行机制:

- 当调用 pause( )方法暂停后,可以通过 start( )方法来恢复播放。
- 当调用 pause( )方法停止后,必须先调用 pause( )方法来使之处于准备状态,然后再通过 start( )方法来开始播放。
- 状态变量 IsStop 的使用:该状态变量用来标记播放状态是否处于停止状态,如果处于停止状态则要重新播放(重新播放的这个过程需要重新载入资源文件)。

任务考核

任务提高:将本程序在实体机中运行,将一 mp3 文件(如 mm. mp3)预先放置于内存卡中的“music”目录中,程序涉及的 mp3 文件的位置定位手机内存卡中的“music”目录中的 mp3 文件,如“music/mm. mp3”。在实体机中调试、运行。

任务二　简易视频播放程序

任务描述

在 Android 手机应用中,播放视频是 Android 手机的主要功能之一。本任务通过一个具体的播放视频程序的实现过程,介绍使用 VideoView 类播放视频文件的具体流程。程序运行界面如图 8. 6 所示。

图 8. 6　使用 VideoView 类播放视频

任务分析

在本任务中,定义了 1 个 VideoView 类和 3 个按钮,VideoView 用于播放视频,而 3 个按钮分别用于控制视频(VideoView 组件)的播放、暂停、停止。

① 单击“播放”按钮后从指定的资源中获取视频文件,并开始播放。

② 单击“暂停”按钮控制视频的暂停播放。

③ 单击“停止”按钮控制视频的停止播放。

本任务中，也通过 MediaController 类控制 VideoView 类的播放控制，如图 8.6 底部灰色部分，为 MediaController 控制部分。

本任务的主要目的通过制作简易视频播放器，熟悉 VideoView 类的开发，重点和难点如下：

(1)熟悉 VideoView 类；
(2)学习 VideoView 类的一些常用方法；
(3)VideoView 类与 MediaController 类的相互关联；
(4)掌握 VideoView 类开发的基本步骤和方法；
(5)掌握 Android 开发播放视频类程序的基本方法。

任务准备

1. Android 系统播放视频方法

制作分析：在 Android 系统中，一般有 2 种方法来播放视频，一种是使用 MediaPlayer 和 SurfaceView 类相结合的方式，另一种是使用 Android 自带的 VideoView 类。二者比较，使用 VideoView 播放视频的好处是简单，因为它实现了 SurfaceView 以及控制方法，需要的时候直接拿来使用就可以了，缺点是灵活性不够。而使用 MediaPlayer 结合 SurfaceView 来播放视频的话，好处是可以更灵活的对其进行自定义，但缺点是难度比较大。Android 系统提供的 VideoView 类在播放视频上更加易用。

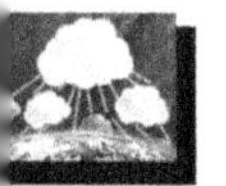

2. VideoView 类

VideoView 类是 Android 提供的专门用来播放视频的类，它位于 android. widget 包下，作用于 ImageView 类似，区别是 ImageView 类用来显示图片，而 VideoView 用来播放视频。

VideoView 类的主要使用方法如表 8.2 所示。

表 8.2 VideoView 类的主要使用方法

方　　法	描述说明
setVideoViewPath(String path)	设置播放视频的路径
setVideoURL(Uri uri)	设置视频的 URI
start( )	开始播放
seekTo(int msec)	定位
pause( )	暂停播放
stop( )	停止播放
setMediaController(MediaController)	设置与 MediaController 类关联

VideoView 类继承自 SurfaceView，并且实现了 MediaPlayerController 这个用于控制媒体播放的接口，通过 VideoView 类的 setMediaController（MediaController）方法可以设置 VideoView 类与 MediaController 类的关联，通过 MediaController 类控制视频播放。

3. **媒体控制面板** MediaController

媒体控制面板 MediaControllers 使得在 VideoView 上有一个用于对媒体播放进行控制的面板，包括快进、快退、播放、暂停按钮以及一个进度条，使得采用 VideoView 对控制播放的视频非常容易。媒体控制面板 MediaController 常用方法如表 8.3 所示。

表 8.3　MediaController 类常用方法

方　　法	描述说明
hide( )	隐藏 MediaController
show( )	显示 MediaController
show(int timeout)	设置 MediaController 显示的时间，以毫秒计算，如果设置为 0 则一直到调用 hide( )时隐藏

4. Android **使用** VideoView **类播放视频步骤**

Android 中使用 VideoView 类播放视频的基本步骤如下：

STEP 1 在布局文件中创建 VideoView 组件；

STEP 2 调用 VideoView 的如下两个方法来加载指定的视频：

① 加载本地视频：setVideoViewPath(String path)加载 path 文件代表的本地视频。

② 加载网络视频：setVideoViewUri(Uri uri)加载 Uri 所对应的网络视频。

STEP 3 视频播放控制：

①通过 MediaController 控制：主要设置 VideoView 与 MediaController 的相互关联。

```
VideoView videoShow;
MediaController mController;
VideoShow. setMediaController ( MController ); /* 设置 VideoView ( ViedoShow ) 与
MediaController(MController)的关联 */
MController. setMediaPlayer ( VideoShow ); /* 设置 MediaController ( MController ) 与
VideoView(VideoShow)的关联 */
```

② 调用 VideoView 的 start( )、stop( )、pause( )方法来控制视频的播放。

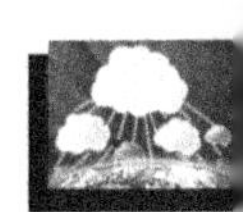

 任务实现

1. **新建** Android **项目**

在 Eclipse 中，新建一 Android 应用程序。

2. 导入视频文件到虚拟 SD 卡

准备视频文件素材：本任务准备的视频文件为“ceshi. 3gp”，在 DDMS 工具中，导入到虚拟 SD 卡中。本任务为导入到 sdcard 目录中，在 Eclipse 中引用的资源路径为“/mnt/sdcard/ceshi. 3gp”。如图 8. 7 所示。

3. 对 Activity 进行布局

1) 编写 strings. xml 定义字符串变量

对布局文件中要引用的字符串变量进行定义，代码如下：

```
<? xml version = "1.0" encoding = "utf - 8"? >
<resources >
    <string name = "app_name" >简易视频播放程序</string >
    <string name = "action_settings" >Settings </string >
    <string name = "play" >播放</string >
    <string name = "pause" >暂停</string >
    <string name = "stop" >停止</string >
</resources >
```

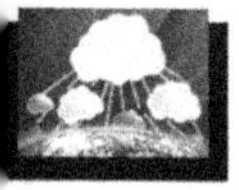

图 8.7　导入资源文件

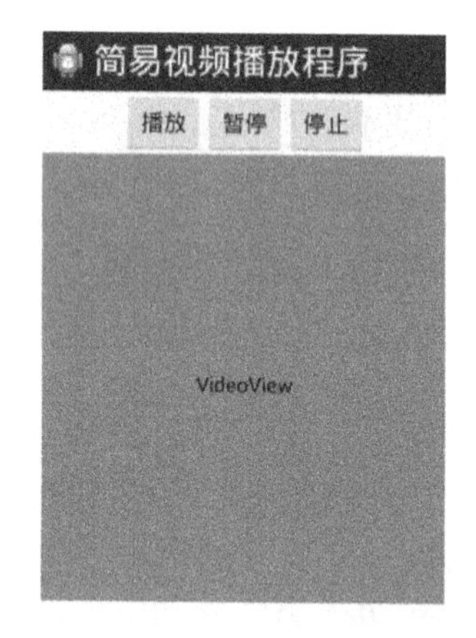

图 8. 8　布局效果图

2) 对 MainActivity 布局：activity_main. xml

界面布局文件：涉及的类有 3 个 Button（控制音乐的播放、暂停、停止）和 1 个 VideoView，综合整体布局设计，考虑用混合布局。布局文件 activity_main. xml 如下，布局效果如图 8. 8 所示。

```
<?xml version = "1.0" encoding = "utf-8"? >
<LinearLayout xmlns: android = "http: //schemas. android. com/apk/res/android"
    android: layout_width = "wrap_content"
    android: layout_height = "wrap_content"
    android: orientation = "vertical" >
    <RelativeLayout
    android: layout_gravity = "center"
    android: layout_width = "wrap_content"
    android: layout_height = "wrap_content"        >
    <Button
        android: id = "@ +id/BtnPlay"
        android: layout_width = "wrap_content"
        android: layout_height = "wrap_content"
        android: text = "@ string/play" / >
    <Button
        android: id = "@ +id/BtnPause"
        android: layout_width = "wrap_content"
        android: layout_height = "wrap_content"
        android: layout_toRightOf = "@ id/BtnPlay"
        android: text = "@ string/pause" / >
    <Button
        android: id = "@ +id/BtnStop"
        android: layout_width = "wrap_content"
        android: layout_height = "wrap_content"
        android: layout_toRightOf = "@ id/BtnPause"
        android: text = "@ string/stop" / >
      </RelativeLayout >
    <VideoView
    android: id = "@ +id/VideoViewShow"
    android: layout_width = "wrap_content"
    android: layout_height = "355dp"        / >
</LinearLayout >
```

4. 主程序功能实现

MainActivity. java 程序文件编写步骤：根据该程序实现的功能以及 UI 界面的设计，先引用相关的类；然后分别声明 UI 界面中的控件变量 Button、VideoView，因为涉及用 Button 类控制视频的播放，所以声明一变量 IsStop 以标志视频播放的状态（是否停止默认值为 False），该变量为 True 代表处于停止状态。具体代码如下：

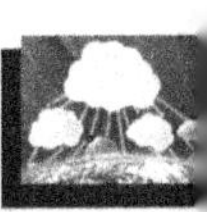

```
package com.example.project8_2;
/*导入相关类*/
import android.os.Bundle;
import android.annotation.SuppressLint;
import android.app.Activity;
import android.view.View;
import android.view.View.OnClickListener;
import android.widget.Button;
import android.widget.MediaController;
import android.widget.Toast;
import android.widget.VideoView;
public class MainActivity extends Activity {
/*定义变量*/
VideoView VideoShow;
MediaController MController;
Button BtnPlay;
Button BtnPause;
Button BtnStop;
Boolean IsStop;
    @SuppressLint("SdCardPath")
    @Override
    protected void onCreate(Bundle savedInstanceState) {
        super.onCreate(savedInstanceState);
        /*载入布局文件*/
        setContentView(R.layout.activity_main);
        /*创建 MediaController 对象*/
        MController = new MediaController(MainActivity.this);
        /*获取布局界面上的 VideoView 组件*/
        VideoShow = (VideoView)findViewById(R.id.VideoViewShow);
        /*获取布局界面上的 Button 组件*/
        BtnPlay = (Button)findViewById(R.id.BtnPlay);
        BtnPause = (Button)findViewById(R.id.BtnPause);
        BtnStop = (Button)findViewById(R.id.BtnStop);
        /*加载 Path 文件的视频*/
        VideoShow.setVideoPath("/mnt/sdcard/ceshi.3gp");
        IsStop = false;
        if(VideoShow! = null)
        {
```

```
/*设置 VideoShow 与 MController 的关联*/
VideoShow.setMediaController(MController);
/*设置 MController 与 VideoShow 的关联*/
MController.setMediaPlayer(VideoShow);
/*让 VideoShow 获取焦点*/
VideoShow.requestFocus();
}
/*视频播放按钮事件*/
BtnPlay.setOnClickListener(new OnClickListener(){
    @Override
    public void onClick(View v) {
        // TODO Auto-generated method stub
        /*有无载入视频*/
        if(VideoShow! =null)
        {
            /*判断视频播放状态*/
            if(IsStop = =false)
            {
            /*视频不处于停止状态*/
            VideoShow.start();
            }
            else
            {
                /*视频处于停止状态，要重新载入视频文件*/
                VideoShow.setVideoPath("/mnt/sdcard/ceshi.3gp");
                VideoShow.start();
                IsStop = false;
            }
            /*Toast 提示视频播放信息*/
                Toast.makeText(MainActivity.this, "视频播放中",
Toast.LENGTH_SHORT).show();
        }
    }
});
/*视频暂停按钮事件*/
BtnPause.setOnClickListener(new OnClickListener(){
    @Override
    public void onClick(View v) {
```

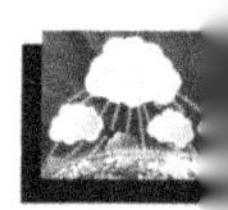

```
                        // TODO Auto - generated method stub
                        if(VideoShow! = null)
                        {
                            /* 暂停播放视频 */
                            VideoShow.pause();
                            /* Toast 提示视频播放信息 */
                              Toast.makeText(MainActivity.this, "视频播放暂停中",
Toast.LENGTH_SHORT).show();
                        }
                    }
                });
                /* 视频停止按钮事件 */
                BtnStop.setOnClickListener(new OnClickListener() {
                    @ Override
                    public void onClick(View v) {
                        // TODO Auto - generated method stub
                        if(VideoShow! = null)
                        {
                            /* 停止播放视频 */
                            VideoShow.stopPlayback();
                            /* 设置视频停止播放标志 */
                            IsStop = true;
                            /* Toast 提示视频播放信息 */
                              Toast.makeText(MainActivity.this, "视频播放停止中",
Toast.LENGTH_SHORT).show();
                        }}});}}
```

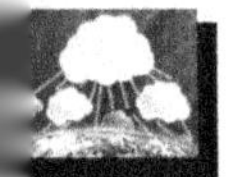

程序说明：

(1)本任务中实现了以下两种方法对视频的播放进行控制：

① 通过 MediaController 类与 VideoView 的相互关联达到对视频的播放进行控制。

② 通过 VideoView 类提供的方法。

(2)调用 VideoView 的 pause( )方法后，视频将暂停播放，这时调用 VideoView 的 start( )方法，视频将继续播放。

(3)调用 VideoView 的 stop( )方法后，视频将停止播放，如果这时直接调用 VideoView 的 start( )方法，程序将报错。报错的原因是因为没有重新载入视频资源文件。所以在本任务中通过一变量 IsStop 来标志视频是否停止播放，如果停止播放，则要重新载入视频资源文件。

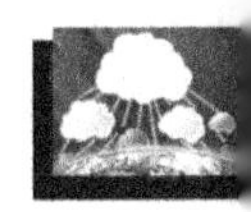

 任务考核

任务提高：将本程序在实体机中运行，将一视频文件(mm. 3gp)预先放置于内存卡中的 video 目录中，程序中涉及的视频文件的位置定位手机内存卡中的“video”目录中的视频文件，如“music/mm. 3gp”。在实体机中调试、运行。

任务三 简易录音功能程序(带播放)

 任务描述

本任务主要是实现简易录音功能，并对其进行播放，主要使用 MediaRecorder 类进行录音，并用 MediaPlayer 类进行播放。程序运行界面如图 8.9 和图 8.10 所示。

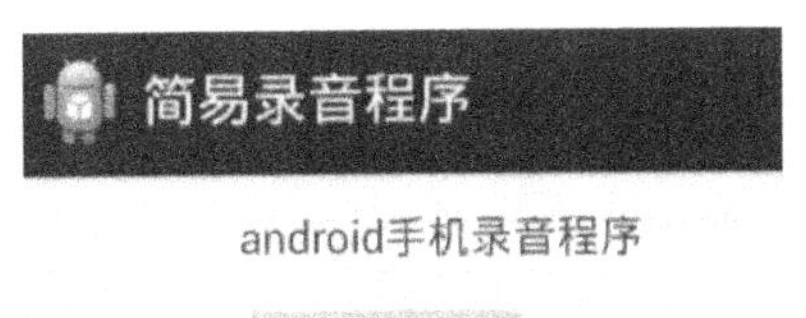

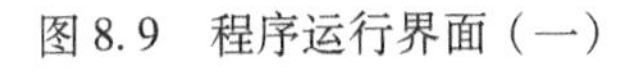

图 8.9 程序运行界面(一)

图 8.10 程序运行界面(二)

 任务分析

在本任务中，定义了 3 个按钮，分别用于控制录音的开始、停止和播放。当单击“开始录音”按钮，开始录音；单击“停止录音”按钮，停止录音，并将录音文件保存(“/mnt/sdcard/ceshi. amr”)；单击“播放录音”按钮，对刚才保存的录音进行播放。

本任务的主要目的是通过录音功能的实现，熟悉 MediaRecorder 类的开发。重点和难点如下：

(1)熟悉 MediaRecorder 类；
(2)学习 MediaRecorder 类的一些常用方法；
(3)掌握 MediaRecorder 类开发的基本步骤和方法；
(4)掌握 Android 开发录音功能类程序的基本方法。

任务准备

1. 任务解析

在播放录音方面，可以用任务一中介绍的MediaPlayer类来播放音频，这个问题不大，本任务的重点和难点是实现录音。

在Android中，提供了MediaRecorder类来实现录音功能。MediaRecorder类主要用来进行媒体采样，包括音频和视频，也就是用来记录音频和视频方法的类，记录的数据一般都是写入到文件里面。因此可以通过MediaRecorder类实现录音频和录视频的功能。

2. MediaRecorder类

使用MediaRecorder前，须利用MediaRecorder类的方法先进行设置，如音频记录的编码格式、音频源等。MediaRecorder类的常用方法如表8.4所示。

表8.4 MediaRecorder类的主要方法

方　法	描述说明
prepar( )	准备录制
start( )	开始录制
stop( )	停止录制
vreset( )	重置MediaRecorder
release( )	释放MediaRecorder占用的资源
setAudioEncoder(int)	设置音频记录的编码格式
setAudoSource(int)	设置音频记录的音频源
setOutoutFormat(int)	设置记录的媒体文件的输出转换格式
setPreviewDisplay(Surface)	设置视频的预览界面
setVideoEncoder(int)	设置视频记录的编码格式
setVideoSource(int)	设置视频记录的视频源
setOutputFile(String)	媒体文件输出路径
setMaxDuration(int)	设置最大记录时长，单位为毫秒

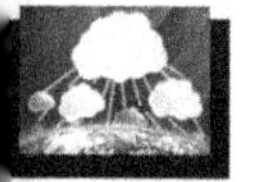

3. 使用MediaRecorder类步骤

在Android中，使用MediaRecorder类实现录音通常分以下6个步骤完成录音：

STEP 1 创建MediaRecorder对象；

```
MediaRecorder mediaRecorder = new MediaRecorder( );
```

STEP 2 设置音频来源（一般为麦克风）；

```
mediaRecorder.setAudioSource(MediaRecorder.AudioSource.MIC);
```

STEP 3 设置音频输出格式；

```
mediaRecorder.setOutputFormat(MediaRecorder.OutputFormat.DEFAULT);
```

STEP 4 设置音频编码方式；

```
mediaRecorder.setAudioEncoder(MediaRecorder.AudioEncoder.DEFAULT);
```

STEP 5 设置输出音频的文件名(路径和文件名)；

```
mediaRecorder.setOutputFile("/mnt/sdcard/ceshi.amr");
```

STEP 6 调用 MediaRecorder 类的 perpare 方法；

```
mediaRecorder.prepare();
```

STEP 7 调用 MediaRecorder 类的 start 方法开始录音；

```
mediaRecorder.start();
```

STEP 8 调用 MediaRecorder 类的 stop 方法停止录音。

```
mediaRecorder.stop();
```

任务实现

1. 新建 Android 项目

在 Eclipse 中，新建一 Android 应用程序。

2. 对 Activity 进行布局

1) 编写 strings.xml 定义字符串变量

对布局文件中要引用的字符串变量进行定义，代码如下：

```
<?xml version="1.0" encoding="utf-8"?>
<resources>
    <string name="app_name">简易录音程序</string>
    <string name="action_settings">Settings</string>
    <string name="showdetails">android 手机录音程序</string>
    <string name="StartRecord">开始录音</string>
    <string name="StopRecord">停止录音</string>
    <string name="PlayRecord">播放录音</string>
</resources>
```

2) 对 MainActivity 布局：activity_main.xml

界面布局文件涉及到的类有 3 个 Button(控制录音的开始、停止、播放)和 1 个用来显

示状态的 VideoView。布局文件 activity_main. xml 代码如下：

```
<? xml version = "1.0" encoding = "utf - 8"?  >
<LinearLayout xmlns: android = "http: //schemas. android. com/apk/res/android"
    android: layout_width = "match_parent"
    android: layout_height = "match_parent"
    android: orientation = "vertical"  >
     <TextView
        android: id = "@ + id/ShowTv"
        android: layout_width = "wrap_content"
        android: layout_height = "wrap_content"
        android: layout_marginTop = "10dp"
        android: layout_marginLeft = "80dp"
        android: text = "@ string/showdetails"  / >
     <Button
        android: id = "@ + id/BtnStart"
        android: layout_width = "wrap_content"
        android: layout_height = "wrap_content"
        android: layout_marginTop = "10dp"
        android: layout_marginLeft = "80dp"
        android: text = "@ string/StartRecord"  / >
     <Button
        android: id = "@ + id/BtnStop"
        android: layout_width = "wrap_content"
        android: layout_height = "wrap_content"
        android: layout_marginTop = "10dp"
        android: layout_marginLeft = "80dp"
        android: enabled = "false"
        android: text = "@ string/StopRecord"  / >
     <Button
        android: id = "@ + id/BtnPlay"
        android: layout_width = "wrap_content"
        android: layout_height = "wrap_content"
        android: layout_marginTop = "10dp"
        android: layout_marginLeft = "80dp"
        android: enabled = "false"
        android: text = "@ string/PlayRecord"  / >
</LinearLayout >
```

3. 主程序功能实现

MainActivity. java 程序文件编写步骤：根据该程序实现的功能以及 UI 界面的设计，先引用相关的类，然后分别声明 UI 界面中的控件变量 TextView、Button，再分别实现开始、停止、播放录音的按钮事件。具体代码如下：

```
package org. dglg. project8_3;
import java. io. IOException;
import android. media. MediaPlayer;
import android. media. MediaRecorder;
import android. os. Bundle;
import android. view. View;
import android. view. View. OnClickListener;
import android. widget. Button;
import android. widget. TextView;
import android. annotation. SuppressLint;
import android. app. Activity;
public class MainActivity extends Activity {
TextView ShowTv;
Button BtnStart;
Button BtnStop;
Button BtnPlay;
MediaPlayer MPlayer;
MediaRecorder mMediaRecorder;
    @ SuppressLint( "SdCardPath" )
    @ Override
    protected void onCreate( Bundle savedInstanceState) {
        super. onCreate( savedInstanceState) ;
        setContentView( R. layout. activity_main) ;
        / * 获取布局界面上的 VideoView 组件 * /
        ShowTv = ( TextView) findViewById( R. id. ShowTv) ;
        BtnStart = ( Button) findViewById( R. id. BtnStart) ;
        BtnStop = ( Button) findViewById( R. id. BtnStop) ;
        BtnPlay = ( Button) findViewById( R. id. BtnPlay) ;
        / * 开始录音按钮事件 * /
        BtnStart. setOnClickListener( new OnClickListener( ) {
            @ Override
            public void onClick( View v) {
                // TODO Auto - generated method stub
```

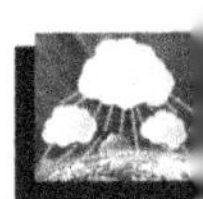

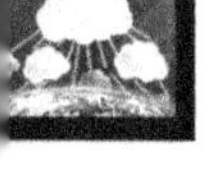

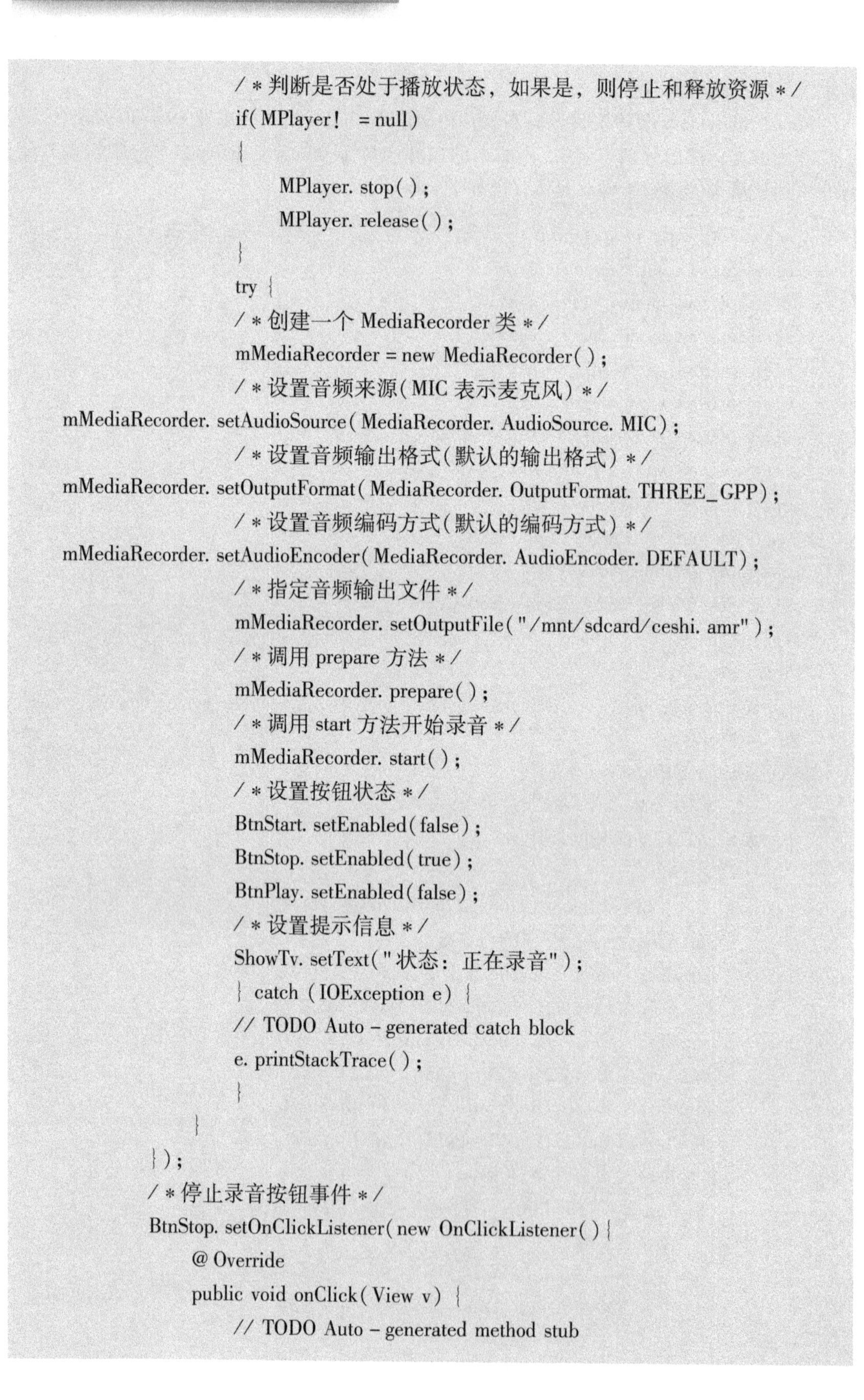

```
                /*判断是否处于播放状态，如果是，则停止和释放资源*/
                if(MPlayer! =null)
                {
                    MPlayer.stop();
                    MPlayer.release();
                }
                try {
                /*创建一个 MediaRecorder 类*/
                mMediaRecorder = new MediaRecorder();
                /*设置音频来源(MIC 表示麦克风)*/
mMediaRecorder.setAudioSource(MediaRecorder.AudioSource.MIC);
                /*设置音频输出格式(默认的输出格式)*/
mMediaRecorder.setOutputFormat(MediaRecorder.OutputFormat.THREE_GPP);
                /*设置音频编码方式(默认的编码方式)*/
mMediaRecorder.setAudioEncoder(MediaRecorder.AudioEncoder.DEFAULT);
                /*指定音频输出文件*/
                mMediaRecorder.setOutputFile("/mnt/sdcard/ceshi.amr");
                /*调用 prepare 方法*/
                mMediaRecorder.prepare();
                /*调用 start 方法开始录音*/
                mMediaRecorder.start();
                /*设置按钮状态*/
                BtnStart.setEnabled(false);
                BtnStop.setEnabled(true);
                BtnPlay.setEnabled(false);
                /*设置提示信息*/
                ShowTv.setText("状态：正在录音");
                } catch (IOException e) {
                // TODO Auto-generated catch block
                e.printStackTrace();
                }
            }
        });
        /*停止录音按钮事件*/
        BtnStop.setOnClickListener(new OnClickListener() {
            @Override
            public void onClick(View v) {
                // TODO Auto-generated method stub
```

```
                /*停止录音和释放资源*/
                mMediaRecorder.stop();
                mMediaRecorder.release();
                /*设置按钮状态*/
                BtnStart.setEnabled(true);
                BtnStop.setEnabled(false);
                BtnPlay.setEnabled(true);
                /*设置提示信息*/
                ShowTv.setText("状态：停止录音");
            }});
        /*播放录音按钮事件*/
        BtnPlay.setOnClickListener(new OnClickListener(){
            @Override
            public void onClick(View v) {
                // TODO Auto-generated method stub
                /*创建 MediaPlayer 对象*/
                MPlayer = new MediaPlayer();
                try {
                /*播放录音*/
                MPlayer.setDataSource("/mnt/sdcard/ceshi.amr");
                MPlayer.setLooping(true);
                MPlayer.prepare();
                MPlayer.start();
                /*设置提示信息*/
                ShowTv.setText("状态：播放录音");
                } catch (IOException e) {
                    // TODO Auto-generated catch block
                    e.printStackTrace();
                }}});   }}
```

程序说明如下：

(1)录音功能涉及录音文件的保存。本任务中，用下面的语句指定音频输出：

```
mMediaRecorder.setOutputFile("/mnt/sdcard/ceshi.amr")
```

该语句指定音频输出路径为“/mnt/sdcard/ceshi.amr”，程序运行后，将在 AVD 中的虚拟 SD 卡中创建名为“ceshi.amr”的音频文件，可在 DDSM 工具中查看，效果如图 8.11 所示，如果要在手机上运行，需注意设置好音频文件的输出路径。

(2)录音播放：运用任务一的知识可对路径为“/mnt/sdcard/ceshi.amr”的音频资源文件进行播放。

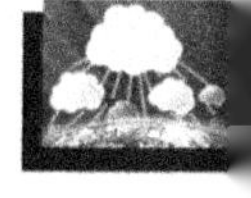

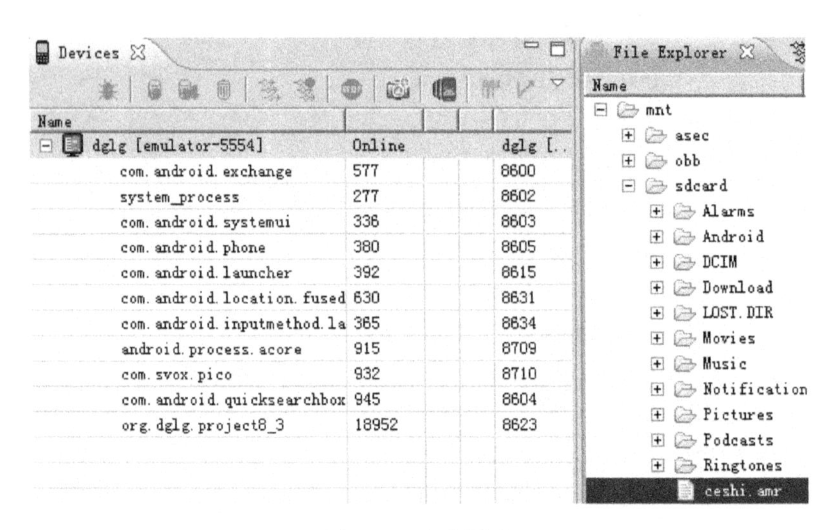

图 8.11　音频输出

(3)按钮状态。因为对录音保存不是采用动态保存，而是指定位置和文件名的，为了避免冲突，设置了按钮状态，防止出现既录音又播放录音的情况。

任务考核

任务提高：将本程序在实体机中调试和运行。

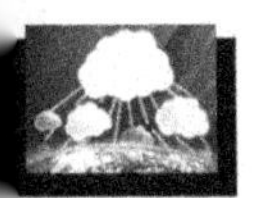

项目九

Android绘画与动画

Android手机开发中也常用到绘图，绘图的基本方法和几种类型的Android动画是Android图像开发的基础。本项目将走进Android的绘图与动画，帮助读者了解Android中绘图与动画的基本知识及应用。

掌握 Android 绘图的基本方法
掌握几种类型的 Android 动画

任务一 Android 简单绘图

任务描述

本任务主要是实现利用 Android 的几个基本对象来实现圆、椭圆、正方形、矩形、多边形、文本等图形的绘制。

任务分析

本任务的主要目的是通过 Android 的几个基本对象来实现圆、椭圆、正方形、矩形、多边形、文本等图形的绘制。重点和难点如下：

(1)了解 Android 绘图的基础知识；
(2)掌握多个 Android 绘图对象的属性和基本用法。

任务准备

1. Android绘图对象

Android绘图的对象包括Color、Paint、Canvas、Path、Drawable和Rect等，其中最基本的三个对象为Color、Paint、Canvas。这些对象都存放在android. graphics包下，使用之前要进行导入。下面一一介绍这些对象。

1) Color（颜色对象）

Color对象相当于现实生活中进行绘画时用的“颜料”。颜色可以用以下几种方式来表示：

(1)使用静态常量：

int color = Color. BLUE;

相应的颜色对应关系如下：

Color. BLACK: // 黑色

Color. BLUE: // 蓝色

Color. CYAN: // 青绿色

Color. DKGRAY: // 灰黑色

Color. YELLOW: // 黄色

Color. GRAY: // 灰色

Color. GREEN: // 绿色

Color. LTGRAY: // 浅灰色

Color. MAGENTA: // 红紫色

Color. RED: // 红色

Color. WHITE: // 白色

Color. TRANSPARENT: // 透明

(2)用color类的静态方法argb(已知ARGB值)：

Android中的颜色用四个数字表示：ARGB分别为透明度(Alpha)、红(Red)、绿(Green)、蓝(Blue)，A、R、G、B一个值可以用一个整数来表示，每个整数8bit(0-255之间)。Alpha值为0时完全透明，为255时完全不透明，具体方法如下所示：

```
int color = Color.argb(A, R, G, B);
```

(3)使用XML资源文件：

定义颜色：

```
<color name="mycolor">#7faaff88</color>
```

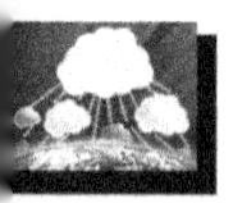

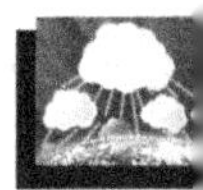

在 xml 文件中定义一个名为 mycolor 的颜色，并可在 Java 代码中调用：

```
int color = getResource().getColor(R.color.mycolor); //返回 mycolor 的资源 ID。
```

注意在创建 color 对象之前，一定要首先需要导入对象：

```
import android.graphics.Color;
```

2)Paint（画笔对象）

Paint 对象相当于现实生活中绘画时用的“画笔”。Paint 类有大量可用的方法，它的方法大体上可以分为两类，一类与图形绘制相关，一类与文本绘制相关，下面列举一部分 Paint 类的常用方法，如表 9.1 所示。

表 9.1　Paint 对象的常用方法

类别	方　法	功　能
图形绘制	setARGB(int a, int r, int g, int b)	设置绘制的颜色，a 代表透明度，r、g、b 代表颜色值
	setAlpha(int a)	设置绘制图形的透明度
	setColor(int color)	设置绘制的颜色，使用颜色值来表示，该颜色值包括透明度和 RGB 颜色
	setAntiAlias(boolean aa)	设置是否使用抗锯齿功能，会消耗较大资源，绘制图形速度会变慢
	setDither(boolean dither)	设定是否使用图像抖动处理，会使绘制出来的图片颜色更加平滑和饱满，图像更加清晰
	setColorFilter(ColorFilter colorfilter)	设置颜色过滤器，可以在绘制颜色时实现不用颜色的变换效果
	setPathEffect(PathEffect effect)	设置绘制路径的效果，如点画线等
	setStyle(Paint.Style style)	设置画笔的样式，如 FILL、FILL_OR_STROKE 或 STROKE
	setStrokeCap(Paint.Cap cap)	当画笔样式为 STROKE 或 FILL_OR_STROKE 时，设置笔刷的图形样式，如圆形样式 Cap.ROUND 或方形样式 Cap.SQUARE
	setStrokeWidth(float width)	当画笔样式为 STROKE 或 FILL_OR_STROKE 时，设置笔刷的粗细度
文本绘制	setFakeBoldText(boolean fakeBoldText)	模拟实现粗体文字
	setTextAlign(Paint.Align align)	设置绘制文字的对齐方向
	setTextScaleX(float scaleX)	设置绘制文字 x 轴的缩放比例，可以实现文字的拉伸的效果

续表 9.1

类别	方　法	功　能
文本绘制	setTextSize(float textSize)	设置绘制文字的字号大小
	setTextSkewX(float skewX)	设置斜体字，skewX 为倾斜弧度
	setTypeface(Typeface typeface)	设置 Typeface 对象，即字体风格，包括粗体、斜体以及衬线体、非衬线体等
	setUnderlineText(boolean underlineText)	设置带有下划线的文字效果
	setStrikeThruText(boolean strikeThruText)	设置带有删除线的效果

Paint 类包含图形的样式、颜色以及绘制任何图形(包括位图、文本和几何图形)所需的其他信息。例如以下代码为使用某种颜色在屏幕上绘图。

```
import android. graphics. Color;
import android. graphics. Paint; //导入 Paint 对象
...
Paint myPaint = new Paint(); //构造 Paint 实例
myPaint. setAlpha(255); //设置透明度为 255
myPaint. setStrokeCap(Paint. Cap . ROUND); //设置笔刷样式为圆形
myPaint. setColor(Color. RED); //使用红色绘图
```

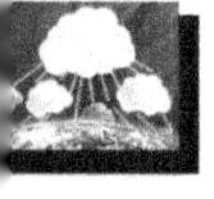

注意在使用 Paint 对象前导入:

```
import android. graphics. Paint;
```

3) Canvas (画布对象)

Canvas 对象相当于现实生活中绘图用的“画布”。

使用 Canvas 类提供的各种方法可以在画布上绘制各类线条、矩形、圆、多边形以及其他各类图形。Canvas 可以绘制的对象具体有以下几种：弧、线(arcs)、填充颜色(argb 和 color)、Bitmap、圆(circle 和 oval)、点(point)、线(line)、矩形(Rect)、图片(Picture)、圆角矩形 (RoundRect)、文本(text)、顶点(Vertices)和路径(path)等。使用 Canvas 对象前要先导入:

```
import android. graphics. Canvas;
```

Canvas 的常用方法如表 9.2 所示。

表 9.2　Canvas 对象的常用方法

方　法	功　能
Canvas()	创建一个空的画布
Canvas(Bitmap bitmap)	以 bitmap 对象创建一个画布，则将内容都绘制在 bitmap 上，bitmap 不得为 null

续表 9.2

方　法	功　　能
Canvas(GL gl)	在绘制 3D 效果时使用，与 OpenGL 有关
drawColor	设置画布的背景色
setBitmap	设置具体的画布
clipRect	设置显示区域，即设置裁剪区
isOpaque	检测是否支持透明
rotate	旋转画布
drawRect(RectF, Paint)	画矩形，第一个参数为图形显示区域，第二个参数为画笔
drawRoundRect (RectF, float, float, Paint)	画圆角矩形，第一个参数为图形显示区域，第二个参数和第三个参数分别是水平圆角半径和垂直圆角半径
drawLine(startX, startY, stopX, stopY, paint)	表示用画笔 paint 从点(startX, startY)到点(stopX, stopY)画一条直线
drawArc(oval, startAngle, sweepAngle, useCenter, paint)	画各类圆弧。oval 为 RectF 类型，即圆弧显示区域；startAngle 和 sweepAngle 为 float 类型，分别表示圆弧起始角度和圆弧度数；3 点钟方向为 0 度；useCenter 设置是否显示圆心；boolean 类型；paint 为画笔
drawCircle (float, float, float, Paint)	用于画圆，前两个参数代表圆心坐标，第三个参数为圆半径，第四个参数是画笔

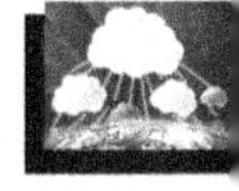

Android 中，屏幕由 Activity 类的对象支配，Activity 类的对象引用 View 类的对象，而 View 类的对象又引用 Canvas 类的对象。通过重写 View. onDraw( )方法，可在指定的画布上绘图。onDraw( )方法唯一参数就是选择在哪一个 Canvas 实例上绘图。

以下是一个通过重写 onDraw( )方法在指定画布 myCanvas 上绘图的例子。

```
import android.app.Activity;
import android.os.Bundle;
import android.view.*;
import android.content.*;
import android.graphics.Canvas;
public class TestCanvas extends Activity {
  @Override
  public void onCreate(Bundle savedInstanceState) {
      super.onCreate(savedInstanceState);
      setContentView(new GraphicsView(this));
  }
  static public class GraphicsView extends View {
      public GraphicsView(Context context) {
```

```
            super(context);
            Canvas myCanvas = new Canvas(); //新建一个画布
            onDraw(myCanvas); //在画布 myCanvas 上绘制
            setBackgroundResource(R.drawable.background); //添加 XML 渐变背景
        }
            @Override
            public void onDraw(Canvas canvas) {
                    //Drawing commands go here
            }
        }
}
```

以上三个对象 Color 对象、Paint 对象、Canvas 对象相结合，就能画出多数基本的图形。除此之外，如果要绘制一些稍微复杂的图形，还要用到其他一些常用对象。

4)其他对象

(1)Path 对象

Path 类包含了一组矢量绘图命令，可以绘制如线条、矩形、曲线等等的矢量图。比如要创建一个圆形的路径，只需要将如下代码添加到上述的 onDraw( )方法中即可在画布 Myanvas 中画出一个圆。其中 addCircle( )方法中的四个参数分别是：圆心的 x 坐标、圆心的 y 坐标、半径(单位是像素)以及绘制方向(CW 是顺时针，CCW 是逆时针)。

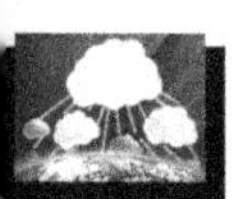

```
        Path circle = new Path();
        circle.addCircle(100, 100, 180, Direction.CCW); //绘制一个圆心坐标在(100, 100)，半径为 180 像素，绘制方向为逆时针的圆。
```

使用前，一样要先导入：

```
import android.graphics.Path;
```

(2)Drawable 对象

Drawable 对象是对图形绘制的一般化抽象。主要针对象位图或纯色这样只用于显示的视觉元素。Drawable 类有很多用于绘制特殊类型图形的子类，包括 BitmapDrawable、ShapDrawable、PictureDrawable、LayerDrawable 等等。可以用三种方法来定义和初始化一个 Drawable 对象：

- 使用保存在项目资源中的一个图片；
- 使用定义 Drawable 对象属性的 XML 文件；
- 使用普通的类构造器。

这几种方法的具体实现在之后的实例中可以看到。

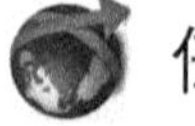

任务实现

1. 新建 Android 项目

在 Eclipse 中，新建一个名为 project 9_1 的 Android 应用，方法可参考项目一，界面如

图 9.1 所示。

2. 布局

先定义一个继承自 View 的类 MyView，再到布局文件 activity_main. xml 中直接进行引用，如图 9.2 所示。

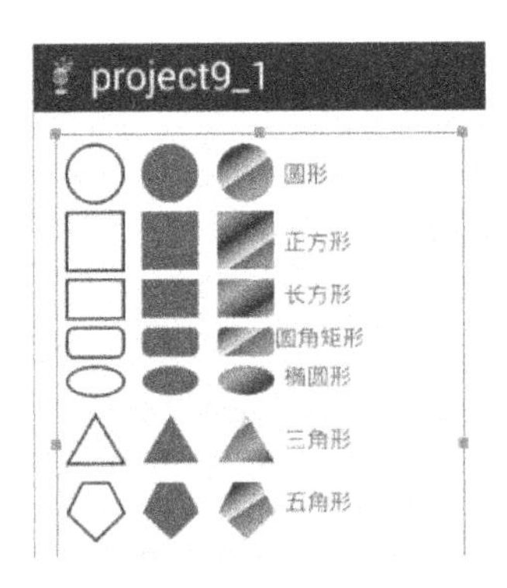

图 9.1　界面设计

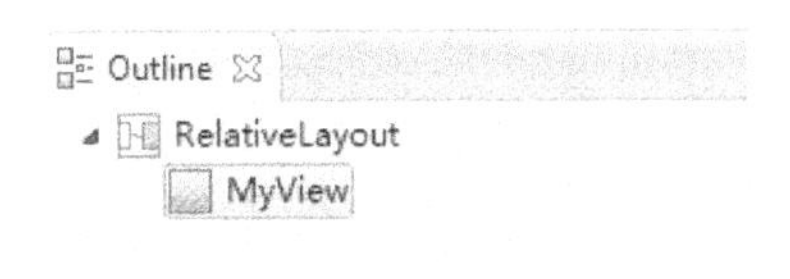

图 9.2　各控件命名

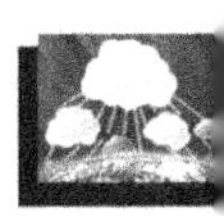

3. 代码

布局文件 activity_main. xml 代码如下：

```
<RelativeLayout xmlns: android = "http: //schemas. android. com/apk/res/android"
    xmlns: tools = "http: //schemas. android. com/tools"
    android: layout_width = "match_parent"
    android: layout_height = "match_parent"
    android: paddingBottom = "@ dimen/activity_vertical_margin"
    android: paddingLeft = "@ dimen/activity_horizontal_margin"
    android: paddingRight = "@ dimen/activity_horizontal_margin"
    android: paddingTop = "@ dimen/activity_vertical_margin"
    tools: context = ". MainActivity"  >
  < org. dglg . project9_1. MyView   //直接引用 MyView 类进行布局
        android: layout_width = "wrap_content"
        android: layout_height = "wrap_content"/ >
</RelativeLayout >
```

类文件 MyView. java 代码如下：

```
package org. dglg. project9_1;
    import android. content. Context;
    import android. graphics. Canvas;
    import android. graphics. Color;
    import android. graphics. LinearGradient;
```

```
import android.graphics.Paint;
import android.graphics.Path;
import android.graphics.RectF;
import android.graphics.Shader;
import android.util.AttributeSet;
import android.view.View;
public class MyView extends View {
    @Override
    protected void onDraw(Canvas canvas) {
        super.onDraw(canvas);
        canvas.drawColor(Color.WHITE);   // 把整张画布绘制成白色
        Paint paint = new Paint();
        paint.setAntiAlias(true);   // 去锯齿
        paint.setColor(Color.BLUE);
        paint.setStyle(Paint.Style.STROKE);
        paint.setStrokeWidth(3);
        canvas.drawCircle(40, 40, 30, paint); // 绘制圆形
        canvas.drawRect(10, 80, 70, 140, paint); // 绘制正方形
        canvas.drawRect(10, 150, 70, 190, paint);   // 绘制矩形
        RectF rel = new RectF(10, 200, 70, 230);
        canvas.drawRoundRect(rel, 15, 15, paint);   // 绘制椭圆矩形
        RectF rel1 = new RectF(10, 240, 70, 270);
        canvas.drawOval(rel1, paint);   // 绘制椭圆
        // 定义一个Path对象，封闭成一个三角形
        Path path1 = new Path();
        path1.moveTo(10, 340);
        path1.lineTo(70, 340);
        path1.lineTo(40, 290);
        path1.close();
        canvas.drawPath(path1, paint);   // 根据Path进行绘制，绘制三角形
        // 定义一个Patch对象，封闭成一个五角形
        Path path2 = new Path();
        path2.moveTo(26, 360);
        path2.lineTo(54, 360);
        path2.lineTo(70, 392);
        path2.lineTo(40, 420);
        path2.lineTo(10, 392);
        path2.close();
```

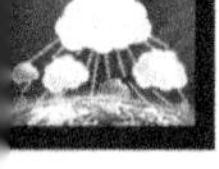

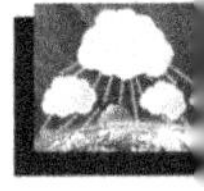

```
// 根据 Path 进行绘制，绘制五角星
canvas. drawPath( path2, paint);
paint. setColor( Color. RED);
paint. setStyle( Paint. Style. FILL);
canvas. drawCircle(120, 40, 30, paint);
canvas. drawCircle(120, 40, 30, paint);        // 绘制正方形
canvas. drawRect(90, 150, 150, 190, paint);      // 绘制矩形
// - - - - - - - - - - - 设置填充风格后绘制 - - - - - - - - - - -
paint. setStyle( Paint. Style. FILL);
paint. setColor( Color. RED);
canvas. drawCircle(120, 40, 30, paint);
canvas. drawRect(90, 80, 150, 140, paint);       // 绘制正方形
canvas. drawRect(90, 150, 150, 190, paint); // 绘制矩形
RectF re2 = new RectF(90, 200, 150, 230);
canvas. drawRoundRect( re2, 15, 15, paint);     // 绘制圆角矩形
RectF re21 = new RectF(90, 240, 150, 270);
canvas. drawOval( re21, paint);     // 绘制椭圆
Path path3 = new Path();
path3. moveTo(90, 340);
path3. lineTo(150, 340);
path3. lineTo(120, 290);
path3. close();
canvas. drawPath( path3, paint); // 绘制三角形
Path path4 = new Path();
path4. moveTo(106, 360);
path4. lineTo(134, 360);
path4. lineTo(150, 392);
path4. lineTo(120, 420);
path4. lineTo(90, 392);
path4. close();
canvas. drawPath( path4, paint);     // 绘制五角形
// - - - - - - - - - - - 设置渐变器后绘制 - - - - - - - - - -
// 为 Paint 设置渐变器
Shader mShader = new LinearGradient(0, 0, 40, 60, new int[] {
    Color. RED, Color. GREEN, Color. BLUE, Color. YELLOW }, null,
    Shader. TileMode. REPEAT);
    paint. setShader( mShader);
    paint. setShadowLayer(45, 10, 10, Color. GRAY);          // 设置阴影
```

```
canvas.drawCircle(200, 40, 30, paint);    // 绘制圆形
canvas.drawRect(170, 80, 230, 140, paint);       // 绘制正方形
canvas.drawRect(170, 150, 230, 190, paint);    // 绘制矩形
RectF re3 = new RectF(170, 200, 230, 230);
canvas.drawRoundRect(re3, 15, 15, paint);        // 绘制圆角矩形
RectF re31 = new RectF(170, 240, 230, 270);
canvas.drawOval(re31, paint); // 绘制椭圆
Path path5 = new Path();
path5.moveTo(170, 340);
path5.lineTo(230, 340);
path5.lineTo(200, 290);
path5.close();
canvas.drawPath(path5, paint);        // 根据 Path 绘制, 绘制三角形
Path path6 = new Path();
path6.moveTo(186, 360);
path6.lineTo(214, 360);
path6.lineTo(230, 392);
path6.lineTo(200, 420);
path6.lineTo(170, 392);
path6.close();
canvas.drawPath(path6, paint);        // 根据 Path 绘制, 绘制五角形
// 设置字符大小后绘制
paint.setTextSize(24);
paint.setShader(null);
// 绘制 7 个文本数据
    canvas.drawText(getResources().getString(R.string.circle), 240, 50,
    paint);
    canvas.drawText(getResources().getString(R.string.square), 240, 120,
    paint);
    canvas.drawText(getResources().getString(R.string.rect), 240, 175,
    paint);
    canvas.drawText(getResources().getString(R.string.round_rect), 230,
    220, paint);
    canvas.drawText(getResources().getString(R.string.oval), 240, 260,
    paint);
    canvas.drawText(getResources().getString(R.string.triangle), 240, 325,
    paint);
```

```
                canvas.drawText(getResources().getString(R.string.pentagon), 240,
                390, paint);
        }
        public MyView(Context context, AttributeSet set) {
            super(context, set);
        }
    }
```

还需要在 string.xml 中添加如下常量定义：

```
<string name="circle">圆形</string>
<string name="square">正方形</string>
<string name="rect">长方形</string>
<string name="round_rect">圆角矩形</string>
<string name="oval">椭圆形</string>
<string name="triangle">三角形</string>
<string name="pentagon">五角形</string>
```

运行效果图如图 9.3 所示。

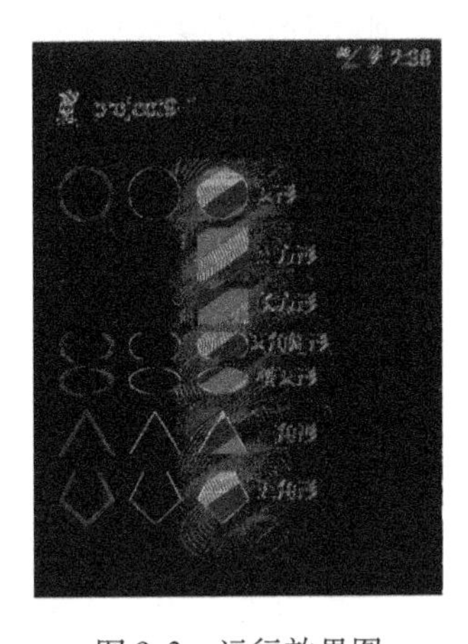

图 9.3 运行效果图

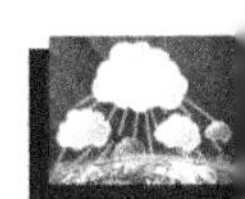

任务二 Android 补间动画实例

任务描述

本任务主要是实现 Android 的几种基本类型动画之一的补间动画。

任务分析

Android 3.0 以前，Android 支持两种动画模式，补间动画(Tween Animation)，帧动画(Frame Animation)，在 Android 3.0 中又引入了一个新的动画系统：属性动画(Property Animation)。本任务的主要目的学习补间动画的几种实现方式，重点和难点如下：

(1)了解 Android 补间动画的基础知识；
(2)掌握 Android 补间动画的几种实现方法。

任务准备

Android 的三种动画类型：

如上所述，Android 动画按实现方法的不同分为三种类型，下面我们分别介绍三种动画特点及实现方式。本节我们要介绍的是第一种补间动画。

补间动画(Tween Animation)，也叫 View Animation，它的原理是给出两个关键帧，通过一些特定算法将给定的属性值在给定的时间内在两个关键帧间渐变。

根据补间动画所变化的内容的不同，如透明度渐变、尺寸伸缩、位置移动及旋转，可将补间动画分为四种，四种补间动画的可分别由以下方式实现，具体见表 9 -3。

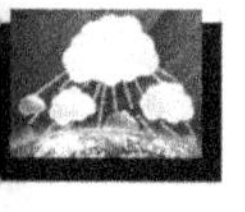

表 9.3　补间动画的四种类型及两种实现方法

渐变透明度动画效果	alpha	AlphaAnimation
渐变尺寸伸缩动画效果	scale	ScaleAnimation
画面转换位置移动动画效果	translate	TranslateAnimation
画面转移旋转动画效果	rotate	RotateAnimation

上面已经讲过，补间动画就是一系列 View 形状的变换，如大小的缩放、透明度的改变、位置的改变以及旋转，动画的定义既可以用 Java 代码定义也可以用 XML 定义。在这里，一般建议用 XML 来定义。

在设置动画时，可以给一个 View 同时设置多个变换的动画，比如从透明至不透明的淡入效果与从小到大的放大效果，这两种动画效果设置同时进行，也可以在一个完成之后再开始另一个。

用 XML 定义的动画放在/res/anim/文件夹内，XML 文件的根元素根据渐变类型的不同设置为 <alpha>(透明度渐变)、<scale>(尺寸大小渐变)、<translate>(位置移动)、<rotate>(旋转)、interpolator 元素或 <set>(表示是以上几个动画的集合，set 可以嵌套)。默认情况下，所有动画是同时运行的，也可以通过 startOffset 属性设置动画开始动作前停顿的时间来控制动画的运行。

可以通过设置 interpolator 属性改变动画渐变的方式，如 AccelerateInterpolator 表示开始时慢，然后逐渐加快。默认为 AccelerateDecelerateInterpolator，表示动画从开始到结束，变化率是先加速后减速的过程。

下面举例说明 XML 文件的定义方法，比如定义一个名为 T_animation . xml 的文件，代码如下所示：

```
<set xmlns: android = "http: //schemas. android. com/apk/res/android" >
        <alpha
            android: fromAlpha = "0. 0"
            android: toAlpha = "1. 0"
            android: duration = "6000"    //设置动画运行时间
            android: startOffset = "3000"/ >    //设置动画动作前停顿时间
</set >
```

以上代码的功能是这个动画的运行总共需要 6 秒钟，等待 3 秒后再播放 3 秒，让目标在 3 秒内透明度从 0(透明)逐渐增加到 1(不透明)。

再比如，定义以下文件来实现对象需要一边做 360 度的旋转一边进行尺寸大小的渐变。代码如下所示：

```
<set xmlns: android = "http: //schemas. android. com/apk/res/android" >
        <rotate
            android: fromDegrees = "0"    //旋转的起始角度
            android: toDegrees = "360"    //旋转的结束角度
            android: pivotX = "50% "    //以图片本身的一半作为 x 轴的坐标
            android: pivotY = "50% "    //以图片本身的一半作为 y 轴的坐标
            android: duration = "2000"/ >
        <scale
            android: pivotX = "50% "    //以图片本身的一半作为 x 轴的坐标
            android: pivotY = "50% "    //以图片本身的一半作为 y 轴的坐标
            android: fromXScale = "0. 0"    //起始 x 尺寸，为 0. 0 说明图片收缩到无
            android: toXScale = "1. 0"    //结束时 x 尺寸，为 1. 0 说明整个图 x 长度
            android: fromYScale = "0. 1"
            android: toYScale = "1. 0"
            android: duration = "2000"/ >
</set >
```

定义好动画的 XML 文件后，可以通过类似下面的代码对指定 View 应用动画。

```
ImageView myImage  =  (ImageView) findViewById (R. id. myImage);
Animation TwAnimation = AnimationUtils. loadAnimation(this, R. anim . T_animation);
myImage. startAnimation(TwAnimation);
```

任务实现

1. 新建 Android 项目

在 Eclipse 中，新建一个名为 project 9_2 的应用，具体方法可参考项目一。

2. 布局

在本例中用的控件为一个图片框，将图片 heart. jpg 加载到图片框中，如图 9. 4。

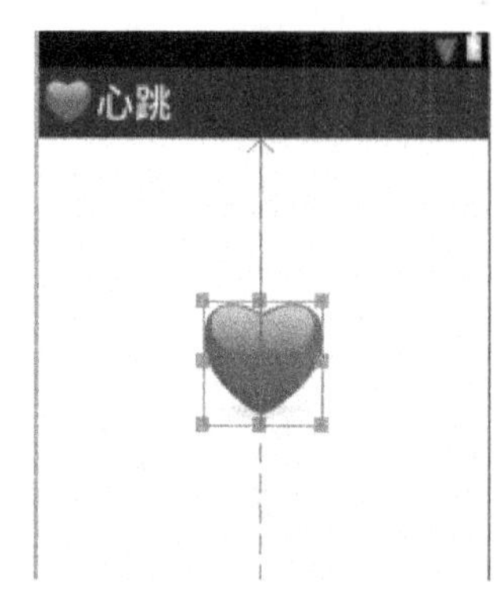

图 9. 4　界面设计

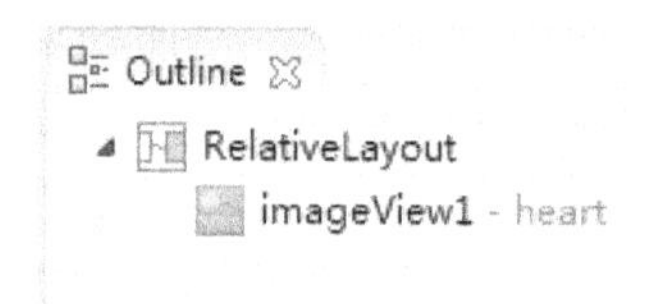

图 9. 5　各控件命名

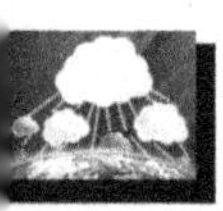

在 res\anim\目录下新建两个 . xml 文件，tolarge. xml 及 tosmall. xml。在这两个文件中用 xml 文件分别将原始图片的尺寸按所要求的比例放大及缩小，具体代码见后续代码。

3. 代码

对 MainActivity 布局：布局文件 activity_main. xml 代码如下：

```
<RelativeLayout xmlns: android = "http: //schemas. android. com/apk/res/android"
    xmlns: tools = "http: //schemas. android. com/tools"
    android: layout_width = "match_parent"
    android: layout_height = "match_parent"
    android: paddingBottom = "@ dimen/activity_vertical_margin"
    android: paddingLeft = "@ dimen/activity_horizontal_margin"
    android: paddingRight = "@ dimen/activity_horizontal_margin"
    android: paddingTop = "@ dimen/activity_vertical_margin"
    tools: context = ". MainActivity"  >
    <ImageView
        android: id = "@ +id/imageView1"
        android: layout_width = "wrap_content"
        android: layout_height = "wrap_content"
        android: layout_alignParentTop = "true"
        android: layout_centerHorizontal = "true"
```

```
        android: layout_marginTop = "93dp"
        android: src = "@drawable/heart"
        android: contentDescription = "@string/app_name" />
</RelativeLayout>
```

文件 tolarge.xml 代码如下，此代码功能为在 500ms 的时间内将对象从原来尺寸的 20% 增大到 100%。

```
<?xml version = "1.0" encoding = "utf-8"?>
<scale xmlns:android = "http://schemas.android.com/apk/res/android"
     android:interpolator = "@android:anim/accelerate_interpolator"
     android:fromXScale = "0.2"
     android:toXScale = "1.0"
     android:fromYScale = "0.2"
     android:toYScale = "1.0"
     android:pivotX = "50%"
     android:pivotY = "50%"
     android:duration = "500">
</scale>
```

文件 tosmall.xml 代码如下，此代码功能为在 500ms 的时间内将对象从原来尺寸的 100% 缩小到 20%。

```
<?xml version = "1.0" encoding = "utf-8"?>
<scale
   xmlns:android = "http://schemas.android.com/apk/res/android"
   android:interpolator = "@android:anim/decelerate_interpolator"
   android:fromXScale = "1.0"
   android:toXScale = "0.2"
   android:fromYScale = "1.0"
   android:toYScale = "0.2"
   android:pivotX = "50%"
   android:pivotY = "50%"
   android:duration = "500">
</scale>
```

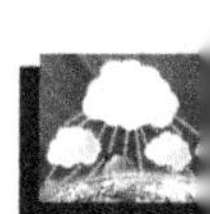

文件 MainActivity.java 代码如下：

```
package org.dglg.project9_2;
import android.app.Activity;
import android.os.Bundle;
```

```
import android.view.Menu;
import android.view.animation.Animation;
import android.view.animation.Animation.AnimationListener;
import android.view.animation.AnimationUtils;
import android.widget.ImageView;
public class MainActivity extends Activity implements AnimationListener {
      private Animation toLargeAnimation;
      private Animation toSmallAnimation;
      private ImageView image;
    @Override
    protected void onCreate(Bundle savedInstanceState) {
      super.onCreate(savedInstanceState);
      setContentView(R.layout.activity_main);
      image = (ImageView) findViewById(R.id.imageView1);
//以下代码将在文件 tolarge.xml 及 tosmall.xml 中定义好的动画运用到程序中
    toLargeAnimation  =  AnimationUtils.loadAnimation(this, R.anim.tolarge);
    toSmallAnimation  =  AnimationUtils.loadAnimation(this, R.anim.tosmall);
    toLargeAnimation.setAnimationListener(this);
    toSmallAnimation.setAnimationListener(this);
    image.startAnimation(toSmallAnimation);
    }
    @Override
    public boolean onCreateOptionsMenu(Menu menu) {
      // Inflate the menu; this adds items to the action bar if it is present.
    getMenuInflater().inflate(R.menu.main, menu);
    return true;
    }
    @Override
    public void onAnimationEnd(Animation animation) {
    if(animation.hashCode() == toLargeAnimation.hashCode()) {
      image.startAnimation(toSmallAnimation);
    } else {
      image.startAnimation(toLargeAnimation);
    }
    }
    @Override
//实现动画从小到大再从大到小不停地变化
    public void onAnimationRepeat(Animation animation) {
```

```
    // TODO Auto - generated method stub
    }
    @ Override
    public void onAnimationStart( Animation animation) {
    // TODO Auto - generated method stub
    }
}
```


运行时可看见一个大小不停变换的心形图案，有心脏跳动的感觉。运行效果如图 9.6 所示。

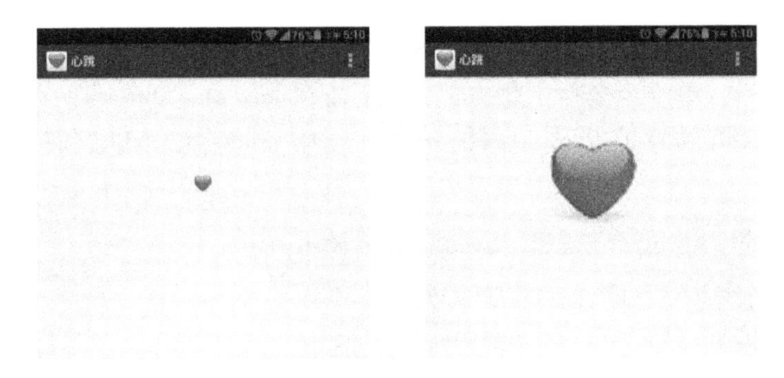

图 9.6　运行效果图

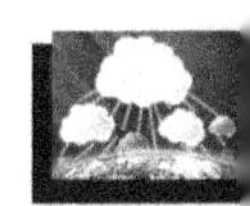

任务三　Android 帧动画实例

任务描述

本任务主要是实现 Android 的几种基本类型动画之一的帧动画。

任务分析

本任务的主要目的是学习几种动画的实现方式之一的帧动画，重点和难点如下：

(1) 了解 Android 帧动画的基础知识；
(2) 掌握 Android 帧动画的实现方法；
(3) 了解属性动画的一般知识。

任务准备

1. 帧动画(Frame animation)

Frame Animation 也称为 Drawable Animation，即帧动画，就如同 GIF 图片，通过一系列

的Drawable依次显示来模拟动画的效果。一般情况下，通过定义一个xml文件的方式来定义动画。例如，如果要定义一个名为anim_drawable.xml的文件来模拟动画效果，定义方式如下：

```
<animation-list xmlns:android="http://schemas.android.com/apk/res/android"
  android:oneshot="true">
   <item android:drawable="@drawable/fly1" android:duration="200"/>
   <item android:drawable="@drawable/fly2" android:duration="200"/>
   <item android:drawable="@drawable/fly3" android:duration="200"/>
</animation-list>
```

注意，在xml文件中，必须以<animation-list>为根元素，以<item>元素表示要轮换显示的图片(此例中fly1、fly2、fly3为需要轮换显示的图片)，duration属性表示各图片要轮换显示的时间，单位为毫秒。此xml文件要放在/res/drawable/目录下。

以下为使用xml文件来形成动画的示例，其中anim_drawable为已经定义好的xml文件。

```
protected void onCreate(Bundle savedInstanceState) {
        // TODO Auto-generated method stub
        super.onCreate(savedInstanceState);
        setContentView(R.layout.main);
        imageView = (ImageView) findViewById(R.id.imageView1);
        imageView.setBackgroundResource(R.drawable.anim_drawable);
        anim = (AnimationDrawable) imageView.getBackground();
    }

    public boolean onTouchEvent(MotionEvent event) {
        if (event.getAction() == MotionEvent.ACTION_DOWN) {
            anim.stop();
            anim.start();
            return true;
        }
        return super.onTouchEvent(event);
    }
```

2. 属性动画(Property Animation)

属性动画在Android 3.0中开始引进，它更改的是对象的实际属性。属性动画和其他类型动画的区别在于：其他动画改变的是对象的绘制效果，而属性动画改变的是对象的实际属性。

比如在补间动画中，它改变的是View的绘制效果，真正的View的属性保持不变。例如，无论你在对话中如何缩放Button的大小，Button的有效点击区域还是没有应用动画时变化前的区域，其位置与大小都不变。而在属性动画中，改变的是对象的实际属性，如Button的缩放中，Button的位置与大小属性值都改变了。而且属性动画不止可以应用于View，还可以应用于任何对象。属性动画只是表示一个值在一段时间内的改变，我们可以自己定义在值改变时所要进行的操作。

在属性动画中，可以对动画应用以下属性：

Duration：动画的持续时间。

TimeInterpolation：属性值的计算方式，如先快后慢。

TypeEvaluator：根据属性的开始、结束值与TimeInterpolation计算出的因子计算出当前时间的属性值。

Repeat Count and behavoir：重复次数与方式，如播放3次、5次、无限循环，可以令动画一直重复，或播放完时再反向播放。

Animation sets：动画集合，即可以同时对一个对象应用几个动画，这些动画可以同时播放也可以对不同动画设置不同开始偏移。

Frame refreash delay：多少时间刷新一次，即每隔多少时间计算一次属性值，默认为10ms，最终刷新时间还受系统进程调度与硬件的影响。

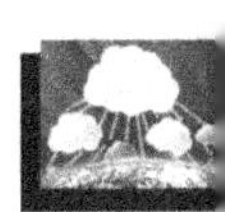

任务实现

1. 新建 Android 项目

在 Eclipse 中，新建一个名为 project 9_3 的应用。此应用在点击“开始”按钮时，动画开始运行，点击“停止”按钮动画停止运行，点击“运行一次”动画运行一次就自动停止。此项目是将12个动作连续的图片连续播放，是典型的帧动画。具体方法可参考项目一。

2. 布局

此项目用到三个按钮控件和一个 View 控件，具体布局及命名如图 9.7 和图 9.8 所示。

图 9.7　界面设计

图 9.8　各控件命名

布局的同时将所用到的如图 9.9 所示的 12 个用于形成动画的图片放置于/res/drawable/目录下。

3. 代码

MainActivity 的布局文件 activity_main. xml 代码如下所示：

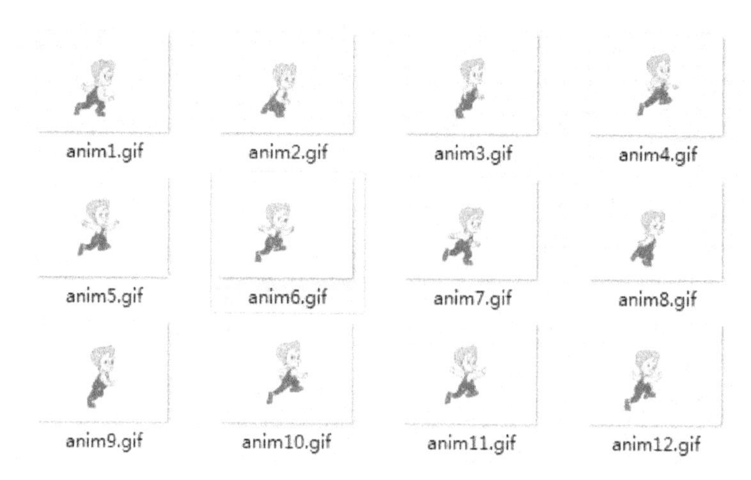

图9.9　用于形成动画的12张图片素材

```
<RelativeLayout xmlns:android="http://schemas.android.com/apk/res/android"
    xmlns:tools="http://schemas.android.com/tools"
    android:layout_width="match_parent"
    android:layout_height="match_parent"
    >

    <Button
        android:id="@+id/btn_start"
        android:layout_width="wrap_content"
        android:layout_height="wrap_content"
        android:layout_alignParentLeft="true"
        android:layout_alignParentTop="true"
        android:text="@string/b_begin"
        tools:ignore="HardcodedText" />

    <Button
        android:id="@+id/btn_stop"
        android:layout_width="wrap_content"
        android:layout_height="wrap_content"
        android:layout_alignParentTop="true"
        android:layout_toRightOf="@+id/btn_start"
        android:text="@string/b_stop"
        tools:ignore="HardcodedText" />
```

```
    <Button
        android:id="@+id/btn_run_once"
        android:layout_width="wrap_content"
        android:layout_height="wrap_content"
        android:layout_alignParentTop="true"
        android:layout_toRightOf="@+id/btn_stop"
        android:text="@string/run_once"
        tools:ignore="HardcodedText" />

    <ImageView
        android:id="@+id/imageView1"
        android:layout_width="300dp"
        android:layout_height="350dp"
        android:layout_alignParentLeft="true"
        android:layout_below="@+id/btn_start"
        android:layout_marginLeft="10dp"
        android:layout_marginTop="18dp"
        tools:ignore="ContentDescription"
        android:contentDescription="@string/app_name" />

</RelativeLayout>
```

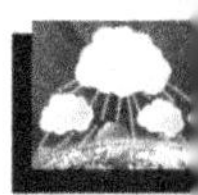

在/res/drawable/目录下建立文件 frame_animation. xml 代码如下：

```
<?xml version="1.0" encoding="utf-8"?>
<animation-list xmlns:android="http://schemas.android.com/apk/res/android"
  android:oneshot="false">
//通过一系列 Drawable 的依次显示来模拟动画的效果
<item android:drawable="@drawable/anim1" android:duration="50" />
  <item android:drawable="@drawable/anim2" android:duration="50" />
  <item android:drawable="@drawable/anim3" android:duration="50" />
<item android:drawable="@drawable/anim4" android:duration="50" />
  <item android:drawable="@drawable/anim5" android:duration="50" />
  <item android:drawable="@drawable/anim6" android:duration="50" />
<item android:drawable="@drawable/anim7" android:duration="50" />
  <item android:drawable="@drawable/anim8" android:duration="50" />
  <item android:drawable="@drawable/anim9" android:duration="50" />
<item android:drawable="@drawable/anim10" android:duration="50" />
  <item android:drawable="@drawable/anim11" android:duration="50" />
```

```
    <item android: drawable = "@drawable/anim12" android: duration = "50" />
</animation - list>
```

文件 MainActivity. java 代码如下:

```
package org. dglg. project9_3;
import android. os. Bundle;
import android. app. Activity;
import android. graphics. drawable. AnimationDrawable;
import android. view. Menu;
import android. view. View;
import android. view. View. OnClickListener;
import android. widget. Button;
import android. widget. ImageView;
public class MainActivity extends Activity implements OnClickListener {

  private Button btn_Start;
  private Button btn_Stop;
  private Button btn_run_once;
  private ImageView imageView;
  private AnimationDrawable animationDrawable;

  @ Override
  protected void onCreate( Bundle savedInstanceState) {
    super. onCreate( savedInstanceState);
    setContentView( R. layout. activity_main);
initWidget( );
    btn_Start. setOnClickListener( this);
    btn_Stop. setOnClickListener( this);
    btn_run_once. setOnClickListener( this);
//以下代码将在文件 frame_animation. xml 中定义好的动画运用到程序中
    imageView. setBackgroundResource( R. drawable. frame_animation);
    Object backgroundObject  =  imageView. getBackground( );
    animationDrawable  =  ( AnimationDrawable) backgroundObject;
  }
  private void initWidget( ) {
    btn_Start = ( Button) findViewById( R. id. btn_start);
    btn_Stop = ( Button) findViewById( R. id. btn_stop);
```

```
        btn_run_once = (Button) findViewById(R. id. btn_run_once);
        imageView = (ImageView) findViewById(R. id. imageView1);
    }

    @ Override
    public boolean onCreateOptionsMenu(Menu menu) {
        // Inflate the menu; this adds items to the action bar if it is present.
        getMenuInflater(). inflate(R. menu. main, menu);
        return true;
    }
    @ Override
    public void onClick(View arg0) {
    switch (arg0. getId()) {
    //点击开始按钮开始动画
case R. id. btn_start:
    animationDrawable. setOneShot(false);
    animationDrawable. stop();
    animationDrawable. start();
    break;
    //点击结束按钮结束动画
    case R. id. btn_stop:
    animationDrawable. stop();
    if (animationDrawable1 ! = null)
    {
    animationDrawable1. stop();
    }
    break;
    //点击运行一次按钮运行一次后结束动画
    case R. id. btn_run_once:
    animationDrawable. setOneShot(true);
    animationDrawable. stop();
    animationDrawable. start();
    break;
    }
    }
}
```

还需要在 string. xml 中添加如下常量定义:

```
<? xml version = "1.0" encoding = "utf-8"? >
<resources>

    <string name = "app_name" >project9_3 </string>
    <string name = "action_settings" >Settings </string>
    <string name = "b_stop" >停止 </string>
    <string name = "b_begin" >开始 </string>
    <string name = "run_once" >执行一次 </string>

</resources>
```

运行效果图如图 9.10 所示，点击“开始”时动画开始，点击“停止”时动画停止运动，点击“运行一次”时只运行一次就停止。

图 9.10　运行效果图

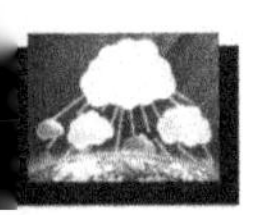

项目十

Android地图服务

地图服务是Android应用的一个重要组成部分，本项目将对百度地图服务进行多方位介绍，帮助读者了解百度地图开发的基本知识。

了解百度地图服务
掌握调用百度地图服务的方法
获取百度地图 API key
掌握百度地图定位方法

任务一　百度地图之显示

任务描述

在 Android 手机应用中，地图服务是手机应用的一个重要组成部分。本任务通过一个百度地图的显示过程，介绍在 Android 手机中显示百度地图的具体流程。

程序运行界面如图 10.1 所示。

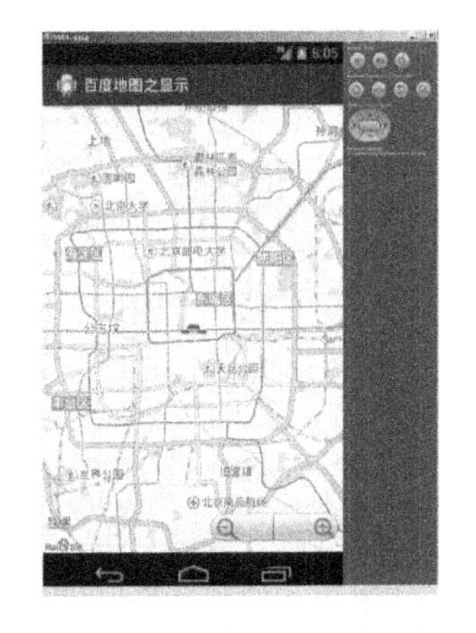

图 10.1　百度地图显示程序

任务分析

在本任务中，定义及引用的只有一个 baiduMap 类，任务的重点是显示百度地图。

本任务的重点和难点如下：

(1)熟悉和了解百度地图SDK及功能；
(2)如何创建百度开发者账号；
(3)如何获取百度地图key；
(4)百度地图的显示。

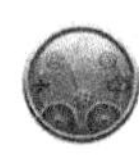

任务准备

1. 百度地图SDK

百度地图Android SDK是一套基于Android 2.1及以上版本设备的应用程序接口。开发者可以使用该套SDK开发适用于Android系统移动设备的地图应用，通过调用地图SDK接口，可以轻松访问百度地图服务和数据，构建功能丰富、交互性强的地图类应用程序。

百度地图Android SDK提供的所有服务是免费的，接口使用无次数限制。但使用前须申请密钥(key)，才可使用百度地图Android SDK。

百度地图SDK主要功能有：

(1)地图：提供地图展示和地图操作功能。地图展示包括普通地图(2D，3D)、卫星图和实时交通图。地图操作包括可通过接口或手势控制来实现地图的点击、双击、长按、缩放、旋转、改变视角等操作。

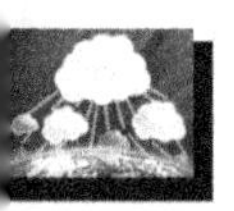

(2)POI检索：POI(Point of Interest)，中文可以翻译为"兴趣点"。在地理信息系统中，一个POI可以是一栋房子、一个商铺、一个邮筒、一个公交站等。其支持周边检索、区域检索和城市内检索。百度地图SDK提供三种类型的POI检索：周边检索、区域检索和城市内检索。

周边检索：以某一点为中心，指定距离为半径，根据用户输入的关键词进行POI检索。

区域检索：在指定矩形区域内根据关键词进行POI检索。

城市内检索：在某一城市内，根据用户输入的关键字进行POI检索。

(3)地理编码：提供地理坐标和地址之间相互转换的能力。

正向地理编码：实现了将中文地址或地名描述转换为地球表面上相应位置的功能。

反向地理编码：将地球表面的地址坐标转换为标准地址的过程。

(4)线路规划：支持公交信息查询、公交换乘查询、驾车线路规划和步行路径检索。

公交信息查询：可对公交详细信息进行查询。

公交换乘查询：根据起点、终点、查询策略，进行线路规划方案。

驾车线路规划：提供不同策略，规划驾车路线，同时支持设置途经点。

步行路径检索：支持步行路径的规划。

(5)定位：采用GPS、WIFI、基站、IP混合定位模式，使用Android定位SDK获取定位信息，使用地图SDK定位图层进行位置展示。

(6)导航：支持百度地图客户端导航和Web页面导航。

2. 获取百度地图API key

(1)百度地图API key的申请地址为http：//lbsyun.baidu.com/apiconsole/key。获取百

度地图 key 前，必须登录百度账号，如图 10.2 所示。如果没有，可免费申请。

图 10.2　登录百度账号

(2)百度账号登录后，会跳转到 API 控制台服务，如图 10.3 所示。

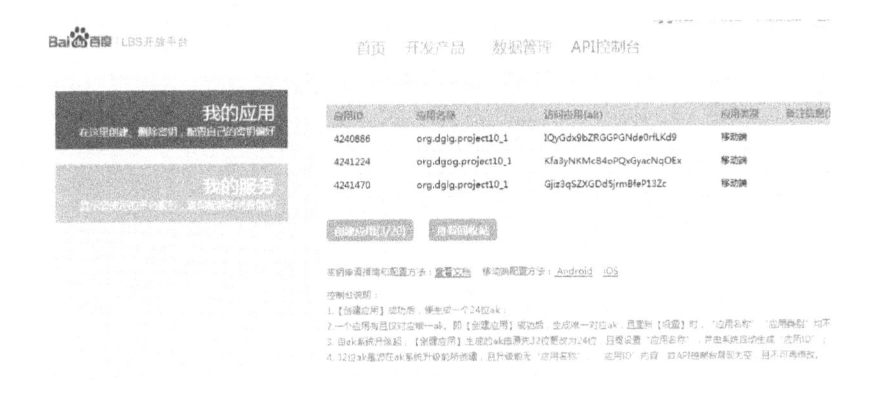

图 10.3　API 控制台

(3)单击图 10.3 中的黄色按钮和“创建应用”按钮，创建应用，如图 10.4 所示。

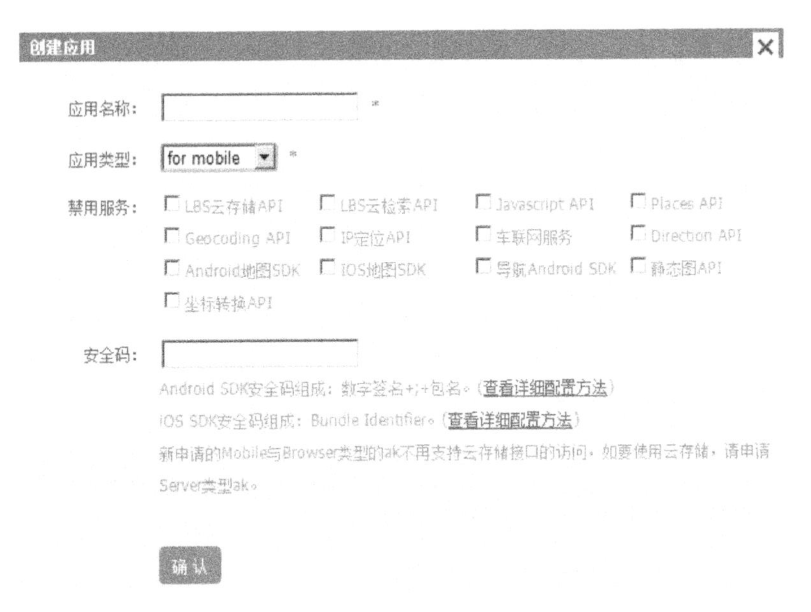

图 10.4　创建应用

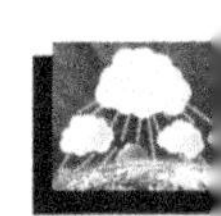

应用配置情况如下。

应用名称：在这输入 Android 应用名称。

应用类型：在此选择 for mobile。

禁用服务：在此可选择一些禁用服务。

安全码：安全码是创建应用的必填项，由数字签名 +“;”+ 包名组成(中间的分号为英文状态下的分号!)。

数字签名：是 Android 签名证书的 SHA1 值，可使用 keytool 工具查看，详细步骤如下：

STEP 1 取得 Android 签名证书文件 keystore 存放路径，可在 Eclipse 中查看到存放路径，单击 Eclipse 中的【windows】菜单中的【preferance】，弹出“preferance”对话框，在其中选择【Android】中的“Build”，可在其中查看到签名证书文件 keystore 的存放路径，如图 10.5 所示(图中选中的文本为 keystore 的存放路径)。

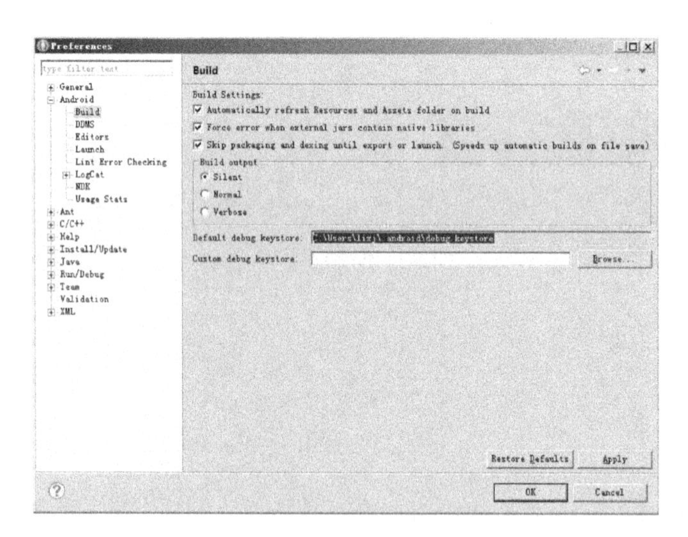

图 10.5　查看签名证书文件路径

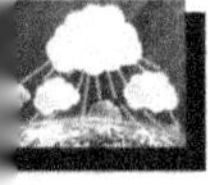

STEP 2 运用 JDK 提供的 keytool 工具查看签名证书文件的认证指纹，启动命令行窗口，在命令行窗口中输入如下命令：

keytool - list - v - keystore <Android keystore 的存储位置>

如：keystore 的存放路径为“C:\Users\lixj\.android\debug.keystore”，则命令为：

keytool - list - v - keystore C:\Users\lixj\.android\debug.keystore

命令输入完成后，会显示“输入密钥库口令”，直接回车，在之后出现的提示信息中可查看到 SHA1 值，如图 10.6 所示，图中的横线部分。

知识拓展

如果运行 keytool 工具时，系统提示“找不到该命令”，则说明未在 PATH 环境变量中定义路径，须增加“%Java_HOME%/bin”路径。

“%Java_HOME%”代表 JDK 的安装路径。

```
C:\windows\system32\CMD.exe
C:\Users\lixj>keytool -list -v -keystore c:\users\lixj\.android\debug.keystore
输入密钥库口令:

*****************  WARNING WARNING WARNING  *****************
* 存储在您的密钥库中的信息的完整性  *
* 尚未经过验证!  为了验证其完整性, *
* 必须提供密钥库口令。                  *
*****************  WARNING WARNING WARNING  *****************

密钥库类型: JKS
密钥库提供方: SUN

您的密钥库包含 1 个条目

别名: androiddebugkey
创建日期: 2013-5-7
条目类型: PrivateKeyEntry
证书链长度: 1
证书[1]:
所有者: CN=Android Debug, O=Android, C=US
发布者: CN=Android Debug, O=Android, C=US
序列号: 48e25019
有效期开始日期: Tue May 07 11:55:25 CST 2013, 截止日期: Thu Apr 30 11:55:25 CST
2043
证书指纹:
         MD5: 70:76:99:1A:0E:84:08:1F:58:9A:74:EF:53:AB:8F:A1
         SHA1: 94:DE:10:56:33:6C:9D:E3:96:3F:14:41:B4:BC:6D:4A:D9:E6:E6:11
         SHA256: 4C:7D:33:38:BC:BD:5D:CF:C5:C0:81:EB:44:9B:6F:94:0A:78:38:40:F4:
AD:3A:FB:AF:4F:B2:F1:0E:D8:A9:E1
         签名算法名称: SHA256withRSA
         版本: 3

扩展:

#1: ObjectId: 2.5.29.14 Criticality=false
SubjectKeyIdentifier [
KeyIdentifier [
0000: C7 BA AD CB 01 A6 45 CD   65 05 EF 29 14 E3 F2 5D  ......E.e..>...]
0010: 59 1C BB 88                                        Y...
```

图 10.6　keytool 工具查看 SHA1 值

STEP 3 运行完上述的步骤后，取得 SHA1 值，再加上“;”和包名则组成完整的安全码。本任务中对应的安全码如下(分号前为 SHA1 值，分号后为包名，本任务的包名为：org. dglg. project10_1)：

94：DE：10：56：33：6C：9D：E3：96：3F：14：41：B4：BC：6D：4A：D9：E6：E6；11；org. dglg. project10_1

STEP 4 输入完图 10.4 创建应用中的应用名称、应用类型、安全码后，单击确认按钮，则在 API 控制台中查看申请到的 Key，如图 10.7 所示(横线部分)。

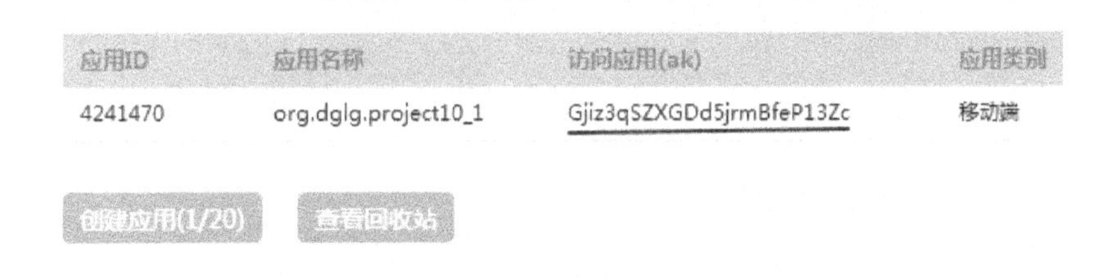

图 10.7　API 控制台 Key 值查看

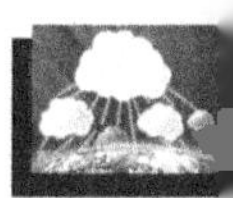

3. 加载百度地图 Android SDK 资源文件

百度地图 Android SDK 目前的版本为 3. 1. 1，进行百度地图 Android 开发前，须在工程中加载 SDK 资源文件，步骤如下：

STEP 1 下载 SDK 资源文件：百度地图 Android SDK 资源文件可免费下载，下载地址为 http：//developer. baidu. com/map/index. php? title = androidsdk/sdkandev - download，

最新的 SDK 资源文件包括：baidumapapi _ v3 _ 1 _ 1. jar、libBaiduMapSDK_v3_1_1. so 和 locSDK_3. 1. jar、liblocSDK3. so(locSDK_3. 1. jar、liblocSDK3. so 为百度定位 SDK 资源文件)。

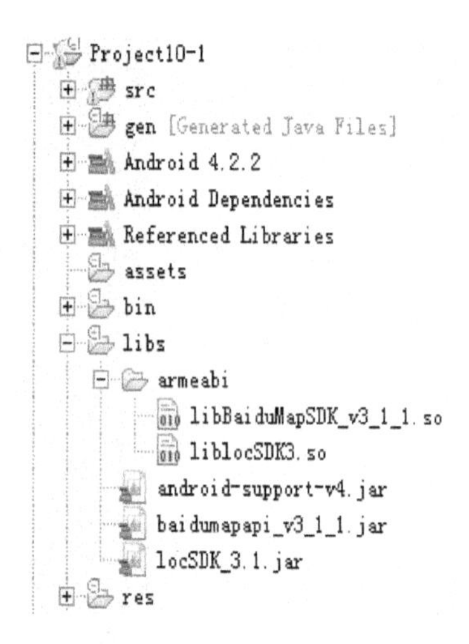

图 10.8　libs 目录结构

STEP 2 将 SDK 资源文件集成到工程：将资源文件中扩展名为 jar 的文件拷贝到工程里的 libs 根目录下，扩展名为 so 的文件拷贝到 libs\armeabi 目录下(如果 libs 里面没有 ameabi 这个目录，则创建)，拷贝完成后的工程目录结构如图 10.8 所示。

STEP 3 armebai 目录添加方法：如果 libs 里面没有 armebai 这个目录，则必须创建这个目录。方法如下：单击 Eclipse 中的【Flie】中的【New】中的【Folder】，如图 10.9 所示。Eclipse 将打开如图 10.10 所示的“New Folder”对话框。

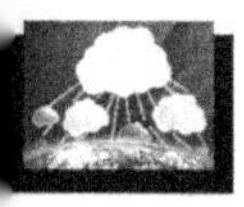

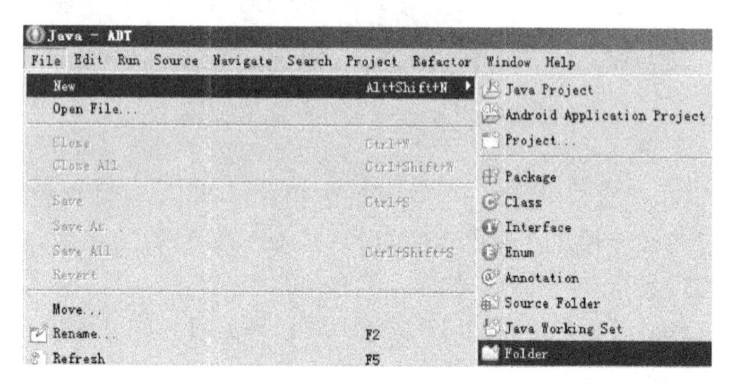

图 10.9　创建目录

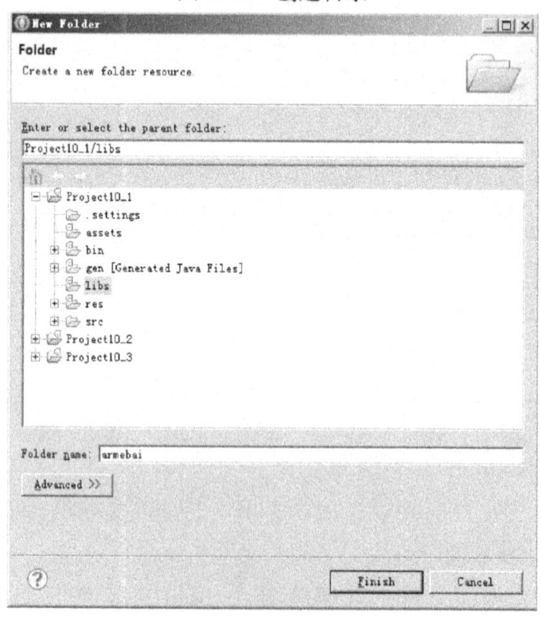

图 10.10　“New Folder”对话框

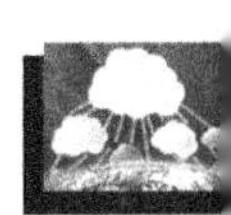

STEP 4 “New Folder”对话框设置：选定要添加 Folder 的位置(本任务为 Project10_1/libs 目录下)，在“Folder name”中输入 Folder 名称 armebai，单击【Finish】按钮完成目录创建。

任务实现

1. 新建 Android 项目

在 Eclipse 中，新建一 Android 应用程序，注意该应用程序的包名，本任务的包名为“org. dglg. project10_1”。

2. 申请百度地图 API key

根据 Android 签名证书的 SHA1 值和本任务的包名组成的安全码申请百度地图 API key。

3. 在 AndroidManifest 中添加 key

在 application 中添加 key，代码如下：

```
<application>
    <meta-data
  android:name="com.baidu.lbsapi.API_KEY"
  android:value="申请到的 key" />
</application>
```

4. 在 AndroidManifest 中添加权限

使用百度地图服务会涉及许多的 Android 权限，以下为百度官方列出的权限：

```
/*允许程序访问账户列表*/
<uses-permission android:name="android.permission.GET_ACCOUNTS" />
/*允许程序请求验证*/
<uses-permission android:name="android.permission.USE_CREDENTIALS" />
/*允许程序管理 AccountManager 中的账户列表*/
<uses-permission android:name="android.permission.MANAGE_ACCOUNTS" />
/*允许一个程序通过账户验证方式访问账户*/
<uses-permission android:name="android.permission.AUTHENTICATE_
ACCOUNTS" />
/*允许程序获取网络信息状态*/
<uses-permission android:name="android.permission.ACCESS_NETWORK_
STATE" />
/*允许程序访问网络连接*/
<uses-permission android:name="android.permission.INTERNET" />
/*允许程序读取设置*/
```

```
    <uses - permission android: name = "com. android. launcher. permission. READ_
SETTINGS" />
    /* 允许程序改变 WiFi 状态 */
    <uses - permission android: name = "android. permission. CHANGE_WIFI_STATE" /
>
    /* 允许程序获取当前 WiFi 接入的状态以及 WLAN 热点的信息 */
    <uses - permission android: name = "android. permission. ACCESS_WIFI_STATE" />
    /* 允许程序写入外部存储 */
    <uses - permission
    android: name = "android. permission. WRITE_EXTERNAL_STORAGE" />
    /* 允许程序收到广播后快速收到下一个广播 */
    <uses - permission android: name = "android. permission. BROADCAST_STICKY" />
    /* 允许程序读写系统设置项 */
    <uses - permission android: name = "android. permission. WRITE_SETTINGS" />
    /* 允许程序访问电话状态 */
    <uses - permission android: name = "android. permission. READ_PHONE_STATE" /
>
```

5. 加载百度地图 Android SDK 资源文件

将百度地图 Android SDK 资源文件加载进工程中。

6. 对 Activity 进行布局

本任务的功能是显示百度地图，所以界面布局文件非常简洁，在布局文件中引用百度地图 API 中的地图控件即可，代码如下：

```
<? xml version = "1.0" encoding = "utf - 8"? >
<LinearLayout xmlns: android = "http: //schemas. android. com/apk/res/android"
    android: layout_width = "fill_parent"
    android: layout_height = "fill_parent"
    android: orientation = "vertical" >
    <com. baidu. mapapi. map. MapView
        android: id = "@ + id/bmapView"
        android: layout_width = "fill_parent"
        android: layout_height = "fill_parent"
        android: clickable = "true" />
</LinearLayout>
```

7. 主程序功能实现

MainActivity. java 程序文件编写步骤：主要获取布局文件中的 MapView 以及实现 MapView 的生命周期。具体代码如下：

```
package org. dglg. project10_1;
import com. baidu. mapapi. SDKInitializer;
import com. baidu. mapapi. map. MapView;
import android. app. Activity;
import android. os. Bundle;
public class MainActivity extends Activity {
MapView mMapView = null;
    @ Override
    protected void onCreate( Bundle savedInstanceState) {
        super. onCreate( savedInstanceState) ;
    /* 在使用 SDK 各组件之前初始化 context 信息，传入 ApplicationContext */
    /* 注意该方法要再 setContentView 方法之前实现   */
        SDKInitializer. initialize( getApplicationContext( ) ) ;
        setContentView( R. layout. activity_main) ;
        /* 获取地图控件 */
        mMapView = (MapView) findViewById( R. id. bmapView) ;
    }
    @ Override
    protected void onDestroy( ) {
        super. onDestroy( ) ;
/* 在 Activity 执行 onDestroy 时，执行 mMapView. onDestroy( )，实现地图生命周期
管理 */
        mMapView. onDestroy( ) ;
    }
    @ Override
    protected void onResume( ) {
        super. onResume( ) ;
/* 在 Activity 执行 onResume 时，执行 mMapView. onResume ( )，实现地图生命周
期管理 */
        mMapView. onResume( ) ;
        }
    @ Override
    protected void onPause( ) {
        super. onPause( ) ;
/* 在 Activity 执行 onPause 时，执行 mMapView. onPause ( )，实现地图生命周期管
理 */
        mMapView. onPause( ) ;
        }
}
```

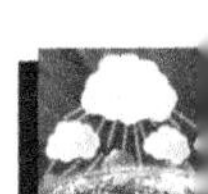

程序说明如下。

使用百度地图进行 Android 开发步骤包括:

① 申请一个与该程序包名对应的百度地图 API key(一个程序申请一个);

② 在工程中载入百度地图 SDK 资源文件;

③ 在 AndroidManifest 中添加 key 和相应权限;

④ 在布局文件中定义百度地图控件;

⑤ 在 Activity 中引用百度地图控件及使用相关类;

⑥ 在 Activity 生命周期中,管理及实现地图生命周期。

任务考核

任务提高:将本程序在实体机中调试并运行。

任务二 百度地图应用

任务描述

本任务是在任务一的基础上进行改进,主要功能是通过经度、纬度查找百度地图,然后在地图上显示标注以标明位置。

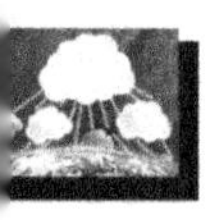

程序运行界面如图 10.11 所示。

图 10.11 百度地图显示程序

任务分析

在本任务中，涉及按经度、纬度查找百度地图 ，并在地图标注(标注覆盖物)。

本任务的重点和难点如下：

(1)进一步熟悉百度地图 SDK 及功能；
(2)了解地图中的经度和纬度；
(3)按经度、纬度查找百度地图；
(4)在百度地图标注覆盖物。

任务准备

1. 经纬度

经纬度是经度与纬度的合成组成一个坐标系统，又称为地理坐标系统。它是一种利用三度空间的球面来定义地球上的空间的球面坐标系统，能够标示地球上的任何一个位置。

1)经线

经线也称子午线，和纬线一样是人类为度量方便而假设出来的辅助线，定义为地球表面连接南北两极的大圆线上的半圆弧。任两根经线的长度相等，相交于南北两极点。每一根经线都有其相对应的数值，称为经度。经线指示南北方向。

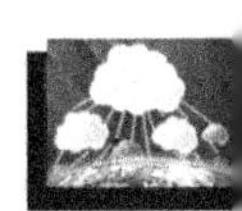

2)纬线

纬线和经线一样是人类为度量方便而假设出来的辅助线，定义为地球表面某点随地球自转所形成的轨迹。任何一根纬线都是圆形而且两两平行。纬线的长度是赤道的周长乘以纬线的纬度的余弦，所以赤道最长，离赤道越远的纬线，周长越短，到了两极就缩为 0。纬线指示东西方向。

3)经度

经度是地球上一个地点离一根被称为本初子午线的南北方向走线以东或以西的度数。

4)纬度

纬度是指某点与地球球心的连线和地球赤道面所成的线面角，其数值在 0 ~ 90 度之间。位于赤道以北的点的纬度叫做北纬，记为 N；位于赤道以南的点的纬度称南纬，记为 S。

5)中国地理位置四至点

最北端：位于东八区，北纬 53 度 33 分，黑龙江省的(漠河县)漠河以北的黑龙江主航道的中心线上。

最南端：位于东八区，北纬 3 度 52 分，南沙群岛的曾母暗沙(八仙暗沙和立地暗沙在曾母暗沙西南，是南海未曾正式公布的地名。虽然它们比曾母暗沙更靠南，但一般认为曾母暗沙是中国最南端)。

最西端：位于东五区，东经 73 度 40 分，一般认为是新疆维吾尔自治区位于乌兹别里山口的乌恰县(真正的西至点在帕米尔高原上，在中国、吉尔吉斯斯坦、塔吉克斯坦三国交界处略南的一座雪山上)。

最东端：位于东九区，东经135度2分30秒，黑龙江和乌苏里江交汇处的黑瞎子岛。

2. 导入地图标注图标

对地图进行标注前，须导入要标注的图标。方法为：直接将该文件复制到项目文件下的res/drawable-xxx中，Android会自动为这份资源在R清单文件中创建一个索引项：R. drawable. file_name，在java类中使用如下语法访问该资源：

[<package_name. >]R. drawable. file_name

在XML代码中则使用如下语法访问该资源：

@[<package_name. >] drawable/file_name

3. 地图标注

在对百度地图进行开发时，可根据自己实际需求，利用标注覆盖物，在地图指定的位置上(利用经度、纬度)添加标注信息。运用百度地图实现该功能代码比较简单，具体实现方法如下：

```
mMapView = (MapView) findViewById(R.id.bmapView); /*获取地图控件 */
mBaiduMap = mMapView.getMap(); /*获取地图 */
LatLng point = new LatLng(38.96, 116.40);   /*根据经度、纬度定义Marker坐标点(代码中纬度为38.96，经度为116.40) */
BitmapDescriptor bitmap = BitmapDescriptorFactory
.fromResource(R.drawable.icon_marka);   /*构建Marker图标*/
OverlayOptions option = new MarkerOptions()
.position(point)
.icon(bitmap);   /*构建MarkerOption，用于在地图上添加Marker*/
mBaiduMap.addOverlay(option); /*在地图上添加Marker，并显示*/
```

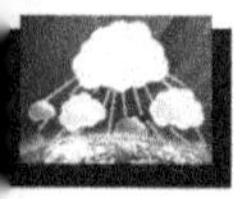

4. 更新地图状态

考虑到手机屏幕大小问题，在地图显示过程中，始终显示的是地图的某一部分。可以用以下方法更新地图状态，使得对应的经度、纬度位置显示在手机屏幕中心：

```
mMapView = (MapView) findViewById(R.id.bmapView); /*获取地图控件 */
mBaiduMap = mMapView.getMap(); /*获取地图 */
/*封装经度、纬度 */
LatLng point = new LatLng(38.96, 116.40);
MapStatus mMapStatus = new MapStatus.Builder()
.target(point)
.zoom(12)
.build();
/*改变地图状态*/
MapStatusUpdate mMapStatusUpdate =
                MapStatusUpdateFactory.newMapStatus(mMapStatus);
mBaiduMap.setMapStatus(mMapStatusUpdate);
```

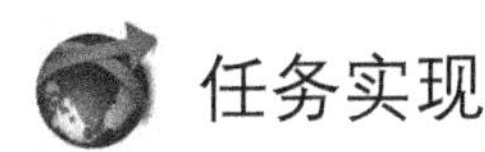

任务实现

1. 新建 Android 项目

在 Eclipse 中，新建一 Android 应用程序，注意该应用程序的包名，本任务的包名为“org. dglg. project10_2”。

2. 申请百度地图 API key

根据 Android 签名证书的 SHA1 值和本任务的包名组成的安全码申请百度地图 API key。

3. 在 AndroidManifest 中添加 key 并添加权限

依照任务一的方法，在 AndroidManifest 中添加 key 和权限。

4. 导入标注图标

将标注图标导入项目中，本任务导入的文件为 pos. gif，在 java 文件中使用 R. drawable. pos 来访问该资源。

5. 加载百度地图 Android SDK 资源文件

将百度地图 Android SDK 资源文件加载进工程中。

6. 对 Activity 进行布局

(1)编写 strings. xml 定义字符串变量：对布局文件中要引用的字符串变量进行定义，代码如下：

```
<? xml version = "1.0" encoding = "utf-8"? >
<resources>
    <string name = "app_name" >百度地图应用</string>
    <string name = "action_settings" >Settings</string>
    <string name = "StrLon" >经度</string>
    <string name = "StrLat" >纬度</string>
    <string name = "StrLonshow" >113.77</string>
    <string name = "StrLatshow" >23.05</string>
    <string name = "StrShow" >查找</string>
</resources>
```

(2)根据本任务实现的功能，对程序进行布局，采用线性混合布局，代码如下：

```
<? xml version = "1.0" encoding = "utf-8"? >
<LinearLayout xmlns:android = "http://schemas.android.com/apk/res/android"
    android:layout_width = "fill_parent"
    android:layout_height = "fill_parent"
    android:orientation = "vertical" >
```

```
    <LinearLayout
        android: layout_width = "match_parent"
        android: layout_height = "wrap_content"
        android: orientation = "horizontal"
        android: gravity = "center"  >
    <TextView
      android: layout_width = "wrap_content"
      android: layout_height = "wrap_content"
      android: text = "@ string/StrLon"/>
    <EditText
            android: id = "@ +id/EtLat"
            android: layout_width = "wrap_content"
            android: layout_height = "wrap_content"
            android: text = "@ string/StrLonshow"  />
  <TextView
      android: layout_width = "wrap_content"
      android: layout_height = "wrap_content"
      android: text = "@ string/StrLat"/>
        <EditText
            android: id = "@ +id/EtLng"
            android: layout_width = "wrap_content"
            android: layout_height = "wrap_content"
            android: text = "@ string/StrLatshow"  />

        <Button
            android: id = "@ +id/BtnShow"
            android: layout_width = "wrap_content"
            android: layout_height = "wrap_content"
            android: text = "@ string/StrShow"  />
    </LinearLayout>
    <com. baidu. mapapi. map. MapView
        android: id = "@ +id/bmapView"
        android: layout_width = "fill_parent"
        android: layout_height = "fill_parent"
        android: clickable = "true"  />
</LinearLayout>
```

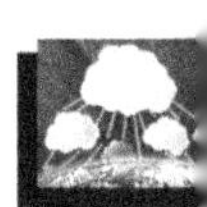

7. 主程序功能实现

MainActivity. java 程序文件编写步骤：主要获取布局文件中的 MapView 以及实现 MapView 的生命周期。具体代码如下：

```
package org. dglg. project10_2;
import com. baidu. mapapi. SDKInitializer;
import com. baidu. mapapi. map. BaiduMap;
import com. baidu. mapapi. map. BitmapDescriptor;
import com. baidu. mapapi. map. BitmapDescriptorFactory;
import com. baidu. mapapi. map. MapStatus;
import com. baidu. mapapi. map. MapStatusUpdate;
import com. baidu. mapapi. map. MapStatusUpdateFactory;
import com. baidu. mapapi. map. MapView;
import com. baidu. mapapi. map. MarkerOptions;
import com. baidu. mapapi. map. OverlayOptions;
import com. baidu. mapapi. model. LatLng;
import android. app. Activity;
import android. os. Bundle;
import android. view. View;
import android. view. View. OnClickListener;
import android. widget. Button;
import android. widget. EditText;
import android. widget. Toast;
public class MainActivity extends Activity {
Button BtnShow;
EditText EtLng;
EditText EtLat;
MapView mMapView = null;
BaiduMap mBaiduMap;
    @ Override
    protected void onCreate( Bundle savedInstanceState) {
        super. onCreate( savedInstanceState) ;
        /*在使用 SDK 各组件之前初始化 context 信息 */
        /*注意该方法要在 setContentView 方法之前实现 */
        SDKInitializer. initialize( getApplicationContext( ) ) ;
        setContentView( R. layout. activity_main) ;
        /*取得布局文件中对象 */
        BtnShow = ( Button) findViewById( R. id. BtnShow) ;
        EtLng = ( EditText) findViewById( R. id. EtLng) ;
```

```
            EtLat = (EditText)findViewById(R. id. EtLat);
            /* 获取地图控件 */
            mMapView = (MapView) findViewById(R. id. bmapView);
            /* 获取地图 */
            mBaiduMap = mMapView. getMap();
            /* 显示按钮事件 */
            BtnShow. setOnClickListener(new OnClickListener() {
    @Override
    public void onClick(View v) {
    // TODO Auto - generated method stub
    /* 获取 EditText 输入的内容 */
    String StrLng = EtLng. getEditableText(). toString(). trim();
    String StrLat = EtLat. getEditableText(). toString(). trim();
    /* 判断 EditText 是否输入内容 */
    if(StrLng. equals("") || StrLat. equals(""))
    {
    Toast. makeText(MainActivity. this, "请输入有效的经度、纬度", Toast. LENGTH_
LONG). show();
    }
    else
    {
      /* 将输入的经度、纬度转换的 double 类型 */
      double DLng = Double. parseDouble(StrLng);
      double DLat = Double. parseDouble(StrLat);
      /* 封装经度、纬度 */
      LatLng point = new LatLng(DLng, DLat);
            MapStatus mMapStatus = new MapStatus. Builder()
            . target(point)
            . zoom(12)
            . build();
            //构建 Marker 图标
            BitmapDescriptor bitmap = BitmapDescriptorFactory
            . fromResource(R. drawable. pos);
            //构建 MarkerOption, 用于在地图上添加 Marker
            OverlayOptions option = new MarkerOptions()
            . position(point)
            . icon(bitmap);
            /* 地图更新 */
```

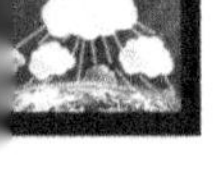

```
            MapStatusUpdate mMapStatusUpdate = MapStatusUpdateFactory.
            newMapStatus(mMapStatus);
            //改变地图状态
            mBaiduMap.setMapStatus(mMapStatusUpdate);
            //在地图上添加 Marker，并显示
            mBaiduMap.clear();
            mBaiduMap.addOverlay(option);
            }
            }
        });
    }
    @Override
    protected void onDestroy() {
        super.onDestroy();
        /*执行 mMapView.onDestroy()，实现地图生命周期管理 */
        mMapView.onDestroy();
    }
    @Override
    protected void onResume() {
        super.onResume();
        /*执行 mMapView.onResume ()，实现地图生命周期管理*/
        mMapView.onResume();
        }
    @Override
    protected void onPause() {
        super.onPause();
        /*执行 mMapView.onPause ()，实现地图生命周期管理 */
        mMapView.onPause();
        }
}
```

程序说明如下。

程序运行流程：

① 检测经度、纬度有没有输入。

② 经度、纬度输入后，按经度、纬度设置地图标注，并更新地图状态，将经度、纬度位置设置为地图中心点。

 任务考核

任务提高：将本程序在实体机中调试并运行。

项目十一

Android网络编程

随着Internet的高度发展，互联网已经发展为有线互联网和移动互联网。Android在网络编程和Internet应用也已经在全世界网络应用中，占据了大壁江山。本项目将走进Android的网络编程，帮助读者了解Android中网络编程和Internet应用的基本知识及应用。

了解 Internet 应用的基本知识
掌握 Android 的网络编程中，实现通过 HTTP 的 HttpURLConnection 方法和 HttpClient 方法访问网络
WebView 组件浏览网页的方法

任务一　利用 HttpURLConnection 访问网络

任务描述

本任务主要是启动 Tomcat 作为服务器，运行用来接收客户端发布的 GET 请求，并且向客户端发回响应的 Java Web 实例，如图 11.1 所示。

```
http://localhost:8081/dglg/HttpURLConnection_GET.jsp
localhost:8081/dglg/HttpURL ×

收到客户端发布的GET信息：
```

图 11.1　服务器端响应收到消息的界面

客户端方面(MainActivity):在编辑文本框中输入要发布的GET请求消息,单击图11.2中的"向服务器发布并返回消息"按钮,将利用HttpURLConnection向Tomcat服务器发布GET请求。在服务器接收到GET请求信息后,通过Java Web实例HttpURLConnection_ GET.jsp做出响应,向客户端发回"收到客户端发布的GET信息:"和客户端发布的GET请求消息,如图11.3所示。

图11.2 MainActivity启动后等待发布消息界面

图11.3 服务器收到消息并向客户端响应的界面

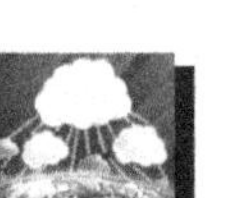

 任务分析

本任务主要是通过HttpURLConnection访问HTTP网络,分为服务器端和客户端的应用。

服务器端要启动Tomcat来接收客户端发送的请求,并通过Java Web实例做出响应。要启动Tomcat就要先安装Tomcat服务器软件,再对其进行设置和启动,这样才能在浏览器中对客户端的GET请求做出回应。

客户端是利用HttpURLConnection向Tomcat服务器发布GET请求,并且显示服务器端对客户端做出回应的内容。

重点和难点如下:

(1)了解HttpURLConnection访问网络的基础知识;

(2)掌握启动Tomcat作为服务器接收客户端发送GET请求的知识;

(3)掌握在服务器接收到请求信息后,通过Java Web实例HttpURLConnection_GET.jsp做出响应的方法。

 任务准备

1. HttpURLConnection 介绍

HttpURLConnection是Android使用HTTP进行网络通信的主要方法之一,此类属于java.net包,能实现HTTP请求的发送和HTTP响应的获取。但是HttpURLConnection对象是不能直接创建的,这是因为HttpURLConnection类是抽象类,不能直接创建对象。因此,要

创建 HttpURLConnection 对象，就要用到 URL 类的 openConnection( )方法来实现。

例如，想要向网站 http：//www. dglg. net 发送 HTTP 请求，可以为此网站创建一个 HttpURLConnection 对象，代码如下：

```
URL url =new URL(“http：//www. dglg. net/”)；
HttpURLConnection urlConnection = (HttpURLConnection)url. openConnection；
```

这时，仅仅是创建一个 HttpURLConnection 对象，还不能连接到网站。但是可以通过 GET 或 POST 请求向网站发送 HTTP 请求。

2. GET 请求和 POST 请求

1)GET 请求

GET 请求是 HttpURLConnection 对象的默认发送请求，发送 GET 请求的方法是在指定的网站地址后面加要发送的参数即可，参数的形式是：？参数名 = 参数值，这里的参数值就包含了要发送的内容，参数可以是多个，多个参数间用逗号隔开。

例如，要指定的网站地址是“http：//localhost：8081/DGLG”，要发送参数名为 getcs，getcs 的参数值是“GET 请求”，那么发送 GET 请求的链接地址是“"http：//localhost：8081/DGLG？ getcs =" +getcs”。

以下用简单的例子，具体说明如何使用 HttpURLConnection 向服务器发送 GET 请求，其步骤如下：

(1)向服务器发送 GET 请求。

通过 send( )方法的编写，向指定的网站地址“http：//localhost：8081/GETPATH”，用参数“？message =" +message”实现发送内容为“GET 请求已发送”的请求，代码如下：

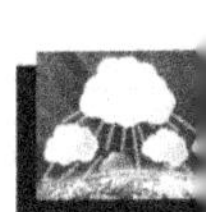

```
public void  send ( ){
        String message ="GET 请求已发送"；
        message = URLEncoder. encode(message，"UTF -8")；//转码
        System. out. println(message)；
String path  ="http：//localhost：8081/GETPATH？ message =" +message;
        URL url  =new URL(path)；
        HttpURLConnection conn  =  (HttpURLConnection)url. openConnection( )；
        conn. setConnectTimeout(5 * 1000)；//设置限制连接的时间
        conn. setRequestMethod("GET")；    //指定 GET 请求方法，不指定则为默认
的请求方法 GET，在这里不用也是可以的。
        InputStream inStream  =  conn. getInputStream( )；
        byte[ ] data  =  StreamTool. readInputStream(inStream)；
        String result =new String(data，"UTF -8")；
        System. out. println(result)；
    }
```

(2)从服务器读取数据。

由于本例 message 的内容只有一行，可以直接读取，代码如下：

```
String message = request.getParameter("message");
```

2)POST 请求

GET 请求只能发送 1024 个字节以内的数据，因此要发送比 1024 个字节大的数据就要用 POST 请求来实现了。发送 POST 请求，要通过 HttpURLConnection 类的方法 setRequestMethod()来指定，代码如下：

```
HttpURLConnection    conn = (HttpURLConnection)url.openConnection();
conn.setRequestMethod("POST");
```

3. send()方法

本任务中，自定义 send()方法，主要实现创建一个 HTTP 连接，将输入的内容通过 GET 请求发送到 Web 服务器，然后获取服务器的处理结果。主要代码片断为：

```
    target = "http://192.168.1.101:8081/DGLG/HttpURLConnection_GET.jsp?content=" + base64(content.getText().toString().trim());    //要访问的 URL 地址和输入的内容
    HttpURLConnection urlConn = (HttpURLConnection) url.openConnection();    //创建一个 HTTP 连接
    InputStreamReader in = new InputStreamReader(  urlConn.getInputStream()); //获得读取的内容
    BufferedReader buffer = new BufferedReader(in); // 获取输入流对象
    String inputcontent = null;
    while ((inputcontent = buffer.readLine()) != null) {//通过循环逐行读取输入流中的内容
    result += inputcontent + "\n";    }
    in.close();    //关闭字符输入流对象
```

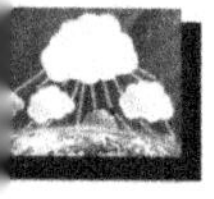

4. Base64 编码

用 Base64 编码解决用 GET 请求传递中文参数时产生乱码的问题。主要代码片断为：

```
    content = Base64.encodeToString(content.getBytes("utf-8"), Base64.DEFAULT);
  //对字符串进行 Base64 编码
    content = URLEncoder.encode(content);    //对字符串进行 URL 编码
```

5. 创建线程和 Handler

创建线程是为了实现 send()发送和获取 GET 消息，主要代码片断为：

```
new Thread(new Runnable() {//创建一个新线程
public void run() {
    send();    //发送文本内容到 Web 服务器
  Message m = handler.obtainMessage(); //获取一个 Message
  handler.sendMessage(m); //发送消息
  }
    }).start(); //开启线程
```

6. Tomcat 服务器的安装

本任务安装 Tomcat 的是 7.0 版本，安装的步骤如下：

(1)运行 Tomcat 的安装文件就会打开标题为“Apache Tomcat setup”对话框，这时单击“Next”按钮，打开“License Agreement”对话框，单击“I Agree”按钮，如图 11.4、图 11.5 所示。

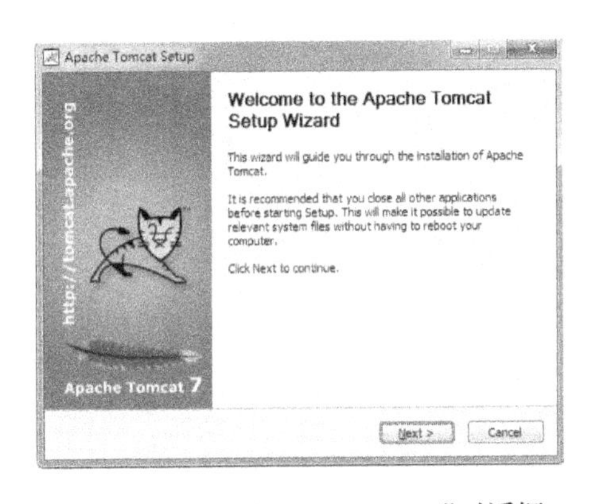

图 11.4 “Apache Tomcat setup”对话框

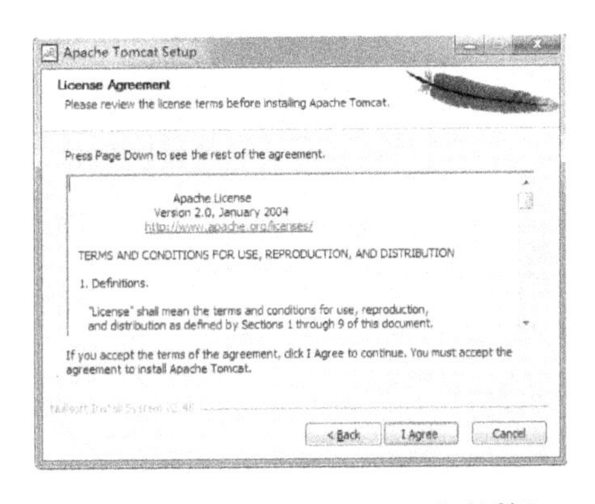

图 11.5 “License Agreement”对话框

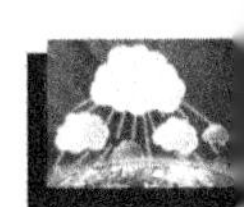

(2)这时就会打开“Choose Components”选择组件对话框，直接单击“Next”按钮，打开“Configuration”设置对话框，这个对话框的参数的设置相当重要，其他参数不变，本任务将“HTTP/1.1 Connector port”的默认值“8080”改为“8081”，这样可以避免与其他应用用同一个连接端口而产生冲突。如图 11.6、图 11.7 所示。

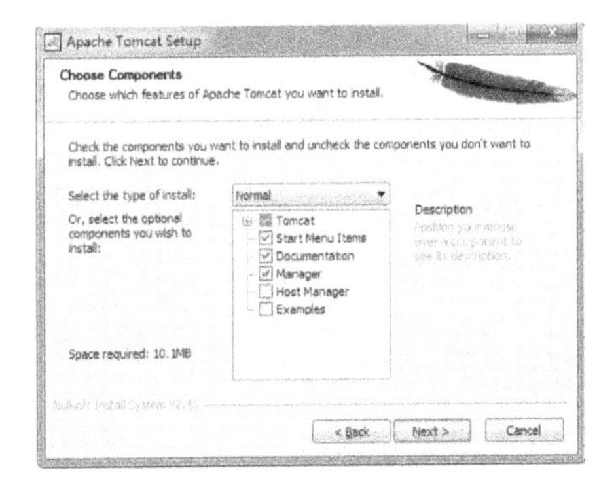

图 11.6 “Choose Components” 对话框

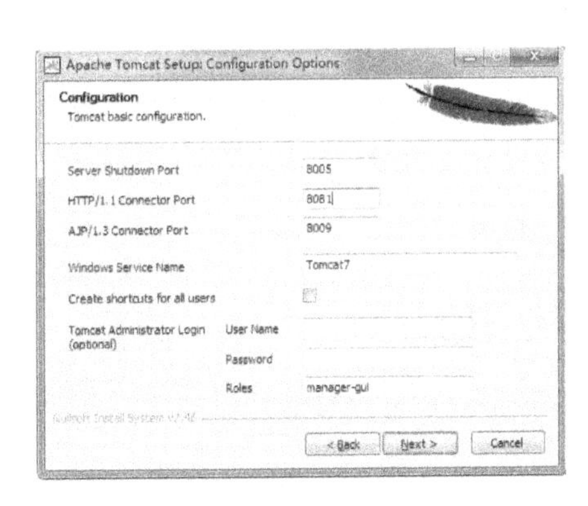

图 11.7 “Configuration” 对话框

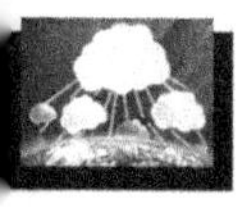

(3)这时就会打开“Java Virtual Machine”Java虚拟机对话框，设置好路径，单击“Next”按钮，打开“Choose Install Location”选择本地安装路径对话框，单击“Install”按钮，即时开始安装。如图11.8、图11.9所示。

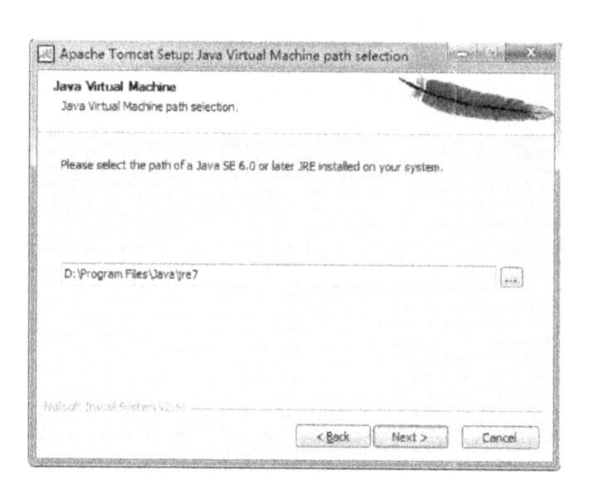

图11.8　“Java Virtual Machine”对话框

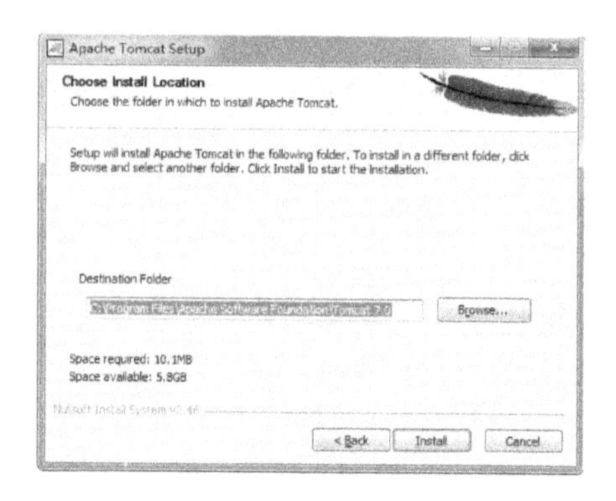

图11.9　“Choose Install Location”对话框

(4)完成安装后就会打开“Completing the Apache Tomcat Setup Wizard”完成安装对话框，图11.10所示。打开IE浏览器输入并打开地址“http://localhost:8081”，如果打开Tomcat7.0运行界面，表示安装成功，如图11.11所示。

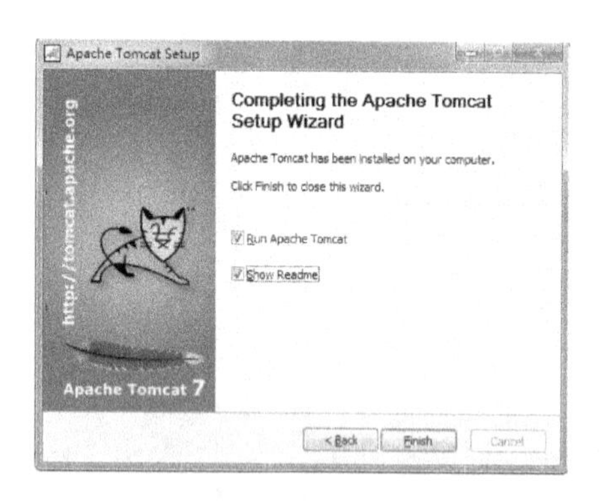

图11.10　完成安装对话框

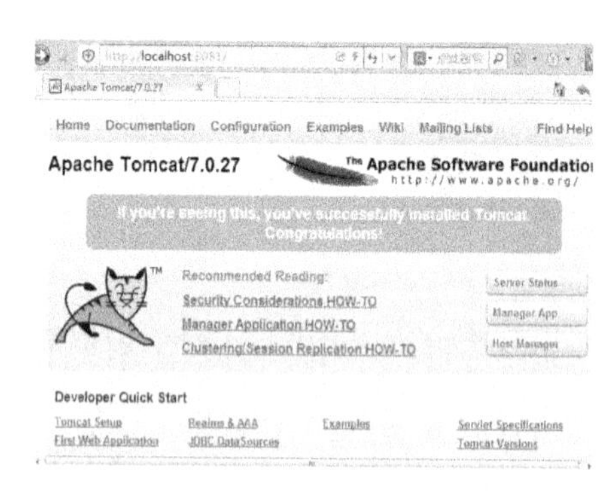

图11.11　运行Tomcat7.0界面

7. 接收客户端发送请求的Java Web实例

服务器端通过Java Web实例HttpURLConnection_GET.jsp来实现，接收客户端发过来的Post请求，并将服务器响应的结果在客户端显示出来。本任务为了运行它，将其保存在“D:\Tomcat 7.0\webapps\dglg”路径下，代码片断如下：

```
content = request.getParameter("content"); //获取输入的GET信息
content = content.replaceAll("%2B","+"); //替换content中的加号，这是由于在进行URL编码时，将+号转换为%2B了
BASE64Decoder decoder = new BASE64Decoder();
content = new String(decoder.decodeBuffer(content),"utf-8");   //进行base64解码
```

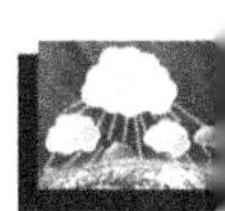

 任务实现

1. 新建 Android 项目

参考项目一创建项目的方法，在 Eclipse 中新建 Android 项目名为 Project11_1 和包为 org. dglg. project11_1 的程序，在新建 Android 项目时会自动创建一个 Activity，本任务中 Activity 的名称是 MainActivity。

2. 对 MainActivity 进行布局

1) 对 MainActivity 的 strings. xml 进行布局

```
<? xml version = "1.0" encoding = "utf -8"? >
<resources >
  <string name = "app_name" >HttpURlConnection 发布 GET 消息 </string >
  <string name = "button" >向服务器发布并返回消息 </string >
</resources >
```

2) 对 MainActivity 的布局文件 activity_main. xml 进行布局

本任务中，MainActivity 的布局文件是 activity_main. xml，用来实现“向服务器发布并返回消息”的布局，代码如下：

```
<? xml version = "1.0" encoding = "utf -8"? >
<LinearLayout xmlns: android = "http: //schemas. android. com/apk/res/android"
    android: layout_width = "fill_parent"
    android: layout_height = "fill_parent"
    android: gravity = "center_horizontal"
    android: orientation = "vertical" >
    <EditText
        android: id = "@ +id/content"
        android: layout_width = "match_parent"
        android: layout_height = "wrap_content" />
    <Button
        android: id = "@ +id/button"
        android: layout_width = "wrap_content"
        android: layout_height = "wrap_content"
        android: text = "@ string/button" />
    <ScrollView
        android: id = "@ +id/scrollView1"
        android: layout_width = "match_parent"
        android: layout_height = "wrap_content"
        android: layout_weight = "1" >
```

```
    <LinearLayout
            android: id = "@ +id/linearLayout1"
            android: layout_width = "match_parent"
            android: layout_height = "match_parent"  >
            <TextView
                android: id = "@ +id/result"
                android: layout_width = "match_parent"
                android: layout_height = "wrap_content"
                android: layout_weight = "1" />
        </LinearLayout>
    </ScrollView>
</LinearLayout>
```

3. 服务器端 Java Web 实例 HttpURLConnection_GET. jsp 的实现

此实例用来接收客户端发过来的 POST 请求，并将服务器响应的结果在客户端显示出来。本任务为了运行它，将其保存在“D:\Tomcat 7.0\webapps\dglg”路径下，代码如下：

```
<%@ page contentType = "text/html; charset = utf - 8" language = "java" import = "sun
. misc. BASE64Decoder"% >
<%
String content = "";
if(request. getParameter("content")! =null){
content = request. getParameter("content"); //获取输入的 GET 信息
content = content. replaceAll("%2B"," +"); //替换 content 中的加号，这是由于在进
行 URL 编码时，将 + 号转换为%2B 了
BASE64Decoder decoder = new BASE64Decoder();
content = new String(decoder. decodeBuffer(content),"utf -8");    //进行 base64 解码
}
% >
<% ="收到客户端发布的 GET 信息:"% >
<% =content% >
```

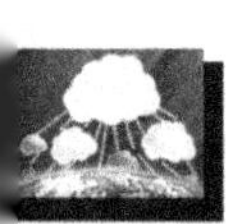

4. 实现 MainActivity 程序功能

完整的代码如下：

```
package org. dglg. project11_1;
    //导入相关的类
    import java. io. BufferedReader;
```

```
import java. io. IOException;
import java. io. InputStreamReader;
import java. io. UnsupportedEncodingException;
import java. net. HttpURLConnection;
import java. net. MalformedURLException;
import java. net. URL;
import java. net. URLEncoder;
import android. app. Activity;
import android. os. Bundle;
import android. os. Handler;
import android. os. Message;
import android. util. Base64;
import android. view. View;
import android. view. View. OnClickListener;
import android. widget. Button;
import android. widget. EditText;
import android. widget. TextView;
import android. widget. Toast;
public class MainActivity extends Activity {
    private EditText content;    //定义一个输入文本内容的编辑框对象
    private Button button;    //定义"向服务器发布并返回消息"按钮对象
    private Handler handler; //定义一个 Handler 对象
    private String result = "";    //定义一个代表显示内容的字符串
    private TextView TV;    //定义一个显示结果的文本框对象
    @ Override
    protected void onCreate( Bundle savedInstanceState) {
        super. onCreate( savedInstanceState) ;
        setContentView( R. layout. main) ;
        content = (EditText) findViewById( R. id. content) ;    //获取输入文本内容
的 EditText 组件
        TV = (TextView) findViewById( R. id. result) ;    //获取显示结果的 TextView
组件
        button = (Button) findViewById( R. id. button) ; //获取"向服务器发布并返
回消息"按钮组件
        //为"向服务器发布并返回消息"按钮添加单击事件监听器
        button. setOnClickListener( new OnClickListener( ) {
            @ Override
            public void onClick( View v) {
```

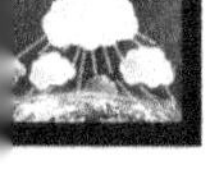

```
                if ("".equals(content.getText().toString())) {
        Toast.makeText(MainActivity.this, "请输入发布的消息!",
        Toast.LENGTH_SHORT).show();     //显示消息提示
                    return;
                }
    // 创建一个新线程，用于发送并读取 GET 信息
        new Thread(new Runnable() {
            public void run() {
            send();     //发送文本内容到 Web 服务器
            Message m = handler.obtainMessage(); // 获取一个 Message
            handler.sendMessage(m); // 发送消息
                    }
                }).start(); // 开启线程
            }
        });
    //创建一个 Handler 对象
        handler = new Handler() {
            @Override
            public void handleMessage(Message msg) {
                if (result != null) {
                    resultTV.setText(result); // 显示获得的结果
                    content.setText("");        //清空文本框
                }
                super.handleMessage(msg);
            }
        };
    }
    public void send() {
    String target = "";
target = "http://192.168.1.101:8081/DGLG/HttpURLConnection_GET.jsp?content="
  +base64(content.getText().toString().trim());    //要访问的URL地址
        URL url;
        try {
            url = new URL(target);
  HttpURLConnection urlConn = (HttpURLConnection) url.openConnection();    //创建
一个 HTTP 连接
  InputStreamReader in = new InputStreamReader(
        urlConn.getInputStream()); // 获得读取的内容
```

```
            BufferedReader buffer = new BufferedReader(in); // 获取输入流对象
            String inputcontent = null;
            //通过循环逐行读取输入流中的内容
                while ((inputcontent = buffer.readLine()) != null) {
                    result += inputcontent + "\n";
                }
                in.close();    //关闭字符输入流对象
                urlConn.disconnect();    //断开连接
            } catch (MalformedURLException e) {
                e.printStackTrace(); //输出异常信息
            } catch (IOException e) {
                e.printStackTrace(); //输出异常信息
            }
        }
        //对字符串进行 Base64 编码
        public String base64(String content){
        try {
    content = Base64.encodeToString(content.getBytes("utf-8"), Base64.DEFAULT);
//对字符串进行 Base64 编码
    content = URLEncoder.encode(content);    //对字符串进行 URL 编码
            } catch (UnsupportedEncodingException e) {
                e.printStackTrace();    //输出异常信息
            }
            return content;
        }
    }
```

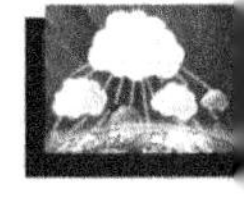

任务考核

1. 参考本任务，安装 Tomcat 7.0 版本。

2. 实现用 Get 方法通过 HttpURLConnection 访问 HTTP 网络。其中服务器端启动 Tomcat 来接收客户端发送的 Get 请求，并通过 Java Web 实例 AnswerGet.jsp 作出响应；客户端的 MainActivity 实现点击“发送”按钮，将在文本框中输入的文本用发送 Get 方法通过 HttpURLConnection 到指定的 Url 地址的服务器端，并由服务器端发回接收后的信息。

任务二 利用 HttpClient 访问网络

任务描述

与任务一相似，同样是向 Tomcat 服务器发布 HTTP 请求，服务器会对客户端发布来的消息向客户端做出响应；差别在于本任务是利用 HttpClient 向服务器发布 POST 请求。

(1)本任务主要是启动 Tomcat 作为服务器，运行用来接收客户端 HttpClient 发布的 POST 请求，并且向客户端发回响应的 Java Web 实例，HttpClient_POST. jsp 文件，如图 11. 12 所示。

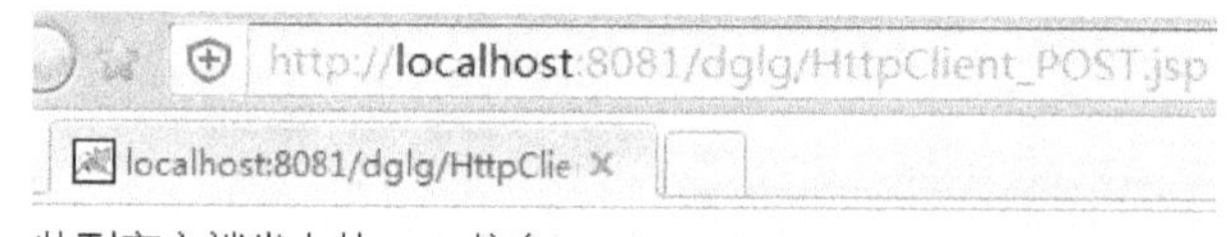

图 11. 12　HttpClient_POST. jsp 响应收到消息的界面

(2)启动客户端的 MainActivity，在编辑文本框中输入要发布的 POST 请求消息，单击图 11. 13 中的"向服务器发布并返回消息"按钮，将利用 HttpClient 向 Tomcat 服务器发布 POST 请求。在服务器接收到 POST 请求信息后，通过 Java Web 实例 HttpClient_POST. jsp 作出响应，向客户端发回"收到客户端发布的 POST 信息:"和客户端发布的 POST 请求消息，如图 11. 14 所示。

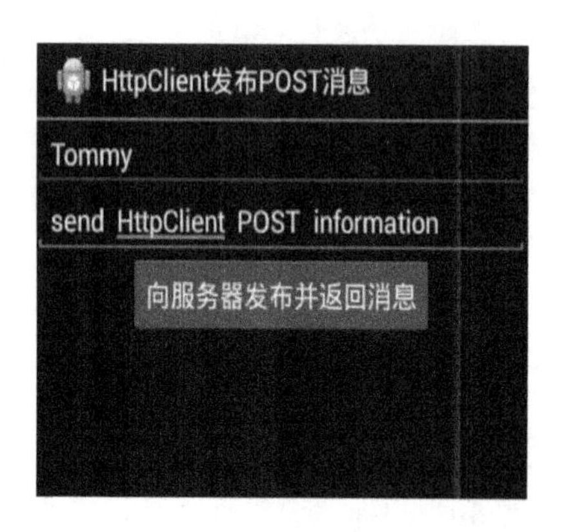

图 11. 13　MainActivity 启动后等待发布消息界面

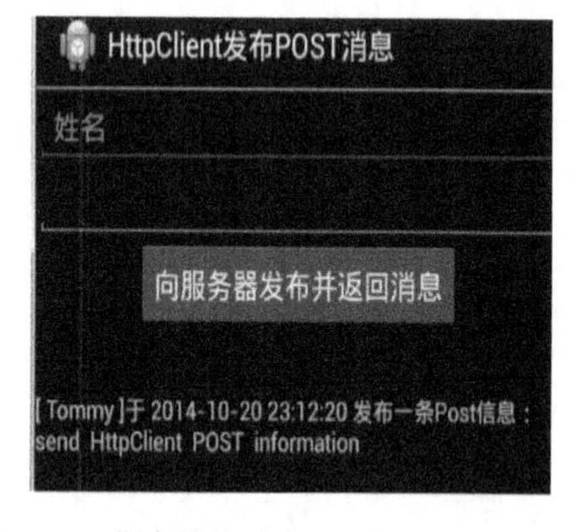

图 11. 14　服务器收到消息并向客户端响应界面

任务分析

本任务主要是通过 HttpClient 访问 HTTP 网络，分为服务器端和客户端的应用。

服务器端要启动 Tomcat 来接收客户端发送的请求，并通过 Java Web 实例 HttpClient_POST. jsp 做出响应。

客户端是利用 HttpURLConnection 向 Tomcat 服务器发布 GET 请求，并且显示服务器端 Java Web 实例 HttpClient_POST. jsp 对客户端做出回应的内容。

重点和难点如下：

(1) 了解 HttpClient 访问网络的基础知识；

(2) 掌握启动 Tomcat 作为服务器接收客户端发送 POST 请求的知识；

(3) 掌握在服务器接收到请求信息后，通过 Java Web 实例 HttpClient_POST. jsp 做出响应的方法。

任务准备

1. HttpClient 介绍

由于集成了 HttpClient，在 Android 中可以直接使用 HttpClient 访问网络，并且能实现复杂的联网操作。

Java 提供的网络访问方法被 HttpClient 封装成了 HttpGet、HttpPost、HttpResponse 类以简化操作，HttpGet、HttpPost、HttpResponse 分别表示发送 GET 请求、发送 POST 请求、处理响应的对象。

2. POST 请求

HttpClient 的 POST 请求是传递复杂数据的请求。其步骤如下：

(1)创建 HttpClient 和 HttpPost 对象。

(2)用 HttpPost 的 setParams()或 setEntity()方法添加请求参数。

(3)用 HttpClient 的 execute()方法发送请求，这时会返回一 HttpResponse 对象。

(4)用 HttpResponse 的 getEntity()方法获取包含服务器响应内容的 HttpEntity 对象，以此对象获取服务器响应内容。

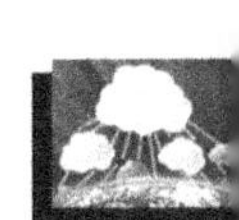

3. send()方法

本任务中，自定义 send()方法，主要实现创建一个 HTTP 连接，将输入的内容通过 HttpClient 的 GET 请求发送到 Web 服务器，然后获取服务器的处理结果。主要代码片断为：

```
String target = "http://192.168.1.66:8081/blog/HttpClient_Post.jsp";   //要提交的目标地址
HttpClient httpclient = new DefaultHttpClient();   //创建 HttpClient 对象
HttpPost httpRequest = new HttpPost(target);   //创建 HttpPost 对象
List<NameValuePair> params = new ArrayList<NameValuePair>(); //将要传递的参数保存到 List 集合中
params.add(new BasicNameValuePair("param", "post"));   //标记参数
params.add(new BasicNameValuePair("content", content.getText().toString())); //内容
```

```
httpRequest.setEntity(new UrlEncodedFormEntity(params, "utf-8")); //设置编码方式
HttpResponse httpResponse = httpclient.execute(httpRequest); //执行 HttpClient 请求
```

4. 编码方式设置和转码

本任务在 send()设置编码方式，代码如下：

```
httpRequest.setEntity(new UrlEncodedFormEntity(params, "utf-8")); //设置编码方式
```

在 HttpClient_POST.jsp 中进行转码，代码如下：

```
content = new String(content.getBytes("iso-8859-1"),"utf-8"); //对内容进行转码
```

5. 创建线程和 Handler

创建线程和 Handler 是为了实现 send()发送和获取 POST 消息，主要代码片断为：

```
new Thread(new Runnable() {
                public void run() {
                send();
                Message m = handler.obtainMessage(); // 获取一个 Message
                handler.sendMessage(m); // 发送消息
                }
            }).start(); // 开启线程
            }
        });
handler = new Handler() {
        @Override
        public void handleMessage(Message msg) {
            if (result != null) {
                TV.setText(result); // 显示获得的结果
                content.setText(""); // 清空内容编辑框
                }
            super.handleMessage(msg);
        }
    };
}
```

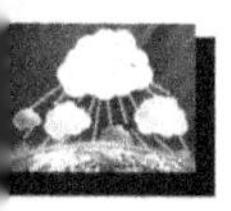

6. 接收客户端发送请求的 Java Web 实例

Java Web 实例 HttpClient_Post.jsp 用来接收客户端发过来的 Post 请求，并将服务器响应的结果在客户端显示出来，在本任务为了运行它，将其保存在"D:\Tomcat 7.0\webapps\dglg"路径下。

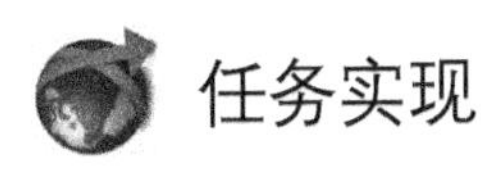

任务实现

1. 新建 Android 项目

参考项目一创建项目的方法，在 Eclipse 中新建 Android 项目名为 Project11_2 和包为 org. dglg. project11_2 的程序，在新建 Android 项目时会自动创建一个 Activity，本任务中 Activity 的名称是 MainActivity。

2. 对 MainActivity 进行布局

1) 对 MainActivity 的 strings. xml 进行布局

```
<? xml version = "1.0" encoding = "utf - 8"? >
<resources >
   <string name = "app_name" >HttpClient 发布 post 消息 </string >
   <string name = "button" >向服务器发布并返回消息 </string >
</resources >
```

2) 对 MainActivity 的布局文件 activity_main. xml 进行布局

本任务中，MainActivity 的布局文件是 activity_main. xml，用来实现“向服务器发布并返回消息”的布局，代码如下：

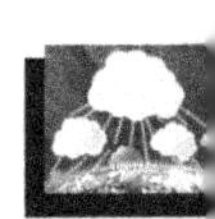

```
<? xml version = "1.0" encoding = "utf - 8"? >
<LinearLayout xmlns: android = "http: //schemas. android. com/apk/res/android"
     android: orientation = "vertical"
     android: gravity = "center_horizontal"
     android: layout_width = "fill_parent"
     android: layout_height = "fill_parent" >
      <EditText
        android: id = "@ +id/content"
        android: layout_height = "wrap_content"
        android: layout_width = "match_parent"
        android: inputType = "textMultiLine"/ >
      <Button
        android: id = "@ +id/button"
        android: layout_width = "wrap_content"
        android: layout_height = "wrap_content"
```

```
            android:text = "@string/button" />
        <ScrollView
            android:id = "@ +id/scrollView1"
            android:layout_width = "match_parent"
            android:layout_height = "wrap_content"
            android:layout_weight = "1"  >
        <LinearLayout
                android:id = "@ +id/linearLayout1"
                android:layout_width = "match_parent"
                android:layout_height = "match_parent"  >
        <TextView
            android:id = "@ +id/result"
            android:layout_width = "match_parent"
            android:layout_height = "wrap_content"
            android:layout_weight = "1" />
        </LinearLayout>
        </ScrollView>
    </LinearLayout>
```

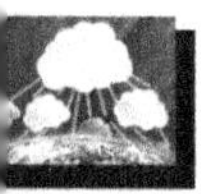

3. 服务器端 Java Web 实例 HttpClient_Post.jsp 的实现

此实例用来接收客户端发过来的 POST 请求，并将服务器响应的结果在收客户端显示出来，在本任务为了运行它，将其保存在“D:\Tomcat 7.0\webapps\dglg”路径下，代码如下：

```
<%@ page contentType = "text/html; charset = utf -8" language = "java" %>
<%
String param = request.getParameter("param"); //获取参数值
if(!"".equals(param) || param! =null){
if("post".equals(param)){
  String content = request.getParameter("content"); //获取输入的 Post 信息
  if(content! =null && name! =null){
  content = new String(content.getBytes("iso -8859 -1"),"utf -8"); //对内容进行转
码
  String date = new java.util.Date().toLocaleString();    //获取系统时间
  out.println("[ " +name +" ]于 " +date +" 发布一条 Post 信息:");
  out.println(content);
    }
  }
}
%>
```

4. 实现 MainActivity 程序功能

完整代码如下：

```
package org.dglg.project11_2;
//导入相关的类
import java.io.IOException;
import java.io.UnsupportedEncodingException;
import java.util.ArrayList;
import java.util.List;
import org.apache.http.HttpResponse;
import org.apache.http.HttpStatus;
import org.apache.http.NameValuePair;
import org.apache.http.client.ClientProtocolException;
import org.apache.http.client.HttpClient;
import org.apache.http.client.entity.UrlEncodedFormEntity;
import org.apache.http.client.methods.HttpPost;
import org.apache.http.impl.client.DefaultHttpClient;
import org.apache.http.message.BasicNameValuePair;
import org.apache.http.util.EntityUtils;
import android.app.Activity;
import android.os.Bundle;
import android.os.Handler;
import android.os.Message;
import android.view.View;
import android.view.View.OnClickListener;
import android.widget.Button;
import android.widget.EditText;
import android.widget.TextView;
import android.widget.Toast;
public class MainActivity extends Activity {
  private EditText content; // 定义一个输入文本内容的编辑框对象
  private Button button; // 定义"向服务器发布并返回消息"按钮对象
  private Handler handler; //定义一个 Handler 对象
  private String result = ""; //定义一个代表显示内容的字符串
  private TextView TV; //定义一个显示结果的文本框对象
  @Override
  protected void onCreate(Bundle savedInstanceState) {
      super.onCreate(savedInstanceState);
```

```
            setContentView(R.layout.activity_main);
        content = (EditText) findViewById(R.id.content); // 获取输入文本内容的
EditText 组件
            TV = (TextView) findViewById(R.id.result); //获取显示结果的 TextView 组
件
            button = (Button) findViewById(R.id.button); //获取"向服务器发布并返回消
息"按钮组件
            // 为"向服务器发布并返回消息"按钮添加单击事件监听器
            button.setOnClickListener(new OnClickListener() {
                @Override
                public void onClick(View v) {
    if ("".equals(content.getText().toString())) {
    Toast.makeText(MainActivity.this, "请输入完整!", Toast.LENGTH_SHORT).show();
            return;
    }
        // 创建一个新线程,用于从网络上获取文件
        new Thread(new Runnable() {
                public void run() {
                send();
                Message m = handler.obtainMessage(); // 获取一个 Message
                handler.sendMessage(m); // 发送消息
                }
            }).start(); // 开启线程
            }
        });
        handler = new Handler() {
            @Override
            public void handleMessage(Message msg) {
                if (result != null) {
                    TV.setText(result); // 显示获得的结果
                    content.setText(""); // 清空内容编辑框
                    }
                super.handleMessage(msg);
            }
        };
    }
    public void send() {
```

```
        String target = "http://192.168.1.66:8081/blog/HttpClient_Post.jsp";
//要提交的目标地址
        HttpClient httpclient = new DefaultHttpClient();    //创建 HttpClient 对象
        HttpPost httpRequest = new HttpPost(target);    //创建 HttpPost 对象
        //将要传递的参数保存到 List 集合中
        List<NameValuePair> params = new ArrayList<NameValuePair>();
        params.add(new BasicNameValuePair("param", "post"));    //标记参数
    params.add(new BasicNameValuePair("content", content.getText().toString()));
//内容
        try {
            httpRequest.setEntity(new UrlEncodedFormEntity(params, "utf-8")); //设置
编码方式
              HttpResponse httpResponse = httpclient.execute(httpRequest);    //执行
HttpClient 请求
            if (httpResponse.getStatusLine().getStatusCode() == HttpStatus.SC_OK){
//如果请求成功
            result += EntityUtils.toString(httpResponse.getEntity());    //获取返回
的字符串
                }else{
                    result = "请求不成功";}
            } catch (UnsupportedEncodingException e1) {
                e1.printStackTrace();    //输出异常信息
            } catch (ClientProtocolException e) {
                e.printStackTrace();    //输出异常信息
            } catch (IOException e) {
                e.printStackTrace();    //输出异常信息
            }
        }
    }
```

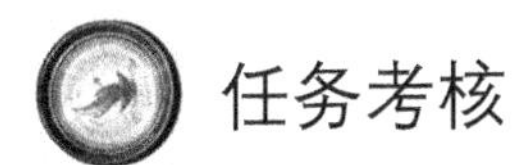

任务考核

实现用 Post 方法 HttpClient 访问 HTTP 网络。其中服务器端启动 Tomcat 来接收客户端发送的 Post 请求，并通过 Java Web 实例 AnswerPost.jsp 做出响应；客户端的 MainActivity 实现点击“发送”按钮，将在文本框中输入的文本用发送 Post 方法通过 HttpClient 到指定的 Url 地址的服务器端，并由服务器端发回接收后的信息。

任务三 实现显示指定城市的天气预报的网页

任务描述

本任务主要是通过打开中国天气网获取指定城市的天气预报信息，在启动MainActivity时，实现在屏幕上显示默认城市的天气预报信息，本任务默认的城市是北京市，如图11.15所示。

单击图11.15中“北京市”、“上海市”、“广州市”按钮中的任意一个按钮，都将在屏幕上显示与按钮名称相应的城市的天气预报信息。本任务中单击的是“广州市”按钮，因此会在屏幕上显示“广州市”的天气预报信息，如图11.16所示。

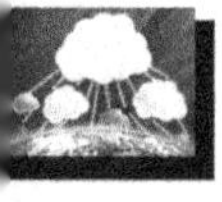

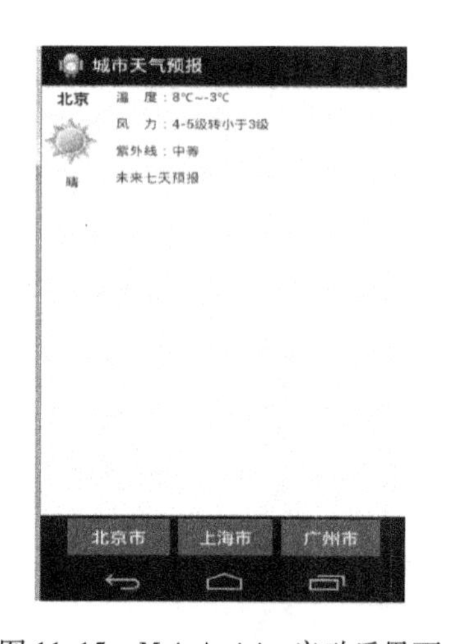

图11.15　MainActivity启动后界面

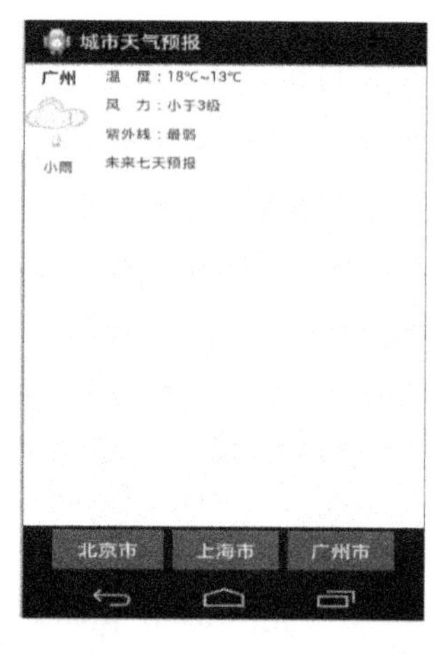

图11.16　单击“广州市”按钮后界面

任务分析

在本任务中，通过添加WebView组件，并允许其使用JavaScript，然后使用WebView组件的setWebChromeClient()方法，使弹出的JavaScript对话框不会被屏蔽掉。

在本任务中，使用到名称为“北京市”、“上海市”、“广州市”等多个按钮，因此要为它们个别添加OnClickListener单击事件监听器，并重写onClick()方法，为不同的按钮事件做出不同的响应，再调用openUrl()方法来获取不同城市的天气预报信息。

由于中国天气网(http://www.weather.com.cn)提供了单个城市24小时的天气预报插件，本任务才得以利用其实现显示指定城市的天气预报信息。

重点和难点如下：

(1)了解WebView组件打开网页的基础知识；
(2)掌握OnClickListener单击事件监听器的使用；
(3)掌握openUrl()方法的应用。

任务准备

1. WebView 组件介绍

在 Android 中，提供了一个使用 WebKit 引擎的内置浏览器，通过这个内置浏览器就能打开网页了，但是要通过 WebView 组件才能使用内置浏览器。

既然 WebView 组件是专门用于打开网页的，那么要如何实现呢？

（1）要在 Java 文件中创建 WebView 组件，或者在布局文件中添加 WebView 组件。一般常用的是在布局文件中添加的方法，可以在布局文件中添加如下代码实现。

```
<WebView
    android: id = "@ +id/webView1"
    android: layout_width = "wrap_content"
    android: layout_height = "0dip"
    android: layout_weight = "0. 92"
    android: focusable = "false" />
```

（2）可以用 WebView 组件的方法执行浏览器操作，WebView 组件的方法如表 11. 1 所示。

表 11. 1 WebView 组件的常用方法

方 法	描 述
loadUrl(String url)	加载指定 URL 对应的网页
loadData(String data, String mimeType, String encoding)	将指定的字符串数据加载到浏览器中
loadDataWithBaseURL(String baseUrl, String data, String mimeType, String encoding, String historyUrl)	基于 URL 加载指定的数据
capturePicture()	创建当前屏幕的快照
goBack()	后退操作，如同浏览器上的后退按钮的功能
goForward()	前进操作，如同浏览器上的前进按钮的功能
stopLoading()	停止加载当前页面
reload()	刷新当前页面

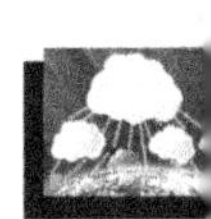

（3）为 MainActivity 的 onCreate()方法获取布局文件中的 WebView 组件，并指定网页的 URL 地址。本任务中，URL 地址是 http://m.weather.com.cn/m/pn12/weather.htm，代码如下：

```
webView = (WebView)findViewById(R.id.webView1);    //获取 WebView 组件
webView.loadUrl("http://m.weather.com.cn/m/pn12/weather.htm"); //设置默认显示的天气预报信息
```

（4）要添加如下代码到 AndroidManifest.xml 文件中，以获得允许访问网络资源权限。

```
<uses-permission android:name="android.permission.INTERNET"/>
```

2. WebView 组件设置 JavaScript 可用

WebView 组件本身不支持 JavaScript，然而许多网页是用 JavaScript 代码编辑的，为了能够浏览这些网页，就要让 WebView 组件支持 JavaScript，通过以下步骤实现。

1）让网页中的大部分 JavaScript 可用

WebView 组件的 WebSettings 对象有一个 setJavaScriptEnable 方法，用这个方法进行设置，可以让 JavaScript 可用。如果定义一个 WebView 组件的名称为 WV，那么对于这个 WV 可以通过以下代码设置让 JavaScript 可用。

```
WV.getSettings().setJavaScriptEnabled(true);    //设置 JavaScript 可用
```

2）显示用 window.alert()方法弹出的对话框

使用 setJavaScriptEnable 方法设置 JavaScript 可用，在显示带有弹出的对话框的网页时，不能显示用 window.alert()方法弹出的对话框，为了能显示此对话框，要用 WebView 组件的 setWebChromeClient()方法来实现，代码如下：

```
WV.setWebChromeClient(new WebChromeClient());
```

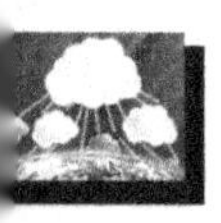

3. OnClickListener 单击事件监听器

由于本任务中用到多个按钮，在单击不同的按钮时，需要对单击事件设置不同的响应。因此，要先让 MainActivity 实现 OnClickListener 接口，以添加单击事件监听器，代码如下：

```
public class MainActivity extends Activity implements OnClickListener {}
```

在获取布局管理器中，在添加按钮的同时添加单击事件监听器。

4. openUrl()方法

本任务中，openUrl()方法是自定义的打开网页获取天气预报信息的方法，可以根据不同的参数，获取不同城市的天气预报信息，代码如下：

```
private void openUrl(String id){
webView.loadUrl("http://m.weather.com.cn/m/pn12/weather.htm?id="+id+"
");//获取显示天气预报信息
}
```

任务实现

1. 新建 Android 项目

参考项目一创建项目的方法，在 Eclipse 中新建 Android 项目名为 Project11_3 和包为 org.dglg.project11_3 的程序，在新建 Android 项目时会自动创建一个 Activity，本任务中

Activity 的名称是 MainActivity。

2. 对 MainActivity 进行布局

1）对 MainActivity 的 strings. xml 进行布局

```
<? xml version = "1.0" encoding = "utf - 8"? >
<resources >
    <string name = "app_name" >城市天气预报</string >
    <string name = "bj" >北京市</string >
    <string name = "sh" >上海市</string >
    <string name = "gz" >广州市</string >
</resources >
```

2）对 MainActivity 的布局文件 activity_main. xml 进行布局

本任务中，MainActivity 的布局文件是 activity_main. xml，用来实现“城市天气预报”的布局，代码如下：

```
<? xml version = "1.0" encoding = "utf - 8"? >
<LinearLayout xmlns: android = "http: //schemas. android. com/apk/res/android"
    android: layout_width = "fill_parent"
    android: layout_height = "match_parent"
    android: gravity = "center_horizontal"
    android: orientation = "vertical" >
  <WebView
      android: id = "@ + id/webView1"
      android: layout_width = "wrap_content"
      android: layout_height = "0dip"
      android: layout_weight = "0. 92"
      android: focusable = "false" / >
  <LinearLayout
      android: orientation = "horizontal"
      android: layout_width = "wrap_content"
      android: layout_height = "wrap_content"
     >
  <Button
      android: id = "@ + id/bj"
      android: layout_width = "107dp"
      android: layout_height = "wrap_content"
      android: text = "@ string/bj" / >
```


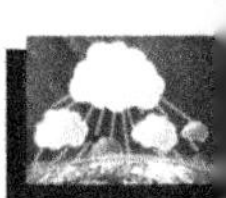

```
    <Button
        android:id="@+id/sh"
        android:layout_width="103dp"
        android:layout_height="wrap_content"
        android:text="@string/sh" />
    <Button
        android:id="@+id/gz"
        android:layout_width="105dp"
        android:layout_height="wrap_content"
        android:text="@string/gz" />
    </LinearLayout>
</LinearLayout>
```

3. 指定允许访问网络资源权限

在 AndroidManifest. xml 文件中，要添加如下代码以获得允许访问网络资源权限。

```
<uses-permission android:name="android.permission.INTERNET"/>
```

AndroidManifest. xml 文件完整的代码如下：

```
<?xml version="1.0" encoding="utf-8"?>
<manifest xmlns:android="http://schemas.android.com/apk/res/android"
    package="org.dglg.project11_3"
    android:versionCode="1"
    android:versionName="1.0" >
    <uses-sdk android:minSdkVersion="17" />
  <uses-permission android:name="android.permission.INTERNET"/>
    <application
        android:icon="@drawable/ic_launcher"
        android:label="@string/app_name" >
        <activity
            android:name=".MainActivity"
            android:label="@string/app_name" >
            <intent-filter>
                <action android:name="android.intent.action.MAIN" />
                <category android:name="android.intent.category.LAUNCHER" />
            </intent-filter>
        </activity>
    </application>
</manifest>
```

4. 实现 MainActivity 程序功能

完整的代码如下：

```
package org. dglg. project11_3;
import android. app. Activity;
import android. os. Bundle;
import android. view. View;
import android. view. View. OnClickListener;
import android. webkit. WebChromeClient;
import android. webkit. WebView;
import android. webkit. WebViewClient;
import android. widget. Button;
public class MainActivity extends Activity implements OnClickListener {
    private WebView webView;    //定义 WebView 组件的对象
    @ Override
    protected void onCreate( Bundle savedInstanceState) {
        super. onCreate( savedInstanceState);
        setContentView( R. layout. main);
        webView = ( WebView) findViewById( R. id. webView1);    //获取 WebView 组件
        webView. getSettings( ). setJavaScriptEnabled( true);    //设置 JavaScript 可用
    webView. setWebChromeClient( new WebChromeClient( ) ); //处理 JavaScript 对话框
        webView. setWebViewClient( new WebViewClient( ) );    //处理各种通知和请求
事件，如果不使用该句代码，将使用内置浏览器访问网页
         webView. loadUrl ( " http://m. weather. com. cn/m/pn12/weather. htm "); //设
置默认显示的天气预报信息
        webView. setInitialScale(57 * 4);    //将网页内容放大 4 倍
    //获取布局管理器中添加的"北京市"、"上海市"、"广州市"按钮
        Button bj = ( Button) findViewById( R. id. bj);
        bj. setOnClickListener( this);
        Button sh = ( Button) findViewById( R. id. sh);
        sh. setOnClickListener( this);
        Button gz = ( Button) findViewById( R. id. gz);
        gz. setOnClickListener( this);
    }
    @ Override
    public void onClick( View view) {
        switch( view. getId( ) ) {
        case R. id. bj:    //单击"北京市"按钮
```

```
            openUrl("101010100T");
            break;
        case R.id.sh:   //单击"上海市"按钮
            openUrl("101020100T");
            break;
        case R.id.gz:   //单击"广州市"按钮
            openUrl("101280101T");
            break;
        }
    }
    //打开网页获取显示天气预报信息的方法,
    private void openUrl(String id){
        webView.loadUrl("http://m.weather.com.cn/m/pn12/weather.htm? id=" + id
+"");
    }
}
```

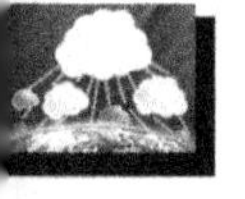

项目十二

Android游戏开发

如今，手机游戏成为越来越多人休闲娱乐不可或缺的物品。由于巨大的需求，手机游戏行业发展迅速，出现了一批著名游戏（如“捕鱼达人”）。手机游戏还能为开发者带来巨大的利润，因此越来越受到重视。我国手机游戏开发起步较晚，还没有真正形成产业链和生态圈，还有很大的发展空间。

试想一下，当几个朋友一起玩的时候有一款游戏能让大家同时在一部手机上玩，不需要太高的技术，休闲而又不缺刺激，这就是本项目任务一的《数字炸弹》。相信很多同学都玩过飞行游戏《雷电》，有没有想过自己也可以开发一款类似的游戏呢？任务二《太空飞行》也许可以帮助到你。

理解 Android 游戏的基本概念
学会 SurfaceView 框架的搭建
学会简单的碰撞检测
学会爆炸动画的处理

任务一　数字炸弹

任务描述

本任务制作一款名叫“数字炸弹”的游戏。游戏规则如下：由游戏主持人或者系统随机抽取 1 ～ 100 中的任意一个整数作为炸弹数，玩家轮流竞猜哪个是炸弹数，猜中炸弹数的人要受惩罚，游戏结束。猜不中的就不断缩小下一个玩家的竞猜范围，例如：炸弹数为 30，玩家 1 猜的是 50，那么玩家 2 的竞猜范围缩小为 1 ～ 50；如果玩家 2 猜的是 20，那么玩家 3 的竞猜范围缩小为 20 ～ 50，直到有人猜中为止。在有人猜中炸弹数后手机播放警报并振动作为提醒。

任务分析

本任务比较简单，运用前面所学过的知识完全可以做出来。思路是这样的，制作两个Activity，Activity1负责设置炸弹数，Activity2负责竞猜。炸弹数利用Intent从Activity1传递到Activity2。主程序设置两个变量minNo和maxNo分别存放竞猜范围的最小值和最大值，利用函数判断炸弹数与玩家竞猜数的大小关系来确定下一轮竞猜的范围：如果玩家竞猜数大于炸弹数，那么maxNo等于玩家竞猜数；如果玩家竞猜数小于炸弹数那么minNo等于玩家竞猜数；如果两者相等则爆炸，游戏结束。

为了让游戏更加完善和健壮，还要对玩家竞猜数进行合法性判断，例如：玩家输入是否为空值、玩家输入是否在竞猜范围内等。

本任务的重点和难点如下：

(1)理解简单Android游戏；
(2)熟悉利用控件开发简单游戏；
(3)学会设置全屏、播放游戏声音等。

任务准备

1. Android 游戏简介

Android游戏就是运行在Android平台上的游戏程序。为了吸引用户，开发者已经推出成千上万、五花八门的Android游戏，大概可以从以下角度去分类。

1)从游戏运行方式分类

单机游戏：此类游戏只能在游戏本身所安装的手机上运行，不需要网络支持，如经典的《扫雷》等。

网络游戏：此类游戏的运行需要无线网络的支持，没有无线网络，游戏就不能持续地运行下去，如《QQ斗地主网络版》等。

2)从游戏内容分类

文字游戏：此类游戏以文字为主，不需要大量的图像、动画和人物，以对文字进行猜想的形式运行，如本任务的游戏《数字炸弹》等。

休闲游戏：一种简单轻松的让玩家休闲放松的游戏，容易上手又无须耗费大量的精力，如《开心消消乐》等。

益智游戏：此类游戏集趣味性与益智性于一体，让玩家在玩游戏的过程中锻炼脑筋的灵活性，如《植物大战僵尸》等。

棋牌游戏：顾名思义，此类游戏以棋牌为主，如《斗地主》和《麻将》等。

射击游戏：此类游戏让玩家控制飞机、坦克等进行射击，如《全民飞机大战》等。

角色扮演游戏：在此类游戏中玩家会扮演游戏中的一个或多个角色，一般来说整个游戏就是一个大型的故事，剧情会根据玩家的不同选择有不同的发展，如《时空猎人》等。

动作冒险游戏：此类游戏通常会设定若干个关卡，玩家控制游戏中的虚拟角色进行冒险，例如与敌人战斗、寻找物品和通过险要地形等，如《神庙逃亡》等。

策略经营游戏：此类游戏让玩家模拟经营一个企业、运营一座城市等，在游戏中设定特殊的任务让玩家完成，如《大富翁》等。

2. 炸弹数的传递的处理

在该任务中，炸弹数是在 Activity1 中设定然后传递到 Activity2 中，这里就涉及两个 Activity 的跳转和参数传递。可以参考前面章节的内容。

3. 设置全屏

为了有更好的用户体验，游戏要设置成全屏，一般设置全屏有两种方法。

- 方法一：修改 AndroidManifest. xml 配置文件，在 <activity> 标签中增加属性：

```
android: theme = "@ android: style/Theme. NoTitleBar. Fullscreen"
```

- 方法二：修改 Activity 的 onCreate( ) 函数，在 super( ) 和 setContentView( ) 两个方法之间加入下面两条语句：

```
this. requestWindowFeature( Window. FEATURE_NO_TITLE);      //去掉标题栏
this. getWindow ( ) . setFlags ( WindowManager. LayoutParams. FLAG _ FULLSCREEN,
WindowManager. LayoutParams. FLAG_FULLSCREEN);               //设置成全屏
```

4. 设置手机振动

在玩家猜中炸弹数后，会让手机振动以提醒，设置振动的方法如下：

在 AndroidManifest. xml 中增加调用振动许可。

```
<uses - permission android: name = "android. permission. VIBRATE" />
```

创建振动对象 vibrator 并调用振动服务 VIBRATOR_SERVICE。

```
Vibrator vibrator = (Vibrator) getSystemService( VIBRATOR_SERVICE);
```

5. 播放声音

在 Android 游戏中，播放声音一般使用 MediaPlayer 和 SoundPool 两个类。前者一般用于播放较大的(大于 1M)持续时间较长的音频，而后者一般用于播放短小的音频，所以在播放游戏背景音乐时一般会选用 MediaPlayer，而播放音效时一般选用 SoundPool。其中 MediaPlayer 的常用方法如表 12. 1 所示。

表 12. 1　MediaPlayer 类的常用方法

方　　法	解　　释
create(Context context, int resid)	通过资源 ID 创建一个多媒体播放实例
pause( )	暂停播放

续表 12.1

方 法	解 释
release()	释放 MediaPlayer 对象和其占用的系统资源
start()	开始播放
stop()	停止播放

通过 MediaPlayer 播放背景音乐的关键代码如下：

```
private MediaPlayer mediaPlayerMs; //声明一个 MediaPlayer 对象
mediaPlayerMs = MediaPlayer.create(this, R.raw.ms); //实例化
mediaPlayerMs.setLooping(true); //设置音乐为循环播放
mediaPlayerMs.start(); //开始播放
```

SoundPool 类的使用由于本任务中没有涉及，大家可以参考其他资料。

任务实现

1. 界面设计

在 Eclipse 中，新建一名为 Porject12_1 的 Android 项目，并建立两个 Activity 分别命名为 Activity1 和 Activity2，都采用相对布局。

把界面所需要的图片资源导入到 res/drawable-hdpi 中。

设置 Activity1 的背景为 bg.bnp，代码如下。

```
android:background = "@drawable/bg"
```

在 Activity1 中放置两个 ImageButton、一个 EditText 和一个 TextView，各项属性如下：

```
<EditText
    android:id = "@+id/edtNo"
    android:layout_width = "wrap_content"
    android:layout_height = "wrap_content"
    android:layout_below = "@+id/tvTitle"
    android:layout_centerHorizontal = "true"
    android:layout_marginTop = "22dp"
    android:ems = "10"
    android:inputType = "number" />
<ImageButton
    android:id = "@+id/btnGo"
    android:layout_width = "wrap_content"
    android:layout_height = "wrap_content"
```

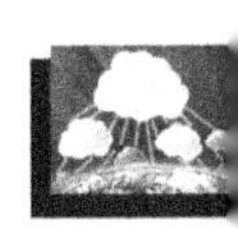

```
        android: layout_alignLeft = "@ +id/edtNo"
        android: layout_below = "@ +id/edtNo"
        android: layout_marginLeft = "38dp"
        android: layout_marginTop = "18dp"
        android: background = "#00000000"
        android: src = "@drawable/okbtn" />
    <TextView
        android: id = "@ +id/tvTitle"
        android: layout_width = "wrap_content"
        android: layout_height = "wrap_content"
        android: layout_alignParentTop = "true"
        android: layout_centerHorizontal = "true"
        android: layout_marginTop = "140dp"
        android: text = "炸弹在 1 ~ 100 之间!"
        android: textColor = "#FF0000"
        android: textSize = "30dp" />
```

默认情况下 ImageButton 会有灰色的背景，要想让其有更好的显示效果，可以把它的背景颜色设成透明，代码如下。

```
android: background = "#00000000"
```

Activity1 设置完成后效果如图 12.1 所示。

图 12.1　Activity1 效果

图 12.2　Activity2 效果

在 Activity2 中放置一个 ImageButton、一个 EditText 和一个 TextView，各项属性如下：

```
<EditText
        android: id = "@ +id/edtNo"
        android: layout_width = "wrap_content"
        android: layout_height = "wrap_content"
        android: layout_below = "@ +id/tvTitle"
        android: layout_centerHorizontal = "true"
        android: layout_marginTop = "22dp"
        android: ems = "10"
        android: inputType = "number" />
    <ImageButton
        android: id = "@ +id/btnGo"
        android: layout_width = "wrap_content"
        android: layout_height = "wrap_content"
        android: layout_alignLeft = "@ +id/edtNo"
        android: layout_below = "@ +id/edtNo"
        android: layout_marginLeft = "38dp"
        android: layout_marginTop = "18dp"
        android: background = "#00000000"
        android: src = "@drawable/okbtn" />
    <TextView
        android: id = "@ +id/tvTitle"
        android: layout_width = "wrap_content"
        android: layout_height = "wrap_content"
        android: layout_alignParentTop = "true"
        android: layout_centerHorizontal = "true"
        android: layout_marginTop = "140dp"
        android: text = "炸弹在1-100之间!"
        android: textColor = "#FF0000"
        android: textSize = "30dp" />
```

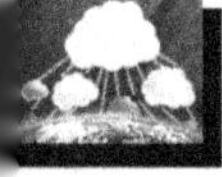

Activity2 设置完成后效果如图 12.2 所示。

2. MainActivity1 代码编写

MainActivity1 主要处理接受炸弹数的输入、检查输入的有效性和传递炸弹数到 MainActivity2。由于游戏可以接受用户自由输入或者电脑随机输入炸弹数两种方式，所以在界面中设置了两个按钮 btnBegin 对应自由输入，btnRBegin 对应电脑随机输入。

btnBegin 监听函数如下：

```
        class MyButtonListener implements OnClickListener {
                @ Override
                  public void onClick( View v) {
                  //获得炸弹数实例
                  edtBoomNo = (EditText)findViewById(R. id. edtBoomNo);
                  //判断玩家是否已经输入
                    if (edtBoomNo. getText(). toString(). length()! =0){
                        if(Integer. parseInt(edtBoomNo. getText(). toString()) > =0
&&Integer.  parseInt(edtBoomNo. getText(). toString()) < =100){
                  //保存炸弹数
                  boomNo = edtBoomNo. getText(). toString();
                  //防止有人可以看到炸弹数，在获得炸弹数后应把输入框清空
                  edtBoomNo. setText("");
                  // 通过 intent 把炸弹数传递给第二个 activity
                        Intent intent = new Intent();
                        intent. putExtra("boomNo", boomNo);
                        intent. setClass(MainActivity1. this, MainActivity2. class);
                        MainActivity1. this. startActivity(intent);
          } else{
            Toast tot;
            tot = Toast. makeText (getApplicationContext (), "请输入 1 ~ 100 的数!",
Toast. LENGTH_LONG);
            tot. setGravity(Gravity. CENTER, 0, 0);
            tot. show();
          }
                          } else{
                          Toast tot;
      tot = Toast. makeText ( getApplicationContext ( ), " 输 入 为 空, 请 输 入!",
Toast. LENGTH_LONG);
                  tot. setGravity(Gravity. CENTER, 0, 0);
                  tot. show();
                        }
                }
        }
```

程序首先接受用户输入并判断输入是否为空，如果为空就输出提醒“输入为空，请输入!”，如果不为空就检查用户是否输入 1 ~ 100 的整数，如果不是就输出提醒“请输入 1 ~ 100 的数!”，经过以上检查后用户的输入就通过 intent 传递到 MainActivity2 并把用户输入框的数据清除。

btnRBegin 监听函数如下:

```
class MyRButtonListener implements OnClickListener {
        @ Override
          public void onClick( View v) {
          //产生 1 ~ 100 的随机整数
              boomNo = String. valueOf( ( int) ( Math. random( )  *  100 + 1) ) ;
          // 通过 intent 把炸弹数传递给第二个 activity
                Intent intent  =  new Intent( ) ;
                intent. putExtra( "boomNo" , boomNo) ;
                intent. setClass( MainActivity1. this, MainActivity2. class) ;
                MainActivity1. this. startActivity( intent) ;
        }
    }
```

程序直接产生 1 ~ 100 的随机整数并通过 intent 传递到 MainActivity2。

3. MainActivity2 代码编写

MainActivity2 主要处理接收 MainActivity1 传递过来的炸弹数、接受玩家输入并检查有效性、判断玩家输入的数据是否为炸弹数并进行相应的操作。玩家输入后点击“输入完毕”按钮 btnGo，btnGo 监听函数如下:

```
class MyButtonListener implements OnClickListener {
          @ Override
          public void onClick( View v) {
            //判断玩家是否已经输入
            if ( edtNo. getText( ). toString( ). length( ) !  =0) {
            //得到玩家输入的数
            int i  =  Integer. parseInt( edtNo. getText( ). toString( ) ) ;
            //判断玩家输入的数是否在最小值和最大值之间
            if( i > = minNo&&i < = maxNo) {
            //得到炸弹数
            int j  =  Integer. parseInt( boomNo) ;
            //如果玩家输入数大于炸弹数，最大值等于玩家输入数
            if( i > j) {
            maxNo = i;
  tvTitle. setText( "炸弹在" + minNo + "和" + maxNo + "之间!" ) ;
  edtNo. setText( "" ) ;
            }
            //如果玩家输入数小于炸弹数，最小值等于玩家输入数
                if( i < j) {
```

```
                minNo = i;
                tvTitle. setText("炸弹" + minNo + "和" + maxNo + "之间!");
                    edtNo. setText("");
                }
                //如果玩家输入数等于炸弹数，设置手机振动并播放警报
                if(i = =j){
                isBoom  =  true;
                //设置振动
        vibrator = (Vibrator)getSystemService(Service. VIBRATOR_SERVICE);
                vibrator. vibrate(new long[]{100, 2000, 100, 4000}, -1);
                //设置声音
                mediaPlayerMs. stop();
                mediaPlayerJd. start();
                tvTitle. setText("你猜中了!");
                //防止用户再点击，把按钮设置为不可用
                btnGo. setClickable(false);
                }
                }else{
        Toast tot;
        tot  =  Toast. makeText(getApplicationContext(), "请输入" + minNo + " - " +
maxNo + "的数!", Toast. LENGTH_LONG);
        tot. setGravity(Gravity. CENTER, 0, 0);
        tot. show();
    }
              }else{
                Toast tot;
         tot  =  Toast. makeText ( getApplicationContext ( ), " 输入为空，请输入!",
Toast. LENGTH_LONG);
        tot. setGravity(Gravity. CENTER, 0, 0);
        tot. show();
              }
              }
          }
```

程序首先定义两个变量 minNo 和 maxNo，分别对应炸弹所在范围的最小值和最大值，在接受用户输入后就判断输入是否为空：如果为空就输出提醒“输入为空，请输入!”，如果不为空就检查用户是否输入 minNo 和 maxNo 之间的整数；如果两者都不是就输出提醒。经过合法性检查后就把玩家输入数与炸弹数进行对比：如果玩家竞猜数大于炸弹数，最大值 maxNo 等于玩家竞猜数；如果玩家竞猜数小于炸弹数，最小值 minNo 等于玩家竞猜数；

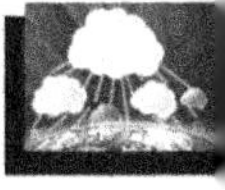

如果玩家竞猜数等于炸弹数，设置手机振动并播放警报。

至此，手机炸弹游戏介绍完毕，我们可以看出游戏开发不一定要很高的技术，有很好的创意也可以开发出受欢迎的好玩的游戏。

任务二 太空飞行

 任务描述

本任务开发一款如图12.3所示的太空飞行游戏，屏幕限制为竖屏，玩家可以通过触屏来控制飞机飞行，如果玩家没有触摸到飞机则不能控制它。在飞行过程中，玩家要靠躲开随机出现的大中小三种石头来赚取游戏积分，躲开大石头得5分、中石头得3分、小石得2分。所有石头的飞行方向为垂直屏幕从上到下，出现位置为屏幕顶部的随机坐标位置。不同种类的石头有不一样的飞行方式，大石头匀速飞行，中石头加速飞行，小石头减速飞行。各种石头的初始速度也不一样，大石头慢，中石头快，小石头最快，并且都有最低速度限制。如果玩家飞机不幸被石头击中会扣减相应的积分，具体是大石头扣50分、中石头扣30分、小石头扣10分，当玩家积分小于0判定失败，游戏结束。为了有更好的用户体验，在飞机与石头碰撞时，要求出现爆炸效果。随着游戏积分的增加游戏难度也会相应提高，提高难度主要靠增加出现的石头批数来体现。游戏每隔30毫秒会出现新批次的石头(一批石头有可能是1个大石头或者2个中石头或者3个小石头，随机决定)。积分、级别和石头批数的关系如表12.2。每次级别和难度的提升都会有文字提示。

图12.3 太空飞行效果

表12.2 积分、级别和石头批数关系表

积　分	级　别	石头批数
<0	失败	0
> =0	1	1
> =30	2	2
> =120	3	3
> =280	4	4

续表 12.2

积　分	级　别	石头批数
> =500	5	5
> =700	6	6
> =1000	获胜	

任务分析

本次任务与上次不同，上次任务中游戏是静态的，没有运动的物体，而这次任务是动态的，所以开发方法就不一样。分析任务描述可以看出这是一款 2D 的飞行游戏，通常这类型的游戏需要很高的执行效率，可以采用 SurfaceView 框架来开发并且用线程来控制游戏的刷新，因此构建一个游戏框架类 GameSurfaceView 负责游戏的总体运行；游戏的两个主要元素是飞机和石头，为飞机构建一个类 Plane，封装了飞机的各种属性、绘图函数和飞机碰撞检测函数等；为石头构建一个类 Rock，封装石头的各种属性、绘图函数和运算函数；由于在飞机与石头碰撞时出现爆炸效果，所以也为爆炸的动画构建一个类 BoomAnimation，封装处理爆炸效果的绘图函数和运算处理。

本任务的重点和难点如下：

(1)理解 SurfaceView 框架；
(2)学会飞行游戏的基本开发方法；
(3)理解游戏中的碰撞检测。

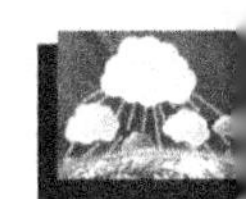

任务准备

1. SurfaceView 游戏框架

SurfaceView 类继承于 View 类，它提供了直接访问内存中的屏幕像素数据的方法，特别适合用来开发要求高执行效率的 2D 游戏。

在使用 SurfaceView 时，必须通过 SurfaceHolder 来访问和控制它，SurfaceHolder 是一个接口，就像是 SurfaceView 的监听器，其常用方法如表 12.3 所示。

表 12.3　SurfaceHolder 常用方法

方　法	说　明
addCallback(SurfaceHolder. Callbackcallback)	为 SurfaceHolder 添加一个 SurfaceHolder. Callback 回调接口
lockCanvas()	获取和锁定 Canvas 对象
lockCanvas(Rect dirty)	获取和锁定 Canvas 对象中指定的 dirty 区块
unlockCanvasAndPost(Canvascanvas)	解锁画布并提交改变

使用SurfaceView类来开发游戏还需要实现SurfaceHolder.Callback接口，该接口必须重写三个函数，代码如下。

```
// SurfaceView第一次创建时响应此函数
surfaceCreated(SurfaceHolder holder)
// SurfaceView状态发生改变时响应此函数
surfaceChanged(SurfaceHolder holder, int format, int width, int height)
//SurfaceView销毁时响应此函数
surfaceDestroyed(SurfaceHolder holder)
```

另外为了使游戏画面能够自动刷新，实现动态效果，我们通常使用线程来对画面进行刷新。

综上所述，一个SurfaceView游戏框架大体如下：

```
public class MySurfaceView extends SurfaceView implements Callback, Runnable {
……//变量声明
public MySurfaceView(Context context) {
    ……// SurfaceView初始化函数
}
@ Override
public void surfaceCreated(SurfaceHolder holder) {
    ……//SurfaceView视图创建，响应此函数
}
@ Override
public void surfaceChanged(SurfaceHolder holder, int format, int width, int height) {
……//SurfaceView视图状态发生改变，响应此函数
}
@ Override
public void surfaceDestroyed(SurfaceHolder holder) {
……//SurfaceView视图消亡时，响应此函数
    }
public void myDraw() {
    try {
    ……//绘图操作
        }
    } catch (Exception e) {
        // TODO: handle exception
    } finally {
    ……
    }
```

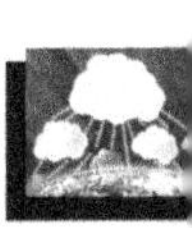

```
    }
    @Override
    public boolean onTouchEvent(MotionEvent event) {
        ……//触屏响应
    }
    public void calc() {
        ……//运算函数
    }
    @Override
    public void run() {
        ……//线程控制
    }
}
```

2. 碰撞检测

从玩家的角度看碰撞检测指的是检测游戏中两个或两个以上的物体是否碰在一起，例如飞机是否撞上石头、子弹是否打中飞机等；从游戏开发者的角度看指的是两个或多个物体所占的屏幕坐标是否相交，像素是否重叠等。下面具体介绍两种简单的碰撞检测。

1）矩形碰撞检测

对于两个矩形的碰撞检测一般采用排除法，只要排除所有两个矩形不相撞的情况，就可以认为这两个矩形是相撞的。假设矩形 1 的顶点坐标为(x1、y1)，宽度为 w1，高度为 h1；矩形 2 的顶点坐标为(x2、y2)，宽度为 w2，高度为 h2，两个矩形不相撞的情况如图 12.4 所示。

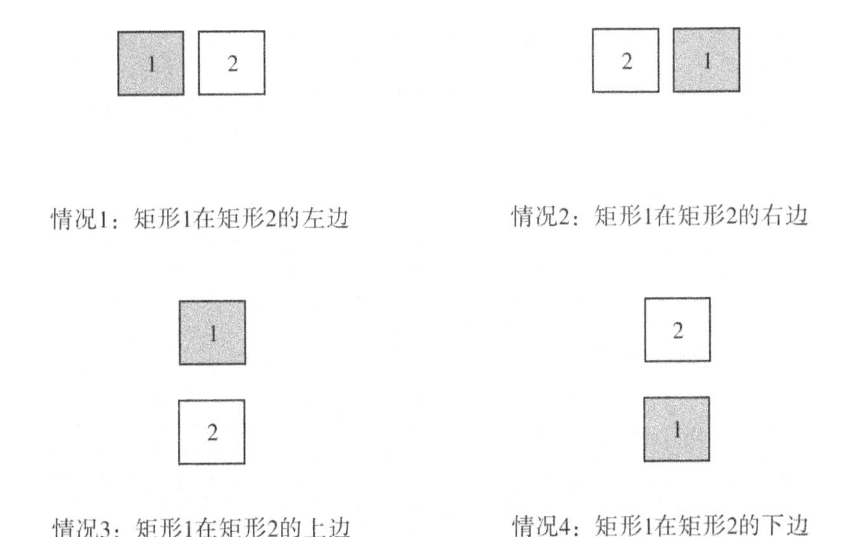

图 12.4　矩形不相撞的四种情况

综上所述，矩形碰撞算法可表示如下：

```
public boolean isCollsionWithRect (int x1, int y1, int w1, int h1, int x2, int y2, int w2,
int h2) {
        if (x1 <= x2 && x1 + w1 <= x2) {
            return false; //情况 1
        } else if (x1 >= x2 && x1 >= x2 + w2) {
            return false; //情况 2
        } else if (y1 <= y2 && y1 + h1 <= y2) {
            return false; //情况 3
        } else if (y1 >= y2 && y1 >= y2 + h2) {
            return false; //情况 4
        }
        return true;
    }
```

2)圆形碰撞检测

两个圆形的碰撞比较简单，只要比较两个圆的圆心之间的距离是否小于等于两个圆的半径之和，如果是则相撞，否则不相撞。假设圆1的圆心坐标为(x1, y1)，半径为r1；圆2的坐标为(x2, y2)，半径为r2，两圆之间的关系如图12.5所示。

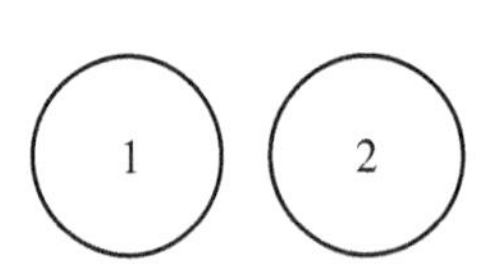

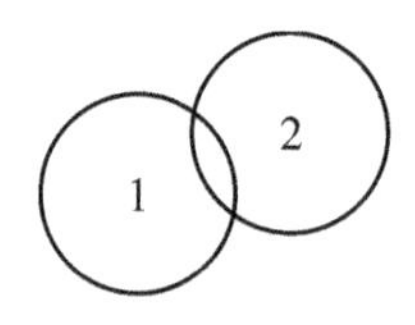

情况1：两圆心距离大于半径之和　　情况2：两圆心距离小于等于半径之和

图12.5　两圆之间的关系

圆形碰撞算法可表示如下：

```
public boolean isCollsionWithCircle(int x1, int y1, int x2, int y2, int r1, int r2){
        /*如果两圆的圆心距小于或等于两圆半径之和则认为发生碰撞，计算方法采
用勾股定理，其中 Math.sqrt:():开平方、Math.pow(double x, double y):X 的 Y 次
方*/
    if(Math.sqrt(Math.pow(x1 - x2, 2) + Math.pow(y1 - y2, 2)) <= r1 + r2){
        return true;
        }
        return false;
    }
```

3）其他碰撞检测

除了以上两种碰撞检测外，还有像素检测、四叉树检测和3D碰撞检测等，读者可以自行查找资料学习。

3. 游戏背景滚动

我们知道所有物体的运动都是相对的，飞行游戏中要模拟飞机在太空中飞行的运动，就必须要让飞机或者背景动起来或两者同时动起来。飞机通常是由玩家控制的，所以它的运动可以通过跟随屏幕上的手指运动来实现。背景的滚动则可以使用同一幅背景图片首尾相接交替播放的方式来实现。在程序编写时可以用同一幅背景图片实例化两个图片变量，例如名叫"BG1"和"BG2"，初始化的时候把它们首尾相接起来，当"BG1"的上边缘滚出屏幕下方时，马上设置它的位置在"BG2"的上面，如此反复形成滚动效果。为了让背景对接更均匀，可以在拼接的时候让它们稍为重叠。图12.6模拟了背景滚动的过程，其中粗线框代表手机屏幕，背景滚动的具体代码请参考任务实现。

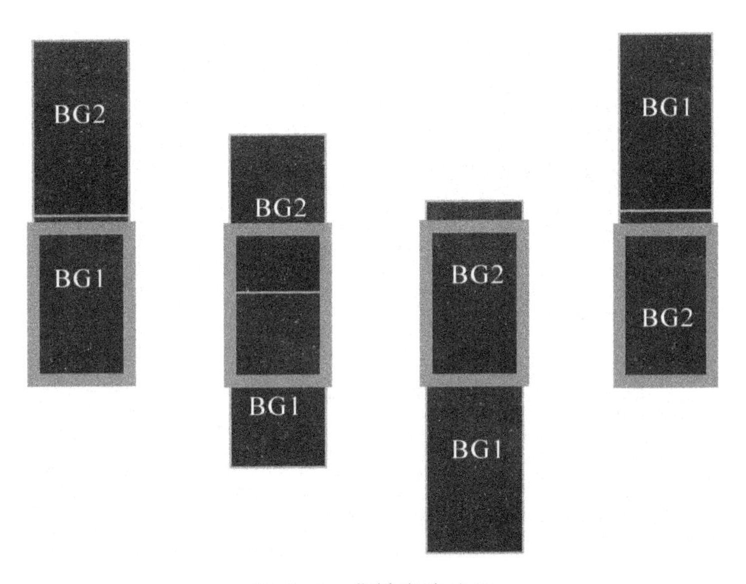

图12.6　背景滚动过程

4. 爆炸动画效果的处理

如图12.7所示，整个爆炸动画所有的帧都放在同一张图片上，可以通过按顺序播放图片上的不同区域来实现爆炸的动画效果。这里要特别注意的是，在处理爆炸图片的时候一定要让每一帧大小相同，这样在编写程序的时候就很容易处理，具体代码请参考任务实现。

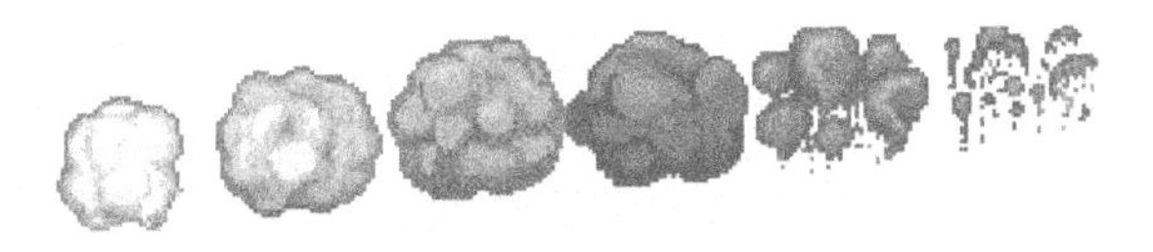

图12.7　爆炸效果图片

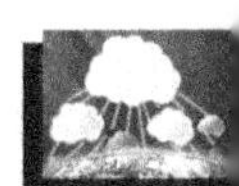

任务实现

1. 游戏框架的搭建

新建一个名为 project12_2 的项目，在该项目中新建一个继承 SurfaceView，名为 GameSurfaceView 的类，并增加 Callback 和 Runnable 接口，各项设置如图 12.8 所示。

New Java Class

Java Class

Create a new Java class.

Source folder: project12-2/src Browse...

Package: com.example.project12_2 Browse...

Enclosing type: Browse...

Name: GameSurfaceView

Modifiers: public default private protected

abstract final static

Superclass: android.view.SurfaceView Browse...

Interfaces: android.view.SurfaceHolder.Callback Add...

java.lang.Runnable Remove

Which method stubs would you like to create?

public static void main(String[] args)

Constructors from superclass

Inherited abstract methods

Do you want to add comments? (Configure templates and default value here)

Generate comments

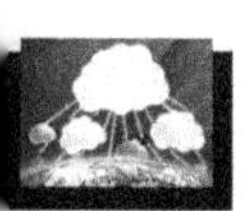

图 12.8　GameSurfaceView 类设置

知识拓展

增加 callback 接口的时候可以点击“Add”按钮，在弹出的对话框中输入“callback”查找，再选择 android. view. SurfaceHolder. Callback。

在点击完成后 Eclipse 会自动创建以下函数，各函数具体作用前文已经有所讲解，这里不再说明。

```
surfaceChanged(SurfaceHolder arg0, int arg1, int arg2, int arg3)
surfaceCreated(SurfaceHolder arg0)
```

```
surfaceDestroyed(SurfaceHolder arg0)
run()
```

除此之外，还要在该类中手动增加绘图函数 myDraw()、触屏监听函数 onTouchEvent(MotionEvent event)、构造函数 GameSurfaceView(Context context)、运算函数 calc()和初始化函数 initGame()等，到此游戏的基本构架就搭建完毕。

2. GameSurfaceView 类

1)游戏状态处理

本游戏一共分为三种状态，分别是游戏中 GAMEING、游戏胜利 GAME_WIN 和游戏结束 GAME_OVER，在程序中定义三个常量来表示。

```
public static final int GAMEING = 1; //游戏中
public static final int GAME_WIN = 2; //游戏胜利
public static final int GAME_OVER = 3; //游戏结束
public static int gameState = GAMEING;
```

当前游戏状态也用一个常量 gameState 来表示，初始状态为 GAMEING。当玩家积分超过 1000，gameState = GAME_WIN；当玩家积分少于 0，gameState = GAME_OVER。在编写游戏的各项功能的时候，要根据 gameState 来判断当前游戏状态，不同状态进行不同处理。

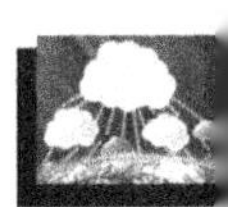

2)积分和游戏级别处理

每当石头超出屏幕，就为玩家增加相应的积分，具体来说：小石头加 1 分、中石头加 3 分、大石头加 5 分，关键代码如下：

```
//当石头飞出屏幕
if (rock. isPassed) {
//根据不同类型的石头加相应的分数
switch (vcRock. elementAt(i). type){
case ROCK_BIG: //大石头加 1 分
      totalPoint + =1;
      break;
      case ROCK_MID: //中石头加 2 分
      totalPoint + =2;
      break;
      case ROCK_SMALL: //小石头加 3 分
            totalPoint + =3;
      break;
   }
}
```

每当玩家飞机与石头碰撞，就扣减玩家相应的分数，具体来说：小石头扣10分、中石头扣30分、大石头扣50分，关键代码如下：

```
if (plane.isCollsionWith(vcRock.elementAt(i))) {
    //发生碰撞后根据石头类型扣减相应分数
    switch (vcRock.elementAt(i).type) {
    case ROCK_BIG:
    totalPoint -=50;
    break;
    case ROCK_MID:
    totalPoint -=30;
    break;
    case ROCK_SMALL:
    totalPoint -=10;
    break;
    }
}
```

游戏级别会随着玩家积分的增加而提升。例如，当游戏积分达到30分，级别就会提升一级，级别的提升主要从生成石头的多少来体现，例如：一开始每次只产生一批石头，而随着级别的提升每次会产生多批石头。

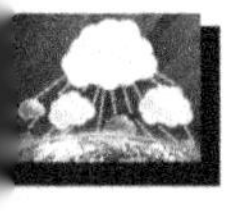

在程序中定义六个常量来表示六个级别。

```
public static final int GRADE_ONE = 1;
public static final int GRADE_TWO = 2;
public static final int GRADE_THREE = 3;
public static final int GRADE_FOUR = 4;
public static final int GRADE_FIVE = 5;
public static final int GRADE_SIX = 6;
private int gameGradeBefore = 0; //上一级别
public static int gameGrade = GRADE_ONE; //当前级别
private boolean isUpGrade = false; //是否升级级别标志
```

每当玩家的积分达到一定的值，游戏会自动升级并出现升级提示。升级关键代码如下：

```
if(totalPoint >=700) { //玩家分数超过700升到第六级
gameGrade = GRADE_SIX;
    if(gameGrade > gameGradeBefore) {//判断是否是第一次达到这个级别
        isUpGrade = true; //设置升级标志
```

```
            gameGradeBefore = gameGrade; //防止重复升级
        }
    }
```

升级提示关键代码如下：

```
if( isUpGrade) {
    //设置画笔
        paint. setTextSize(100);
        paint. setColor( (Color. RED) );
            //每两次游戏循环闪烁一次
            if (upGradeFlash % 2 = = 0) {
          canvas. drawText("升级啦！ 难度增加!", screenW/8, screenH/2, paint);
            }
        }
```

升级提示的关键运算代码如下：

```
    if (isUpGrade) {
                upGradeFlash + +; //升级提示闪烁时间计数器
                if (upGradeFlash > = upGradeFlashTime) {
                //闪烁过后，初始化
                isUpGrade = false;
                upGradeFlash = 0;
            }
        }
```

3) 绘图函数

在游戏中，主要绘制的图形有背景、玩家飞机、不同种类的石头和爆炸效果等，而所有绘图都直接或间接由 MyDraw( ) 函数完成。关键代码如下：

```
    //锁定画布
    canvas = sfh. lockCanvas();
    if (canvas ! = null) {
    canvas. drawColor(Color. WHITE);
    //绘图函数根据游戏状态不同进行不同绘制
    switch (gameState) {
    case GAMEING:
    //游戏背景绘制
        省略
    //飞机绘制
```


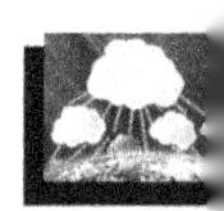

```
plane. draw(canvas, paint);
//石头绘制
for (int i = 0; i < vcRock. size(); i++) {
vcRock. elementAt(i). draw(canvas, paint);
}
//爆炸效果绘制
for (int i = 0; i < vcBoomAnimation. size(); i++) {
vcBoomAnimation. elementAt(i). draw(canvas, paint);
}
//设置画笔字体和颜色
paint. setTextSize(80);
paint. setColor((Color. RED));
//显示得分
canvas. drawText(String. valueOf(totalPoint), 100, 100, paint);
//处理升级提示
if(isUpGrade){
//此处上文已介绍，省略
}
break;
case GAME_WIN:
//绘制游戏胜利背景
canvas. drawBitmap(bmpGameWin, 0, 0, paint);
break;
case GAME_OVER:
//绘制游戏结束背景
canvas. drawBitmap(bmpGameOver, 0, 0, paint);
break;
}
}
```

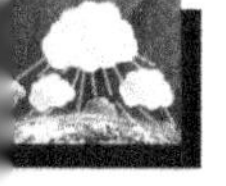

由于石头和爆炸效果每次绘制不止一个图片对象，所以使用 Vector 容器来动态存放。每次创建新的对象时就把它存入容器中，绘制时就从容器中抽出，当对象出屏或绘制完毕时就把它从容器中删除以节省系统资源。

4)石头的生成

游戏会根据不同级别随机生成类型不同、数量不等的石头，每个石头出现的坐标也是随机生成的，关键代码如下：

```
count++; //计数器
if (count % createRockTime == 0) {
```

```
//每次随机生成石头各类
for(int j =0; j< gameGrade; j++){
//随机生成类型
    rockType = (int)(Math.random() * 3+1);
    //石头出现的随机 X 坐标
    int x;
    //不同类型石头出现的位置和数量不一样
    switch (rockType){
        case ROCK_BIG:
        //大石头一次一个
        x = random.nextInt(screenW );
        vcRock.addElement(new Rock(bmpRockBig, 1, x, -40));
        break;
        case ROCK_MID:
        //中石头一次两个
        for(int t = 1; t< =2; t++){
        x = random.nextInt(screenW );
        vcRock.addElement(new Rock(bmpRockMid, 2, x, -20));
        }
        break;
        case ROCK_SMALL:
        //小石头一次三个
        for(int t = 1; t< =3; t++){
        x = random.nextInt(screenW );
        vcRock.addElement(new Rock(bmpRockSmall, 3, x, -10));
        }
        break;
        }
  }
}
```

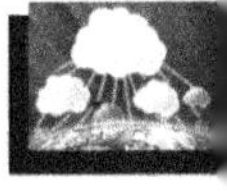

5)爆炸效果的处理

只要玩家飞机与石头相撞，就向爆炸容器增加一个新的爆炸对象，如果爆炸对象已经绘制完毕，就把它从爆炸容器中删除，关键代码如下：

爆炸效果增加：

```
for (int i = 0; i < vcRock.size(); i++) {
//碰撞检测
if (plane.isCollsionWith(vcRock.elementAt(i))) {
```

```
    //添加爆炸效果
    vcBoomAnimation.add(new BoomAnimation(bmpBoomAnimation, vcRock.elementAt(i)
.x, vcRock.elementAt(i).y, 7)); }
    }
```

爆炸效果处理:

```
for (int i = 0; i < vcBoomAnimation.size(); i++) {
BoomAnimation boomAnimation = vcBoomAnimation.elementAt(i);
    if (boomAnimation.isEnd) {
    //播放完毕的从容器中删除
    vcBoomAnimation.removeElementAt(i);
    } else {
//处理爆炸效果运算
    vcBoomAnimation.elementAt(i).calc();
    }
}
```

6)背景滚动处理

如前面所述，背景是由两张图片首尾相接滚动播放来实现的，关键代码如下:

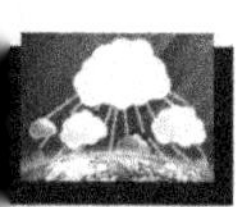

定义变量:

```
//定义两个图片变量，引用的是同一张背景图片资源
private Bitmap bmpBG1;
private Bitmap bmpBG2;
//两个背景坐标
private int bg1x, bg1y;
private int bg2x, bg2y;
//背景滚动速度
private int speed = 5;
```

在初始化函数 initGame()中实例化图片变量并设定好两个图片变量的位置。

```
bmpBG1 = BitmapFactory.decodeResource(res, R.drawable.background);
bmpBG2 = BitmapFactory.decodeResource(res, R.drawable.background);
//设定位置，第一张图片填充满整个屏幕
bg1y = -Math.abs(bmpBG1.getHeight() - GameSurfaceView.screenH);
//第二张背景图紧接在第一张背景的上方稍有重合(20)
bg2y = bg1y - bmpBG1.getHeight() + 20;
```

在绘图函数 myDraw()中绘制背景图片。

```
canvas.drawBitmap(bmpBG1, bg1x, bg1y, paint);
canvas.drawBitmap(bmpBG2, bg2x, bg2y, paint);
```

在运算函数 calc()中设置图片的运动。

```
//设置滚动速度
bg1y += speed;
bg2y += speed;
//当第一张图片的 Y 坐标超出屏幕，立即将其坐标设置到第二张图片的上方
    if (bg1y > GameSurfaceView.screenH) {
        bg1y = bg2y - bmpBG1.getHeight() + 20;
    }
//当第二张图片的 Y 坐标超出屏幕，立即将其坐标设置到第一张图片的上方
    if (bg2y > GameSurfaceView.screenH) {
        bg2y = bg1y - bmpBG1.getHeight() + 20;
    }
```

3. Plane 类

1)飞机的绘图函数

飞机的绘图分成两种，一是正常飞行状态，二是与石头碰撞后闪烁状态。关键代码如下：

```
    public void draw(Canvas canvas, Paint paint) {
        if (isCollision) {//当发生碰撞，让飞机闪烁
        if (flashCount % 2 == 0) {//每 2 次游戏循环，绘制一次飞机
            canvas.drawBitmap(bmpPlane, x, y, paint);
        }
        } else {//过了闪烁时间，直接绘制飞机
            canvas.drawBitmap(bmpPlane, x, y, paint);
        }
    }
```

2)飞机的运算函数

飞机的运算函数主要处理飞机闪烁的时间控制，代码如下：

```
public void calc() {
        //处理闪烁状态
        if (isCollision) {
        //计时器开始计时
        flashCount++;
```

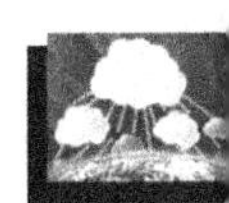

```
                if (flashCount  > = flashTime) {
                    //闪烁时间过后，初始化
                    isCollision  =  false;
                    flashCount  =  0;
                }
            }
        }
```

3)飞机的屏幕监听函数

玩家通过手指触摸来移动飞机避开石头，在编写代码的时候首先要检查玩家手指是否点中飞机，如果否就不能移动。当玩家手指点中飞机并移动起来，就计算每次游戏循环玩家手指移动的距离，让飞机跟随手指移动，关键代码如下：

```
    public boolean isCollsionWith(Rock rock) {//用矩形碰撞检测
                int x2  =  rock. x;
                int y2  =  rock. y;
                int w2  =  rock. rockW;
                int h2  =  rock. rockH;
                if (x  > =  x2 && x  > =  x2  +  w2) {
                    return false;
                } else if (x  < =  x2 && x  +  bmpPlane. getWidth()  < =  x2) {
                    return false;
                } else if (y  > =  y2 && y  > =  y2  +  h2) {
                    return false;
                } else if (y  < =  y2 && y  +  bmpPlane. getHeight()  < =  y2) {
                    return false;
                }
                //发生碰撞并进入闪烁状态
                isCollision  =  true;
                    return true;
    }
```

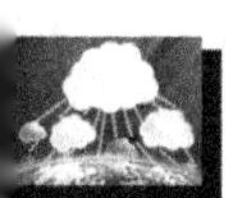

4. BoomAnimation 类

1)构造函数

在构造函数中，就必须手工指明所用的图片有多少帧(frameCount)，然后计算出每一帧的长和宽以便绘图时使用，关键代码如下：

```
public BoomAnimation(Bitmap bmpBoom, int x, int y, int frameCount) {
    this.bmpBoom = bmpBoom;
    this.x = x;
    this.y = y;
    this.frameCount = frameCount;
    //每帧宽度等于图片总宽除以总帧数
    this.boomW = bmpBoom.getWidth() / frameCount;
    this.boomH = bmpBoom.getHeight();
}
```

2)爆炸效果绘制函数

每一次游戏循环只会绘制一帧的爆炸图片，为了不影响其他游戏元素的绘制，在绘制爆炸图片之前必须把当前的画布保存，当绘制结束后再把画布恢复，关键代码如下：

```
public void draw(Canvas canvas, Paint paint) {
        canvas.save();
        //设置裁剪区域
        canvas.clipRect(x, y, x + boomW, y + boomH);
        //绘制画片
        canvas.drawBitmap(bmpBoom, x - playingFrame * boomW, y, paint);
        canvas.restore();
}
```

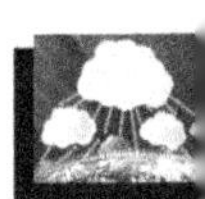

3)爆炸效果的运算函数

只要爆炸效果没有绘制完，就继续逐帧地绘制，如果绘制完了，就设置一个标志(isEnd)，关键代码如下：

```
public void calc() {
    if (playingFrame < frameCount) {
        playingFrame++;
    } else {
        isEnd = true;
    }
}
```

5. Rock 类

1)构造函数

在构造函数中，除了定义石头的各个属性外还为不同类型的石头定义了不同的随机初始飞行速度，代码如下：

```
public Rock(Bitmap bmpRock, int rockType, int x, int y) {
        this.bmpRock = bmpRock; //获得图片
        //获得石头图片的长度和宽度
        this.rockW = bmpRock.getWidth();
        this.rockH = bmpRock.getHeight();
        //石头类型
        this.rockType = rockType;
        //石头出现的坐标
        this.x = x;
        this.y = y;
        //不同的类型有不一样的初始运动速度
        switch (rockType) {
        case ROCK_BIG:
            //大石头初始速度在10到15之间
            speed = (int)(Math.random() * 15 + 10);
            break;
        case ROCK_MID:
            //中石头初始速度在20到25之间
            speed = (int)(Math.random() * 25 + 20);
            break;
        case ROCK_SMALL:
            //小石头初始速度在40到60之间
            speed = (int)(Math.random() * 60 + 40);
            break;
            }
        }
```

2)石头运算函数

该函数提供了各类型石头的不同飞行方式和石头出屏判断，代码如下：

```
public void calc() {
    //不同石头飞行方式不一样
    switch (rockType) {
    case ROCK_BIG:
    if (isPassed == false) {
        //大石头匀速飞行
        y += speed;
        if (y >= GameSurfaceView.screenH) {
```

```
                isPassed = true;
            }
        }
        break;
    case ROCK_MID:
        if (isPassed == false) {
            //中石头加速飞行
            speed += 1;
            y += speed;
            if (y >= GameSurfaceView.screenH) {
                isPassed = true;
            }
        }
        break;
    case ROCK_SMALL:
        if (isPassed == false) {
            //小石头减速飞行，最低速度 30
            if(speed >=31){
            speed -= 1;
            }
            y += speed;
            if (y >= GameSurfaceView.screenH) {
                isPassed = true;
            }
        }
        break;
    }
}
```

到此，太空飞行游戏的基本制作方法已经介绍完毕。我们发现，其实该类型的游戏实现起来无非就是在手机屏幕上不断地循环刷新图片和文字，让玩家感觉整个游戏是动起来的。为了让程序结构更加清晰，每一个类中都会有两种函数，一个是绘图函数，负责绘制图片和文字；另一个是运算函数，负责各种动作和逻辑的运算。游戏刷新时，运算函数的运算结果由绘图函数呈现给玩家。

 任务考核

请修改任务二太空飞行游戏的功能，使得每次升级游戏中的石头出现频率和飞行速度整体加快，让游戏难度更大更刺激，答案可参考网络资源。

项目十三

学生实习安全管理系统（APP）

本项目工作情景的目标是让学生学会开发一个相对完整的APP，学生实习安全管理系统APP，主要任务有四个：它们是登录功能实现，动态生成首页界面模块，学生上下班签到模块，学生求救和求救跟踪的模块。

学会提炼抽象方法，设计公共工具类
掌握学会百度地图开发
掌握基于 Android 平台的网络通信开发的方法

项目概述　学生实习安全管理系统 APP 概述

1. 系统简介

职业院校实习生实训是重要的教学环节，是理论联系实际，培养学生独立工作能力的重要途径。为使实习学生实训工作达到预期目的，保证实习学生实训工作有领导、有计划、有组织地进行，针对目前的社会治安和交通安全情况，建立一个安全管理系统非常必要。

学生实习安全管理系统是基于手机客户端与服务端的架构下的信息管理系统。利用 Android 平台开发手机客户端，Java 开发服务器端。系统有三个角色，分别是老师、学生和管理员。学生端可以实现上下班签到、发送求救信息、信息管理等功能。老师端可以接收学生求救信息、查看学生上下班签到情况等功能。管理员端可以管理学生老师数据、信息数据、学生上下班签到数据、学生求救信息等功能。

1）开发背景

学生顶岗实习是学校教育的重要组成部分，从实习开始，学生就具有既是“学生”也是“准员工”的特殊双重身份，实习期间学校仍然是承担管理职能的重要主体。同时，由于实习时间较长，实习期间的安全问题日益突出，如何加强实习生的安全管理是每个学校亟待解决的现实问题。

针对实习生管理工作的特点和流程，应用移动互联网技术开发出一套实习安全管理系统是非常有用的。该系统应该包括考勤管理、SOS、位置信息、信息沟通、告警等主要功能，为学校管理者和实习生之间提供即时沟通、知识宣贯、信息交流和智能管理的平台，开拓了一条实习安全管理手段新通道。

2)系统结构图

实习安全管理系统结构图如图13.1所示。

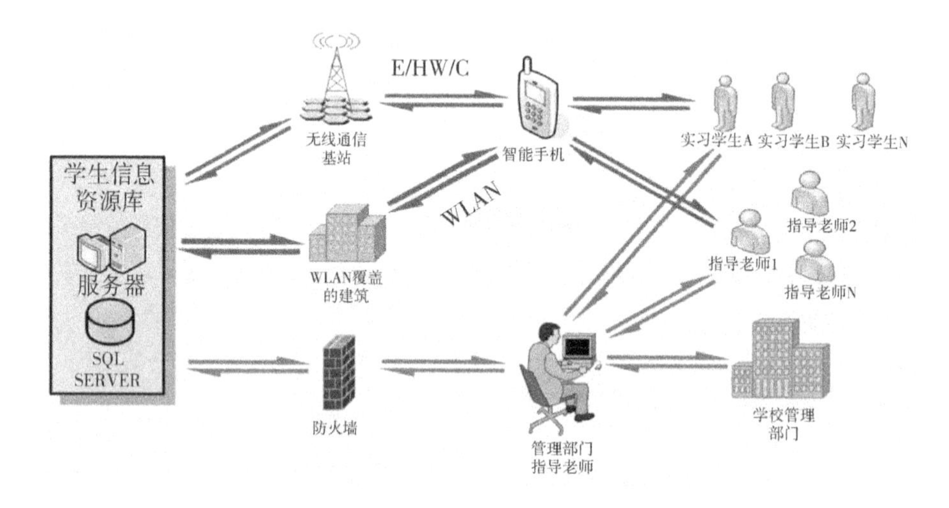

图13.1　实习安全管理系统结构图

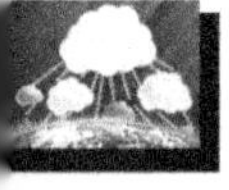

3)系统功能介绍

系统使用主体(三类角色)：实习学生、指导老师和学生管理者。

- 实习学生，系统的主体用户：考勤登记、SOS、信息中心、求救定位等。
- 指导老师：考勤记录、SOS、信息中心、告警(一级告警、二级告警)。
- 学校管理者：SOS、信息中心、一级告警。

4)系统主要功能

(1)实习学生APP功能模块。

- 考勤登记：方便指导老师掌握实习学生在岗情况，了解实习学生安全状况。
- SOS：提供SOS紧急求救功能，地理位置信息及运动轨迹直观展示，方便搜救。
- 地图：地理位置登记，方便找附近同学，方便交流和沟通。
- 信息中心：增加学校、实习学生的信息渠道。

(2)指导老师APP功能模块。

- 考勤记录：完成考勤日常管理，及时掌握实习生在岗情况和安全状况。
- SOS：及时、准确地与相关管理部门、企业和实习生家长沟通，拓宽搜救通道，提高异常事件处理的效率。
- 信息中心：提供即时的信息交流平台，采用群发、分组发、定时发送等信息传播方式。

(3)学校管理者和智能信息库。

- 考勤记录：完成考勤日常管理，及时掌握实习生在岗情况和安全状况。
- SOS：及时、准确地与相关管理部门、企业和实习生家长沟通，拓宽搜救通道，提高异常事件处理的效率。
- 信息中心：提供即时的信息交流平台，采用群发、分组发、定时发送等信息传播方式。
- 告警信息自动采集：提高重点工作的关注度，减轻指导老师工作量，提供工作效率。
- SOS 信息地理信息呈现：运动轨迹展现，救援信息联动，提供遇险者(实习学生)的安全系数。

2. 工程结构说明

工程结构说明如图 13.2 所示。

```
stAndroid2.1
  Android 4.2.2
  Android Private Libraries
  src
    com.eshine.android
      common
      mapdemo.baidu
      st
        attend
        extendedView
        message
        sos
        user
    config
```

图 13.2　工程结构图

1) 系统工具包

系统需要的工具和方法主要由 common、mapdemo. baidu、st 三个包提供。

common 主要是一些公共的模块，例如登录、访问网络获取数据、常用的工具类等。mapdeomo. baidu 是完成百度地图的相关工具类，是最重要的包，app 的功能实现主要是在这个包目录下。st 包的子包主要包括 attend、message、sos，分别是考勤模块、信息中心模块、求救模块。

2) config 的说明

配置各个功能的访问连接，以 key = value 的形式保存，如图 13.3 所示。

- web_url：访问该实习管理系统的网站网址。
- queryStudent：查询实习学生的情况对应的功能模块。
- queryClasses：查询实习学生所在的班级等。

```
web_url=http://14.23.103.106:1080/st2.0
#web_url=http://www.dglg.net:8080/st
queryStudent=/android/queryStudent/studentCtrl.do?action=quer
queryClasses=/android/queryClasses/classCtrl.do?action=queryC
confirmIsWorkOff=/android/isWorkOff/attendCtrl.do?action=conf
isOnDuty=/android/isAttend/attendCtrl.do?action=isOnDuty
classAttend=/android/classAttend/attendCtrl.do?action=queryCl
sendSos=/android/sos/sosCtrl.do?action=send
```

图 13.3　config 文件键值对

任务一　登录模块的实现

任务描述

登录模块是应用系统的窗口，是系统安全的保证。本任务就是实现用户登录功能。完成本任务主要就是完成两点：一是检查客户端的网络链接；二是完成检测后，如何在客户端和服务器端分别存储用户登录信息。

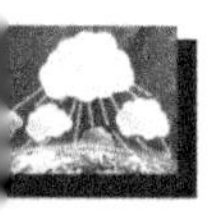

任务分析

本任务是实现用户登录如图 13.4 所示。当用户输入登录账号和密码时，该用户手机端立刻与服务器尝试网络连接，网络连接正常则进行登录用户信息存储等工作。因此，要完成此任务，需要思考如何实现与服务器网络连接检测、用户登录信息存储过程以及 APP 版本号显示和更新。

图 13.4　登录 Activity

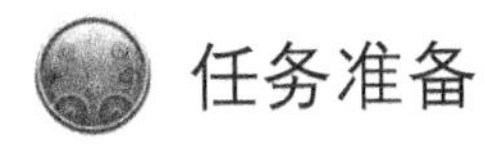

任务准备

实现该任务，需要编写下列工具类及方法。

1. HttpConnectionUtil 网络访问工具类

(1) public synchronized static HttpClient getHttpClient() 获取浏览器。

(2) public static HttpUriRequest getRequest(String url, Map < String, String > params, HttpMethod method) 根据 URI 请求服务器的应答。

(3) public static String getString(HttpPost post) 从服务器获得字符串。

(4) public static String getUrl(String webContentPath, String controller, String action) 用于返回完整的 URL 请求地址字符串。

(5) public static HttpPost getPost(String uri, List < BasicNameValuePair > params) 获得 POST 请求对象、URI 请求地址，也可以带参数，params 如果为 null，则不添加由 BasicNameValue 封装的参数。

(6) public static String getString(String uri, int requestLimit) 请求服务器，返回字符串，URI 字符串形式的请求地址，requestLimit 最多允许的请求失败次数。

(7) public static Object syncConnect(final String url, final HttpMethod method, final HttpConnectionCallback callback) 同步访问网络方法。其中 HttpMethod 为 GET 或者 POST。

(8) public static void asyncConnect(final String url, final Map < String, String > params, final HttpMethod method, final HttpConnectionCallback callback) 同步访问网络方法。其中 HttpMethod 为 GET 或者 POST。

2. LoginSvr 登录服务类

(1) public String submit(Context context, ICmd cmd, LoginCmd cmd2)

(2) public static void getDynamicPass(String userCode, HttpConnectionCallback < Object > callback)

3. DBHelper 数据库管理类

4. Config 阅读 config. properties 文件工具类

5. public final class Constants 保存用户信息(常量)类

6. public class DialogUtil 对话框工具类

(1) public static void showDialog(final Context ctx, String msg, boolean closeSelf) 定义一个显示消息的对话框，closeSelf 为是否关闭当前界面。

(2) public static void showDialog(Context ctx, View view) 定义一个显示指定组件的对话框。

(3) public static void showDialog(Context ctx, String title, String msg) 定义一个有标题、有消息处理的对话框。

(4) public static void showTips(Context ctx, String message) 显示 message 文本对话框。

7. public class StringUtil 字符串处理工具类

(1) public static String compress(String str) 压缩字符串 str。

（2）public static String uncompress(String str) 对字符串 str 解压缩。

8. public class SharedPreferencesUtil 键值对处理工具类

（1）public void putValue(String key, String value) 向 SharedPreferences 中注入数据。

（2）public String getValue(String key) 根据 key 获取对应的 Value。

（3）public void clear( ) 清除 SharedPreferences 中的数据。

任务实现

1. 布局文件

布局文件比较长，主要是 6 个 TextView，分别用来显示用户账号、用户密码提示、APP 版本号、开发者和版权所有等信息。2 个 EditText 用于用户登录前输入账号和密码。

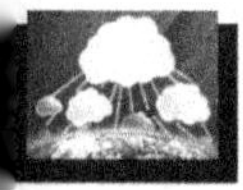

```
<?xml version="1.0" encoding="utf-8"?>
<LinearLayout xmlns:android="http://schemas.android.com/apk/res/android"
    android:layout_width="match_parent"
    android:layout_height="match_parent"
    android:background="@drawable/login_bj"
    android:gravity="center|center_horizontal"
    android:orientation="vertical" >
    <LinearLayout
        android:layout_width="480dp"
        android:layout_height="700dp"
        android:gravity="center|center_horizontal"
        android:orientation="vertical" >
        <TextView
            android:layout_width="310dp"
            android:layout_height="wrap_content"
            android:background="@drawable/logo" />
        <LinearLayout
            android:layout_width="290dp"
            android:layout_height="wrap_content"
            android:focusable="true"
            android:focusableInTouchMode="true"
            android:orientation="horizontal" >
            <TextView
                android:layout_width="wrap_content"
                android:layout_height="40dp"
```

```
            android: layout_gravity = " center_vertical"
            android: layout_marginTop = "20dp"
            android: text = "用户名: "
            android: textSize = "18dp" />
        <EditText
            android: id = "@ +id/login_et_name"
            android: layout_width = "fill_parent"
            android: layout_height = "40dp"
            android: layout_marginLeft = "10dp"
            android: layout_marginTop = "30dp"
            android: paddingLeft = "5dp"
            android: background = "@ drawable/shape"
            android: hint = "用户名..." />
    </LinearLayout >
    < LinearLayout
        android: layout_width = "290dp"
        android: layout_height = " wrap_content"
        android: orientation = " horizontal" >
        < TextView
            android: layout_width = " wrap_content"
            android: layout_height = "40dp"
            android: layout_marginTop = "20dp"
            android: text = "密码: "
            android: textSize = "18dp" />
        < EditText
            android: id = "@ +id/login_et_password"
            android: layout_width = "fill_parent"
            android: layout_height = "40dp"
            android: layout_marginLeft = "10dp"
            android: layout_marginTop = "20dp"
            android: background = "@ drawable/shape"
            android: hint = "密码..."
            android: paddingLeft = "5dp"
            android: password = "true" />
    </LinearLayout >
    < LinearLayout
        android: layout_width = "290dp"
        android: layout_height = " wrap_content"
```

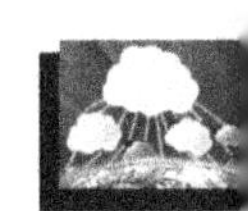

```
        android: layout_marginBottom = "10dp"
        android: layout_marginLeft = "2dp"
        android: layout_marginTop = "20dp"
        android: orientation = "horizontal"  >
    </LinearLayout>
    <Button
        android: id = "@ +id/login_btn"
        android: layout_width = "290dp"
        android: layout_height = "40dp"
        android: layout_marginLeft = "30dp"
        android: layout_marginRight = "30dp"
        android: background = "@drawable/btn_login"
        android: textColor = "#ffffff"  />
    <TextView
        android: layout_width = "fill_parent"
        android: layout_height = "wrap_content"
        android: layout_marginTop = "30dp"
        android: gravity = "center | center_horizontal"
        android: text = "@string/app_version"
        android: textColor = "#979994"
        android: textSize = "13dp"  />
    <TextView
        android: id = "@ +id/verName"
        android: layout_width = "fill_parent"
        android: layout_height = "wrap_content"
        android: layout_marginTop = "8dp"
        android: gravity = "center | center_horizontal"
        android: text = "东莞 * *学校"
        android: textColor = "#979994"
        android: textSize = "13dp"  />
  <TextView
        android: id = "@ +id/verName"
        android: layout_width = "fill_parent"
        android: layout_height = "wrap_content"
        android: layout_marginTop = "8dp"
        android: gravity = "center | center_horizontal"
        android: text = "广州 * *电子科技有限公司 版权所有"
        android: textColor = "#979994"
```

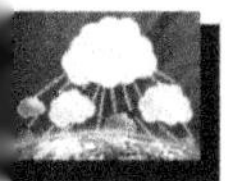

```
            android: textSize = "13dp" />
    </LinearLayout>
</LinearLayout>
```

2. 登录过程处理

1)生成个性的界面

利用 requestWindowFeature(Window.FEATURE_NO_TITLE)方法生成没有标题的界面，再调用 getView()方法获取页面的所有控件。其中该方法的主要代码为

```
        btnLogin = (Button) findViewById(R.id.login_btn);
        edtvName = (EditText) findViewById(R.id.login_et_name);
        edtvPassword = (EditText) findViewById(R.id.login_et_password);
        verName = (TextView) findViewById(R.id.verName);
```

2)检查手机是否已经联网

```
public void isNetworkAvailable(Context context) {
        ConnectivityManager cm = (ConnectivityManager) context
                .getSystemService(Context.CONNECTIVITY_SERVICE);
        if (cm == null) {
        } else {
          NetworkInfo networkinfo = cm.getActivityNetworkInfo();
            if (networkinfo == null || !networkinfo.isAvailable()) {
              show = new AlertDialog.Builder(LoginActivity.this)
              .setTitle("网络设置提示")
              .setMessage("网络不可用，是否现在设置网络?")
              .setPositiveButton("设置", new
DialogInterface.OnClickListener() {   public void
onClick(DialogInterface dialog, int which) {Intent   intent = new
Intent(android.provider.Settings.ACTION_WIRELESS_SETTINGS);
      startActivity(intent);
      }})
              .setNegativeButton("取消", null)
              .show();
            }}}
```

上述代码首先判断是否有 Context.CONNECTIVITY_SERVICE 的服务，有则启动一个 Activity，否则给出“网络不可以用，是否现在设置网络?”的提示。

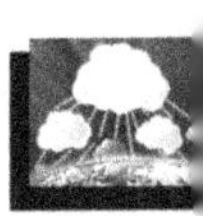

3)登录按钮监听

```
btnLogin. setOnClickListener( new OnClickListener( ) {
        public void onClick( View arg0) {
            String userCode = edtvName. getText( ). toString( );
            String pass = edtvPassword. getText( ). toString( );
            if( userCode = = null || userCode. equals( "" ) ) {
        DialogUtil. showTips( LoginActivity. this, "请输入用户名" );
            } else if( pass = = null || pass. equals( "" ) ) {
        DialogUtil. showTips( LoginActivity. this, "请输入密码" );
            } else{
                btnLogin. setEnabled( false );
                if( progressDialog = = null)
progressDialog = ProgressDialog. show( LoginActivity. this, "请稍等... ", "正在验证用户名和密码... ", true, true);
                else {
if( progressDialog. isShowing( ) )progressDialog. cancel( );
                    progressDialog. show( );
                }
                Thread th = new Thread( loginRun);
                th. start( );
            }
        }
    });
```

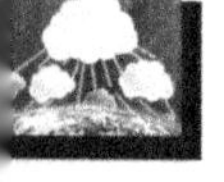

登录按钮监听利用 loginRun 线程处理登录。

```
Runnable loginRun = new Runnable( ) {
    public void run( ) {
        String name = edtvName. getText( ). toString( );
        String pass = edtvPassword. getText( ). toString( );
        LoginCmd cmd = new LoginCmd( );
        cmd. setUserName( name);
        cmd. setPassWord( pass);
        String msg = loginSvr. submit( getApplicationContext( ), cmd, cmd);
//发送登录请求给服务端，包含登录者的账号和密码
        // 提示登录成功
        if ( msg. contains( "连接超时" ) ) {
            // 新启动的线程无法访问 Activity 里的组件
```

```
                // 所以需要通过 Handler 发送信息
                Message msg1 = new Message();
                msg1.what = LoginActivity.LOGINIDENTIFIER;
                msg1.arg1 = 1;
                myHandler.sendMessage(msg1); // 发送消息
                return;
            }else if(msg.contains("异常") || msg.contains("失败")){
                Message msg1 = new Message();
                msg1.what = LoginActivity.LOGINIDENTIFIER;
                msg1.arg1 = 3;
                // 发送消息
                myHandler.sendMessage(msg1);
            }else{
                Message msg1 = new Message();
                msg1.what = LoginActivity.LOGINIDENTIFIER;
                msg1.arg1 = 2;
                msg1.obj = msg; // 传 Obj 对象
                // 发送消息
                myHandler.sendMessage(msg1);
            }


        }
    };
```

代码首先利用 loginSvr. submit(getApplicationContext(), cmd, cmd)发送登录请求给服务端，包含登录者的账号和密码。服务端返回消息 msg。如果登录成功，则 msg1. arg1 =2，msg1. arg1 =1 或者 3(网络访问异常，比如连接超时，其他异常)。然后利用 myHandler. 把消息发送到消息队列当中。

4)登录 loginActivity 跳转到首页的处理

```
    public void setHandler() {
    myHandler = new Handler() {
    public void handleMessage(Message msg) {// 如果该消息是本程序所发送的
                progressDialog.cancel();
                btnLogin.setEnabled(true);
                if (msg.what == LoginActivity.LOGINIDENTIFIER) {
                    if (msg.arg1 == 1) {  // 链接网络超时
                    String name = edtvName.getText().toString();
```

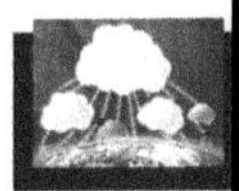

```
                    String pass = edtvPassword. getText( ). toString( );
spu. putValue( SharedPreferencesUtil. STR_USERCODE, name);
spu. putValue( SharedPreferencesUtil. STR_PASSWORD, pass);
     Toast. makeText( LoginActivity. this, "网络超时，请重试!", 5000). show( );
                    } else if (msg. arg1 = = 2) {
               JSONObject obj;
                    try {
                         obj = new JSONObject( (String) msg. obj);
                         String state = obj. getString( "MSG" );
                         String userCode = obj. getString( "userCode" );
                         String userName = obj. getString( "userName" );
                         String sessionID = obj. getString( "sessionID" );
                         String role = obj. getString( "role" );
                         JSONArray ay = obj. getJSONArray( "list" );
                         System. out. println( ay. length( ) );
                         String[ ] func = new String[ ay. length( ) ];
                         for( int i = 0; i < ay. length( ); i + + ) {
                              JSONObject j = ay. getJSONObject( i );
                              String funcCode = j. getString( "funcCode" );
                              func[ i ] = funcCode;
                              Log. d( "func", func[ i ] ); }
          LoginCmd. sessionID = sessionID;
          LoginCmd. userCode = userCode;
          LoginCmd. userName = userName;
          LoginCmd. role = role;
spu. putValue( SharedPreferencesUtil. STR_USERNAME, userName);
          Log. d( "userCode:", userCode);
          Log. d( "sessionID", LoginCmd. sessionID);
          Toast. makeText( LoginActivity. this, state, 5000). show( );
                              if (state. contains( "成功" ) ) {
          Constants. userCode = edtvName. getText( ). toString( );
          String name = edtvName. getText( ). toString( );
          String pass = edtvPassword. getText( ). toString( );
          spu. putValue( SharedPreferencesUtil. STR_USERCODE, name);
          spu. putValue( SharedPreferencesUtil. STR_PASSWORD, pass);
          //将用户名和功能保存到手机数据库
          StringBuilder sb = new StringBuilder( );
```

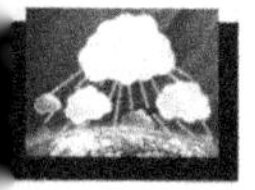

```
JSONArray array = obj. getJSONArray("list");
        for(int i =0; i < array. length(); i + +){
        JSONObject j = array. getJSONObject(i);
        String funcCode = j. getString("funcCode");
                sb. append(funcCode);
                if(i < array. length() -1){
                sb. append(",");
            }
        }
    pass = edtvPassword. getText(). toString();
    Cursor cursor = db. rawQuery("select * from userfunc where
usercode = ? and pass = ?", new String[]{userCode, pass});
    if(cursor! =null&&cursor. getCount() >0){
    db. execSQL("update userfunc set func = ? where usercode = ?
and pass = ?", new Object[]{sb, userCode, pass});
    }else{
    db. execSQL("insert into userfunc values( ? , ?,?)", new
Object[]{userCode, pass, sb});    }
  Intent i = new Intent(getApplicationContext(),   HomeActivity. class);
  i. setFlags(Intent. FLAG_ACTIVITY_CLEAR_TOP); // 注意本行的 FLAG 设置
  i. putExtra("appFuncs", func);
  i. putExtra("appFuncs", (String)msg. obj);
        startActivity(i);
            }
        } catch (JSONException e) {
      e. printStackTrace();
      }
        }else if(msg. arg1 = = 3){
  String name = edtvName. getText(). toString();
  String pass = edtvPassword. getText(). toString();
  spu. putValue(SharedPreferencesUtil. STR_USERCODE, name);
  spu. putValue(SharedPreferencesUtil. STR_PASSWORD, pass);
  Toast. makeText(LoginActivity. this, "请检查网络连接!", 3000). show();
        }
      }
    }
  };
}
```

代码从 loginActivity 跳转到首页的处理的依据是 msg. arg1。当 msg. arg1 = = 1 或者 msg. arg1 = =3 时，网络出问题。当 msg. arg1 = = 2 时，就可以跳转到 HomeActivity。同时，把用户登录的信息保存到数据库 db 的 userfunc 表中和以键值对的形式存储在 SharedPreferencesUtil 的变量中。另外，用户类型等信息存储在 appFuncs 中，通过 intent 传递到 HomeActivity。

任务二　动态生成主页

任务描述

本系统有三个角色的用户，分别是学生、教师和管理员。本任务就是根据三个角色登录系统能动态地生成不同的首页。如图 13.5 管理员角色首页，图 13.6 教师角色首页，图 13.7 学生角色首页。

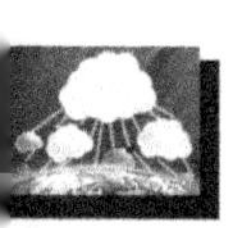

图 13.5　管理员首页

图 13.6　教师首页

任务分析

动态生成首页的布局主要表现在动态添加辅助 SOS、信息中心、上班登记、下班登记、预警和地图六个按钮，还要动态生成最新 APP 版本号，添加包含用户账号的标题条，“更多...”按钮。因此，如何实现上述功能是完成本任务的关键。

图 13.7　学生首页

任务准备

实现该任务，需要理解和编写下列工具类及方法。

(1) public class HomeSvr 登录模块逻辑处理工具类；
(2) private void createButton(View b) 菜单按钮的创建；
(3) public class MyLocationListenner 百度地图工具类。

任务实现

1. 布局文件设计

```
<LinearLayout xmlns: android = "http: //schemas. android. com/apk/res/android"
    android: id = "@ +id/home_root"
    android: orientation = "vertical"
    android: layout_width = "fill_parent"
    android: layout_height = "fill_parent" >
    <include layout = "@ layout/frame_home_header_layout"/ >
```

```
        <ImageView
        android: layout_width = "fill_parent"
        android: layout_height = "wrap_content"
        android: scaleType = "centerCrop"
        android: src = "@drawable/tpbar"
        android: adjustViewBounds = "true"/>
    <ScrollView
                android: id = "@+id/my_scrollview"
                android: layout_width = "match_parent"
                android: layout_height = "fill_parent"
                android: scrollbars = "vertical"  >
                <LinearLayout
                    android: id = "@+id/home_layout"
                    android: layout_width = "fill_parent"
                    android: layout_height = "fill_parent"
                    android: orientation = "vertical" >
                </LinearLayout>
        </ScrollView>
<include layout = "@layout/frame_home_footer_layout"/>
</LinearLayout>
```

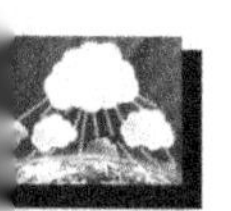

该文件仅仅是显示如图13.8的布局。首页其他的东西需要代码添加。

图13.8　空的首页布局文件视图

2. 动态生成首页过程

1) 登录用户账户显示

```
TextView frameTitleV = (TextView) findViewById(R.id.frame_tv_title);
frameTitleV.setText(LoginCmd.userName);
```

在标题栏里显示用户名。

2) Other 按钮的处理

Other 按钮显示"更多..."，点击后出现"修改密码"，"关于"和"退出"的菜单项。

```
public void onClick(View arg0) {
        final MenuAdapter adapter = new MenuAdapter(menus);
        new AlertDialog.Builder(HomeActivity.this)
        .setAdapter(adapter, new  DialogInterface.OnClickListener()
{   public void onClick(DialogInterface dialog, int which) {
                switch (which) {
                case 0:
        Intent i = new Intent(HomeActivity.this, ModifyPassActivity.class);
            startActivity(i);
                    break;
                case 1:
                    about();
                    break;
                case 2:
                    appExit();
                    break;
                }   }}).show();
            }});
```

上述监听程序利用 switch 语句串联三个菜单项。当 which 为 0 时，跳转到密码修改 ModifyPassActivity。which 为 1 时，调用 about()方法，which 为 2 时，退出本系统。

3. APP 版本号的生成

各用户类型的首页均需生成版本号及版本名称。如果用户当前的版本不是最新，则执行以下步骤。

1) 从服务器里获得版本号

```
verCode = context.getPackageManager().getPackageInfo(
        "com.eshine.android", 0).versionCode;
```

2）从服务器中获得版本名称

```
public String getVerName(Context context) {
        String verName = "";
        try {
            verName = context.getPackageManager().getPackageInfo(
                    "com.eshine.android", 0).versionName;
        } catch (NameNotFoundException e) {
            Log.e("版本名称获取异常", e.getMessage());    }
        return verName;    }
```

3）更新版本代码

```
public void doNewVersionUpdate() {
        int verCode = this.getVerCode(this);
        String verName = this.getVerName(this);
        StringBuffer sb = new StringBuffer();
        sb.append("你目前的版本:");
        sb.append(verName);
        sb.append(", 新版本:");
        sb.append(newVerName);
        sb.append(", \n是否更新");
        Dialog dialog = new AlertDialog.Builder(this)
                .setTitle("软件更新")
                .setMessage(sb.toString())
                .setPositiveButton("更新", new DialogInterface.OnClickListener()
{
        public void onClick(DialogInterface dialog, int which)
{   downDlg = new ProgressDialog(HomeActivity.this);
downDlg.setMessage("正在下载，请稍后...");
downDlg.setProgressStyle(ProgressDialog.STYLE_SPINNER);
String url = Constants.servicePath   +"/setup/uploadCtrl.do? action=download";
                        System.out.println(url);
                        downFile(url);
                    }
                }).setNegativeButton("暂不更新",
                        new DialogInterface.OnClickListener() {
                public void onClick(DialogInterface dialog,
                                    int which) {
```

```
                            dialog.cancel();
                        }
                    }).create();
        dialog.show();
    }
```

在消息提示框中如果选择更新版本，则调用 downFile( )方法依据 URL 的地址下载 APP 新版本。否则，暂时不更新。

4) 下载新版本 APP

```
public void downFile(final String url) {
        downDlg.show();
        new Thread() {
            public void run() {
                HttpClient client = HttpConnectionUtil.getHttpClient();
                HttpGet get = new HttpGet(url);
                HttpResponse response;
                try {
                    response = client.execute(get);
                    HttpEntity entity = response.getEntity();
                    long length = entity.getContentLength();
                    InputStream is = entity.getContent();
                    FileOutputStream fileOutputStream = null;
                    if (is != null) {
                        File file = new File(
    Environment.getExternalStorageDirectory(), UPDATE_SERVERAPK);
                    fileOutputStream = new FileOutputStream(file);
                    byte[] b = new byte[1024];
                    int charb = -1;
                    int count = 0;
                    while ((charb = is.read(b)) != -1) {
                        fileOutputStream.write(b, 0, charb);
                        count += charb;
                        }
                    }
                    fileOutputStream.flush();
                    if (fileOutputStream != null) {
                        fileOutputStream.close();
```

```
                    }
                    down();
                } catch (Exception e) {
                    e.printStackTrace();
                }
            }
        }.start();}
```

采用Http通信的方式，利用文件读写的方式下载APP版本。

5)安装新版本APP

```
public void update() {
    Intent intent = new Intent(Intent.ACTION_VIEW);
    intent.setDataAndType(Uri.fromFile(new File(Environment
    .getExternalStorageDirectory(), UPDATE_SERVERAPK)),
            "application/vnd.android.package-archive");
    startActivity(intent);
}
```

加载版本号和版本名称时，系统会弹出一个对话框提示用户是否为最新版本，若是，则不更新版本，否则更新版本。若是更新版本，则下载并安装新版本。

4. 是否有新信息

1)创建一个Notification通知用户有新信息

```
private void addNotificaction() {
    Notification notification = new Notification();

    notification.icon = R.drawable.txt_file;
    notification.tickerText = "有新信息，请查看";
    notification.defaults = Notification.DEFAULT_SOUND;
    android.media.AudioManager.ADJUST_LOWER;
    nm.notify(119, notification);
}
```

2)显示未读信息的数量

```
private void queryNewMessage() {
    progressDialog = new ProgressDialog(HomeActivity.this);
    Map<String, String> params = new HashMap<String, String>();
    params.put("role", LoginCmd.role);
```

```
        params.put("userCode", LoginCmd.userCode);
        MsgSve.queryNewMessage(params, new NetCallback() {
            public Object execute(String response) {
                Message msg = new Message();
                msg.arg1 = 3;
                msg.obj = response;
                queryHandler.sendMessage(msg);
                return null;
            }
```

如有新信息来，会在手机最上边的状态栏的图标并伴有声音提示。

5. 不同按钮的动态生成

不同类别的用户登录有不同的 Intent，依据不同 Intent 可以动态生成不同首页。

1)获取登录 Activity 传过来的 intent-appFuncs

```
    Intent intent = getIntent();
    String[] msg = intent.getStringArrayExtra("appFuncs");
```

2)转换 msg 为 StringBuilder 格式，提取角色类型并保存在 funcCode

```
    final StringBuilder sb = new StringBuilder();
        for (int i = 0; i < msg.length; i++) {
            String funcCode = msg[i];
        if (funcCode.equals("APPMAP01") || funcCode.equals("APPMAP02")
                    || funcCode.equals("APPMAP03")) {
                sb.append(funcCode + " ");
            }
```

3)生成铺助 SOS 按钮

```
    if (funcCode.equals("sos")) {
        ++funcount;
        Button btn_sos = new Button(getApplicationContext());
    btn_sos.setBackgroundResource(R.drawable.btn_sos);
        createButton(btn_sos);
        btn_sos.setOnClickListener(new View.OnClickListener() {
            public void onClick(View view) {
        if (LoginCmd.sessionID == null) {
Toast.makeText(HomeActivity.this, 该功能在离线状态下无法使用", 1).show();
```

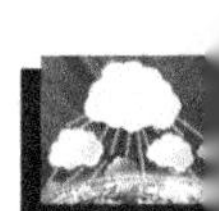

```
                        } else {
Intent i = new Intent(HomeActivity.this, SosSendActivity.class);
            startActivity(i);
                    }}});
            }
```

生成SOS按钮后，编写它的监听器，跳转到SosSendActivity。

4)预警按钮的生成

```
if (funcCode.equals("warning")) {
                ++funcount;
    Button btn_inverse = new Button(getApplicationContext());
        createButton(btn_inverse);
btn_inverse.setBackgroundResource(R.drawable.btn_warning);
btn_inverse.setOnClickListener(new View.OnClickListener() {
    public void onClick(View v) {
        if (LoginCmd.sessionID == null) {
Toast.makeText(HomeActivity.this,"该功能在离线状态下无法使用",
1000).show();
                    } else {
    Intent i = new Intent(HomeActivity.this, WarningActivity.class);
        startActivity(i);}}});
            }
```

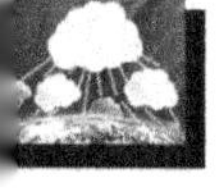

生成预警按钮后，编写它的监听器，跳转到WarningActivity。

5)设置按钮的生成

```
if (funcCode.equals("setting")) {
 ++funcount;
Button btn_map = new Button(getApplicationContext());
btn_map.setBackgroundResource(R.drawable.btn_setting);
        createButton(btn_map);
btn_map.setOnClickListener(new View.OnClickListener() {
        public void onClick(View v) {
final MenuAdapter adapter = new MenuAdapter(menus);
        new lertDialog.Builder(HomeActivity.this)
        .setAdapter(adapter, new DialogInterface.OnClickListener() {
    public void onClick(DialogInterface dialog, int which) {
            switch (which) {
```

```
                    case 0:
        Intent i = new Intent(HomeActivity.this,
    ModifyPassActivity.class);
                    startActivity(i);
                    break;
                    case 1:
                    about();
                    break;
                    case 2:
                    appExit();
                    break;
                    }}}).show();
                        }});
```

生成设置按钮后，编写它的监听器，分别处理密码重设、about()和退出系统。

6)SOS 查询按钮的生成

```
  if (funcCode.equals("sos_list")) {
      ++funcount;
      Button btn_sos_list = new Button(getApplicationContext());
      btn_sos_list.setBackgroundResource(R.drawable.sos_list_menu);
          createButton(btn_sos_list);
          btn_sos_list.setOnClickListener(new View.OnClickListener() {
                      public void onClick(View v) {
                      if (LoginCmd.sessionID == null) {
          Toast.makeText(HomeActivity.this,"该功能在离线状态下无法使用", 1000)
.show();
                              } else {
          Intent i = new Intent(HomeActivity.this, SosQueryActivity.class);
                  startActivity(i);
                              }}});
```

生成 SOS 查询按钮后，编写它的监听器，跳转到 SosQueryActivity。

7) 地图按钮的生成

```
    if(funcCode.equals("sos_locus")) {
        ++funcount;
        Button btn_map = new Button(getApplicationContext());
    btn_map.setBackgroundResource(R.drawable.btn_maps);
```


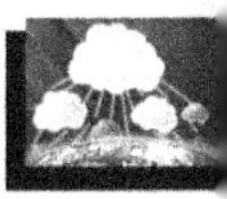

```
        createButton(btn_map);
        btn_map.setOnClickListener(new View.OnClickListener() {
                public void onClick(View v) {
                    if (LoginCmd.sessionID == null) {
    Toast.makeText(HomeActivity.this,"该功能在离线状态下无法使用", 1000).show
();
                    } else {
    Intent i = new Intent(HomeActivity.this, SosLocusActivity.class);
    i.putExtra("MAP_MODE", SosLocusActivity.MAP_MODE_MULTE);
                    startActivity(i);    }
                    }
            });
```

生成地图按钮后，编写它的监听器，跳转到SosLocusActivity。

8）登录者不是学生则跳转到预警处理界面

```
    if(LoginCmd.isStudent() == false){
    Intent i = new Intent(HomeActivity.this, WarningActivity2.class);
            startActivity(i);}
```

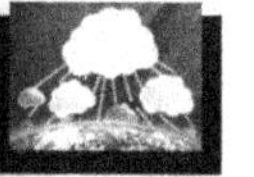

6. 首页按下返回键的处理

```
public boolean onKeyDown(int keyCode, KeyEvent event) {
if (keyCode == KeyEvent.KEYCODE_BACK && event.getRepeatCount() == 0) {
appExit();
            return true;
        } else {
            return super.onKeyDown(keyCode, event);
        }
    }
```

按下手机的返回键，则退出本系统，释放系统占用内存资源。

任务三　上下班签到模块的实现

任务描述

本任务是完成手机客户端上下班签到的开发。学生通过手机实现上下班签到，老师可以及时了解实习学生上下班的情况。系统把签到信息存储在服务器端，老师通过手机端可以查询服务器端的签到信息。

任务分析

上下班签到关键是如何实现把实习学生签到信息上传到服务器的数据库，同时便于教师查阅。解决的问题一是完成数据的上传，问题二是上传成功后如何体现已经完成签到。

任务准备

上下班签到功能的实现主要是完成信息从手机端到服务器端的推送，所以需要编写下列工具类及方法。

1. private void sendAttendRequest( ) 传送数据到服务器

2. public class AttendSve 上传数据工具类

(1) public static void attend(Map < String, String > params, HttpConnectionCallback < Object > callback) 传送数据给服务器，服务器处理完后回调

(2) public static void queryAttend(Map < String, String > params, HttpConnectionCallback < Object > callback) 查询考勤信息，完成后回调

(3) public static void queryAbsent(Map < String, String > params, HttpConnectionCallback < Object > callback) 查询缺勤信息，完成后回调

(4) public static void queryAbsentDetail(Map < String, String > params, HttpConnectionCallback < Object > callback) 查询缺勤详细信息，完成后回调。

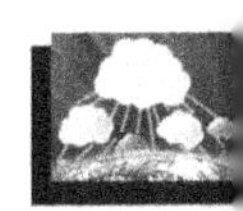

任务实现

1. 布局文件设计

```
<LinearLayout xmlns:android="http://schemas.android.com/apk/res/android"
    android:id="@+id/home_root"
    android:orientation="vertical"
    android:layout_width="fill_parent"
    android:layout_height="fill_parent">
  <include layout="@layout/frame_home_header_layout"/>
      <ImageView
    android:layout_width="fill_parent"
    android:layout_height="wrap_content"
    android:scaleType="centerCrop"
    android:src="@drawable/tpbar"
    android:adjustViewBounds="true"/>
```

```
        <ScrollView
                android: id = "@ +id/my_scrollview"
                android: layout_width = "match_parent"
                android: layout_height = "fill_parent"
                android: scrollbars = "vertical"  >
                <LinearLayout
                    android: id = "@ +id/home_layout"
                    android: layout_width = "fill_parent"
                    android: layout_height = "fill_parent"
                    android: orientation = "vertical" >
                </LinearLayout>
    </ScrollView>
<include layout = "@ layout/frame_home_footer_layout"/>
</LinearLayout>
```

该文件仅仅是显示如图 13.9 的布局。首页其他的东西需要添加代码。

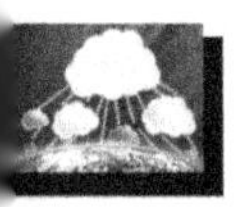

图 13.9　空的首页布局文件视图

2. 上下班签到处理过程

上下班签到主要完成学生实习上下班信息登记，包括学生姓名、实习地址包括百度地图上的经纬度及 POI 地址。完成签到登记后，登记按钮背景色会发生变化，提示用户已经签到。

1）添加上下班登记按钮

其中上班登记按钮的生成代码如下：

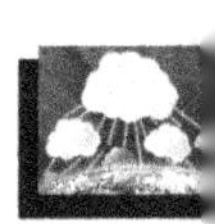

```
if (funcCode.equals("workon")) {
    ++funcount;
    workOnBtn = new Button(getApplicationContext());
    workOnBtn.setBackgroundResource(R.drawable.work_on_yet);
    createButton(workOnBtn);
    workOnBtn.setOnClickListener(new View.OnClickListener() {
        public void onClick(View v) {
            attendType = "0";
            if (locAddress == null || "".equals(locAddress)) {
                Toast.makeText(HomeActivity.this,"获取位置失败，请重新考勤", 2000).show();
            } else {
                sendAttendRequest();
            }
        }
    });
```

上班登记按钮监听器调用 sendAttendRequest( )实现签到处理。

而下班登记按钮的生成代码如下：

```
if (funcCode.equals("workoff")) {
    ++funcount;
    workOffBtn = new Button(getApplicationContext());
    workOffBtn.setBackgroundResource(R.drawable.work_off_yet);
    createButton(workOffBtn);
    workOffBtn.setOnClickListener(new View.OnClickListener() {
        public void onClick(View v) {
            attendType = "1";
            if (locAddress == null || "".equals(locAddress)) {
                Toast.makeText(HomeActivity.this,"获取位置失败，请重新考勤", 2000).show();
            } else {
                sendAttendRequest();
            } }});
```

下班登记按钮监听器调用 sendAttendRequest( )实现签到处理。

2)上下班登记信息处理

登记信息需要存储在服务器的数据里，完成此功能的是 sendAttendRequest( )。它的主要代码如下：

```
            String lon = locData.longitude + "";
                String lat = locData.latitude + "";
                Map<String, String> params = new HashMap<String, String>();
                params.put("attend.address", locAddress);
                params.put("userCode", LoginCmd.userCode);
                params.put("attend.userName", LoginCmd.userName);
                params.put("attend.lon", lon);
                params.put("attend.lat", lat);
                params.put("type", attendType);
AttendSve.attend(params, new HttpConnectionCallback<Object>() {
                    public Object execute(String response) {
                        if (response.contains("成功")) {
                            Message msg = new Message();
                            msg.arg1 = 1;
                            msg.obj = response;
                            attendHandler.handleMessage(msg);
                        } else {
                            Message msg = new Message();
                            msg.arg1 = 0;
                        msg.obj = response;
                            attendHandler.handleMessage(msg);
                        }
                        return null;
                    }
});
```

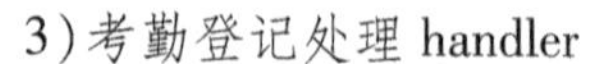

3)考勤登记处理 handler

```
Handler attendHandler = new Handler() {
    public void handleMessage(Message msg) {
        super.handleMessage(msg);
        progressDialog.dismiss();
        JSONObject json = new JSONObject(msg.obj + "");
            String response = json.getString("MSG");
            Looper.prepare();
Toast.makeText(HomeActivity.this, response, Toast.LENGTH_LONG).show();
            if(response.contains("成功")){
                if(response.contains("上班")){
```

```
                        UpdateThread updateThread = new UpdateThread( );
                        updateThread. setAttendType( WORK_ON) ;
                        updateThread. setState( STATE_DONE) ;
                        updateThread. start( ) ;
                    } else if( response. contains( "下班" ) ) {
                        UpdateThread updateThread = new UpdateThread( );
                        updateThread. setAttendType( WORK_OFF) ;
                        updateThread. setState( STATE_DONE) ;
                        updateThread. start( ) ;
                    }
            }
            Looper. loop( ) ;
```

3. 签到完成后，签到按钮背景色改变处理

上下班签到后，上下班按钮会变化背景颜色。具体由下列代码完成。

```
private Handler uiUpdateHandler = new Handler( ) {
        public void handleMessage( Message msg) {
            if (msg. what = = STATE_YET) {
                if( msg. arg1 = = WORK_ON) {
    workOnBtn. setBackgroundResource( R. drawable. work_on_yet) ;
                } else {
    workOffBtn. setBackgroundResource( R. drawable. work_off_yet) ;
                }
            } else if( msg. what = = STATE_DONE) {
                if( msg. arg1 = = WORK_ON) {
  workOnBtn. setBackgroundResource( R. drawable. work_on_finish) ;
                } else {
    workOffBtn. setBackgroundResource( R. drawable. work_off_finish) ;
                }
        }
    }
};
```

其中 R. drawable. work_on_yet，R. drawable. work_off_yet，R. drawable. work_on_finish，R. drawable. work_off_finish 分别是上下班签到前后使用的背景资源。

任务四　学生求救及求救跟踪模块的实现

任务描述

学生外出实习一旦遇到危险，就可以通过客户端发出求救信息，老师手机端立刻有信息提示，提示声可以是特别的铃声。学生发出求救信息后，百度地图间隔5秒定位一次求救学生的位置，从发出求救信息到取消求救信息这段时间内，系统会在百度地图上描绘求救学生的活动轨迹，这样就能更快更方便实施救助。

任务分析

完成本任务的关键是求救信息发送和求救者位置定位。因此，需要解决如下问题即可：

（1）学生端的求救信息如何发送到老师手机端；

（2）求救学生的地理位置如何确定；

（3）求救学生从求救信息发出开始到求救成功（或者取消求救）这段时间，求救学生的所在位置的活动轨迹如何实现。

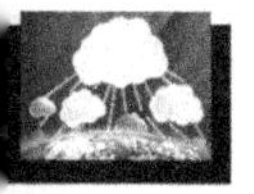

任务准备

本任务实现的关键是地图定位和求救信息在手机之间的传送，因此需要编写下列工具。

1. public class SosService **求救信息服务类**

（1）public static void sendSos（Map <String，String>
params，HttpConnectionCallback <Object> callback）发送求救信息方法。

（2）public static void cancelSos（Map <String，String>
params，HttpConnectionCallback <Object> callback）取消求救信息方法。

（3）public static void hasSos（Map <String，String>
params，HttpConnectionCallback <Object> callback）是否有求救信息方法。

（4）public static void querySos（Map <String，String>
params，HttpConnectionCallback <Object> callback）查询求救信息方法。

（5）public static void addSosLocus（Map <String，String>
params，HttpConnectionCallback <Object> callback）添加求救者位置的方法。

（6）public static void querySosLocus（Map <String，String>
params，HttpConnectionCallback <Object> callback）查询求救者位置的方法。

2. public class MyLocationListenner　**定位SDK监听服务类**

3. public class SendSosService sos 定位后台服务类

(1) private void sendLocationRequest(BDLocation location) 间隔时间内发送位置信息给服务器。

(2) private void initBaiduTools() 百度地图的初始化

4. public class SosWebSocketClientService 服务器端求救信息处理类

任务实现

1. 布局文件

1) 布局文件代码

布局文件代码如下:

```
<? xml version = "1.0" encoding = "utf - 8"? >
<LinearLayout xmlns: android = "http: //schemas. android. com/apk/res/android"
    android: layout_width = "match_parent"
    android: layout_height = "match_parent"
    android: orientation = "vertical" >
     <LinearLayout
        android: id = "@ + id/send_sos_main"
        android: layout_width = "match_parent"
        android: layout_height = "wrap_content"
        android: background = "#D000"
        android: orientation = "vertical" >
         <LinearLayout style = "@ style/ActionBarCompat" >
             <TextView
                 android: layout_width = "fill_parent"
                 android: layout_height = "wrap_content"
                 android: focusable = "true"
                 android: gravity = "center"
                 android: paddingTop = "10dip"
                 android: text = "辅助 SOS"
                 android: textColor = "#ffffff"
          android: textSize = "18sp" / >
         </LinearLayout >
         <RelativeLayout
            android: layout_width = "fill_parent"
            android: layout_height = "fill_parent" >
             <LinearLayout
```

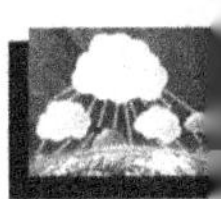

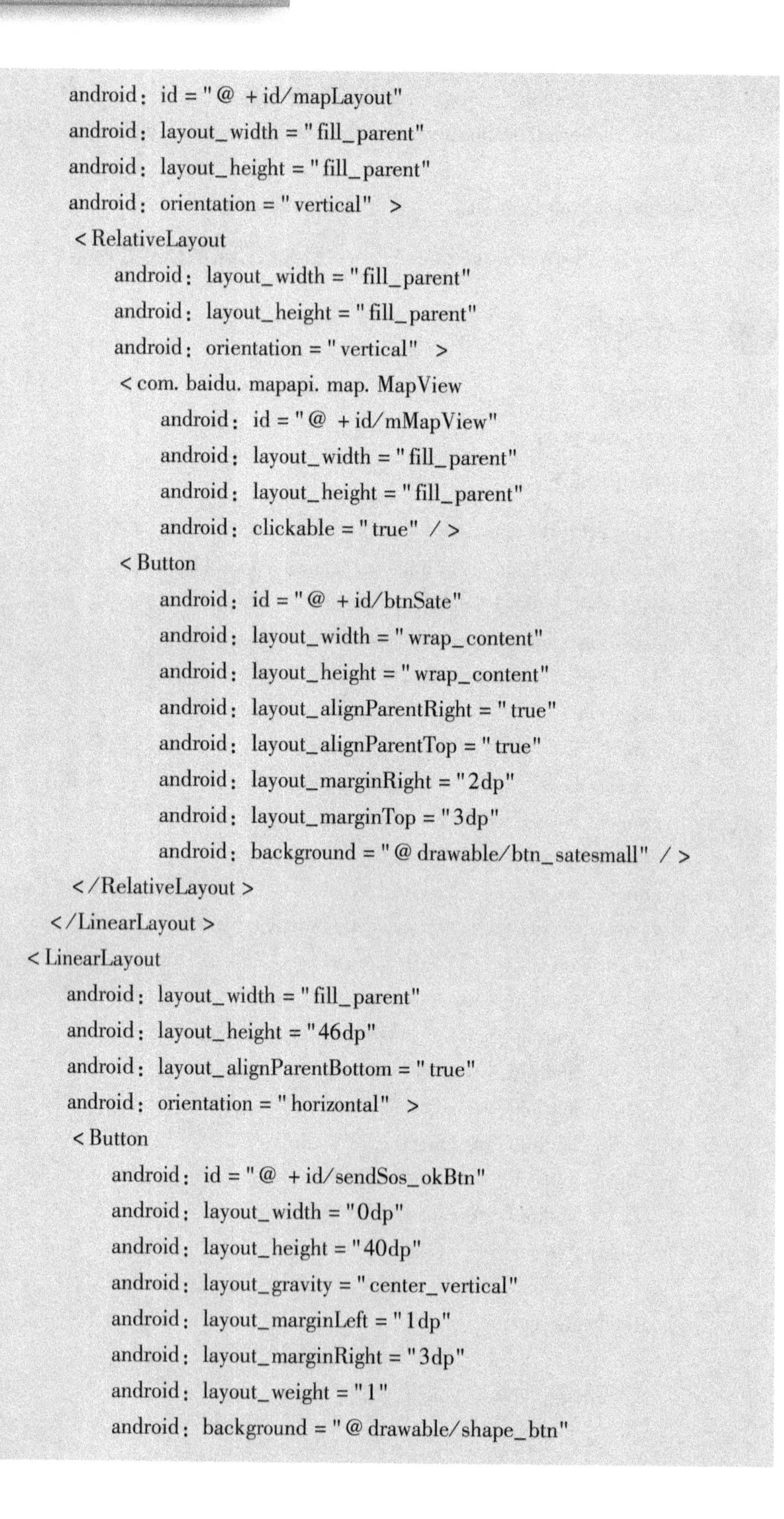

```
            android: id = "@ +id/mapLayout"
            android: layout_width = "fill_parent"
            android: layout_height = "fill_parent"
            android: orientation = "vertical" >
            <RelativeLayout
                android: layout_width = "fill_parent"
                android: layout_height = "fill_parent"
                android: orientation = "vertical" >
                <com. baidu. mapapi. map. MapView
                    android: id = "@ +id/mMapView"
                    android: layout_width = "fill_parent"
                    android: layout_height = "fill_parent"
                    android: clickable = "true" />
                <Button
                    android: id = "@ +id/btnSate"
                    android: layout_width = "wrap_content"
                    android: layout_height = "wrap_content"
                    android: layout_alignParentRight = "true"
                    android: layout_alignParentTop = "true"
                    android: layout_marginRight = "2dp"
                    android: layout_marginTop = "3dp"
                    android: background = "@drawable/btn_satesmall" />
            </RelativeLayout>
        </LinearLayout>
    <LinearLayout
        android: layout_width = "fill_parent"
        android: layout_height = "46dp"
        android: layout_alignParentBottom = "true"
        android: orientation = "horizontal" >
        <Button
            android: id = "@ +id/sendSos_okBtn"
            android: layout_width = "0dp"
            android: layout_height = "40dp"
            android: layout_gravity = "center_vertical"
            android: layout_marginLeft = "1dp"
            android: layout_marginRight = "3dp"
            android: layout_weight = "1"
            android: background = "@drawable/shape_btn"
```

```
                android: text = "发送求助" />
            <Button
                android: id = "@ +id/sendSos_cancelBtn"
                android: layout_width = "0dp"
                android: layout_height = "40dp"
                android: layout_gravity = "center_vertical"
                android: layout_marginLeft = "1dp"
                android: layout_marginRight = "3dp"
                android: layout_weight = "1"
                android: background = "@ drawable/shape_btn"
                android: text = "取消求助" />
        </LinearLayout>
      </RelativeLayout>
    </LinearLayout>
</LinearLayout>
```

其中 com. baidu. mapapi. map. MapView 为百度地图控件，其他的控件是常用控件中的 Button、TextView 等控件。

2) 布局视图

布局视图如图 13. 10 所示。

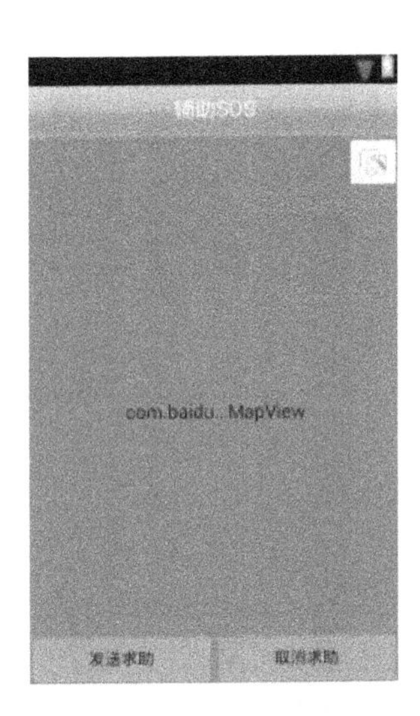

图 13. 10　求救布局视图

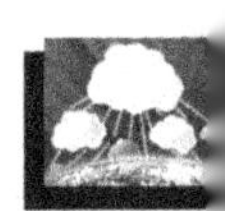

2. 求救及求救信息处理

1) 百度地图初始化

初始化百度地图，设置地图可以点击，默认的位置是东莞市政府。

```
    MyLocationListenner myListener = new MyLocationListenner();
        DemoApplication app = (DemoApplication)this.getApplication();
        if (app.mBMapManager == null) {
        app.mBMapManager = new BMapManager(this);
        app.mBMapManager.init(DemoApplication.strKey, new
    DemoApplication.MyGeneralListener());
  mMapView = (MapView) findViewById(R.id.mMapView);
        mMapController = mMapView.getController();
        mMapController.enableClick(true);
        mMapController.setZoom(17);
  GeoPoint point = new GeoPoint((int)(23.026994 * 1E6), (int)(113.758231 *
1E6));
        mMapController.setCenter(point);
        mMapView.setBuiltInZoomControls(false);
```

2)定位求救者当前位置

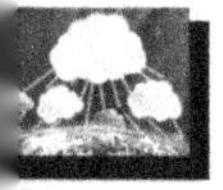

```
        mLocClient = new LocationClient(this);
        locData = new LocationData();
        mLocClient.registerLocationListener(myListener);
        LocationClientOption option = new LocationClientOption();
            option.setOpenGps(true); //打开gps
            option.setCoorType("bd09ll");        //设置坐标类型
            option.setScanSpan(5000);
        mLocClient.setLocOption(option);
        mLocClient.start();
    // 定位图层初始化
    locationOverlay = new MyLocationOverlay(mMapView);
    // 设置定位数据
    locationOverlay.setData(locData);
    // 添加定位图层
    mMapView.getOverlays().add(locationOverlay);
    locationOverlay.enableCompass();
    // 修改定位数据后刷新图层生效
    mMapView.refresh();
```

3)求救信息发送

```
sendSosOkBtn. setOnClickListener( new OnClickListener( ) {
                public void onClick( View arg0) {
AlertDialog. Builder builder = new AlertDialog. Builder(
                            SosSendActivity. this) ;
                        builder. setTitle( "您好，您确定需要发送 SOS 信号吗?") ;
builder. setPositiveButton( "确定",
                            new DialogInterface. OnClickListener( ) {
public void onClick( DialogInterface dialog, int whichButton) {
    progressDialog = ProgressDialog. show(
    SosSendActivity. this, "请稍等 . . . ","正在提交数据 . . . ", true) ; //对话框确
定按钮监听代码
        Thread th = new Thread( sendRun) ; //启动发送求救线程
        th. start( ) ;
    }
}) ;
builder. setNegativeButton( "取消",
                            new DialogInterface. OnClickListener( ) {
public void onClick( DialogInterface dialog, int whichButton) {}
                            }) ;
                builder. create( ). show( ) ;
        }
    }) ;
}
```

4)求救信息发送线程

```
Runnable sendRun = new Runnable( ) {
    public void run( ) {
    String userCode = LoginCmd. userCode; //记录用户信息(求救人信息)
    String type = "0"; // 发送中
    String lon = locData. longitude + "";
        String lat = locData. latitude + "";
        Map <String, String > params = new HashMap <String, String >( ) ;
        params. put( "sos. userCode" , userCode) ;
        params. put( "sos. userName" , LoginCmd. userName) ;
        params. put( "sos. type" , type) ;
        params. put( "sos. sendContent" , "" ) ;
```

```
        params.put("sos.address", button.getText().toString());
        //百度地图当前地址
        params.put("sos.lon", lon);
        params.put("sos.lat", lat);
        SosService.sendSos(params, new
HttpConnectionCallback<Object>() {
            //发送信息给服务器端，服务器保存求救人信息和地址后，回调
        public Object execute(String response) {//回调方法处理代码
                if (response.contains("成功")) {
                Message msg = new Message(); //将回调的信息交给msg
                    msg.arg1 = 1; //成功是1
                    msg.obj = response;
                    sendHandler.sendMessage(msg);
                } else {
                    Message msg = new Message();
                    msg.arg1 = 0; //失败是0
                    msg.obj = response;
                    sendHandler.sendMessage(msg);
                }
                return null;
        }
});
```

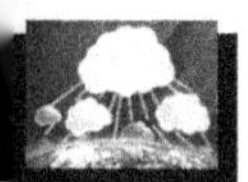

5)求救信息消息处理 handler

```
    Handler sendHandler = new Handler() {
    public void handleMessage(Message msg) {
        super.handleMessage(msg);
        progressDialog.dismiss(); //取消滚动条
        try {
        JSONObject json = new JSONObject(String.valueOf(msg.obj)); //把 msg
转换成 json 格式
            String response = json.getString("MSG"); //得到 msg 的值
            id = json.getString("ID");
            if (response.contains("成功")) {
                LoginCmd.hasSos = "true"; //求救状态成功
                sendSosOkBtn.setEnabled(false);
                sendSosCancelBtn.setEnabled(true);
```

```
                startService( new Intent( SosSendActivity. this,
        SendSosService. class) ); //后台启动地图定位服务并发送到服务器
    bindService( new Intent( SosSendActivity. this, SendSosService. class),
        sosServiceConn, Context. BIND_AUTO_CREATE);
    //绑定学生求救定位服务
    spu. putValue( SharedPreferencesUtil. STR_SOS_STATE, "true"); //保存求救状态
    json. getString( "mobile") ! = null && json. getString( "mobile"). length( ) > 0) {
    SmsManager manager = SmsManager. getDefault( );    //发送求救短信给老师
    ArrayList < String > list = manager. divideMessage( json. getString( "sms") );    //因为
一条短信有字数限制, 因此要将长短信拆分
                for( String text: list) {
manager. sendTextMessage( json. getString( "mobile"), null, text, null, null);
                    }
            }
    }
    DialogUtil. showDialog( SosSendActivity. this, response, false);
```

6)反编码：通过坐标点检索求救者详细地址及周边 POI

```
    public void onGetAddrResult( MKAddrInfo res, int error) {
            if (error ! = 0) {
            String str = String. format( "错误号:% d", error);
    Toast. makeText( SosSendActivity. this, str, Toast. LENGTH_LONG)
                    . show( );
                return;    }
            if (res. type = = MKAddrInfo. MK_REVERSEGEOCODE) {
                String strInfo = res. strAddr;
                if (! strInfo. equals( button. getText( ) ) ) {
if ( !"". equals( button. getText( ) ) || button. getText( ). length( ) > 0)
                    mMapView. removeView( button);
                    button. setText( strInfo);
                    layoutParam = new MapView. LayoutParams(
                    MapView. LayoutParams. WRAP_CONTENT,
    MapView. LayoutParams. WRAP_CONTENT, ptCenter, 0,
    -32, MapView. LayoutParams. BOTTOM_CENTER);
    mMapView. addView( button, layoutParam);
                }
            }
```

```
        }
    });
```

本系统是一个完整的基于C/S的管理系统，其中C端是手机端，S端是服务器端。本书配套的网络资源有完整的APP程序，程序代码可以在Android SDK2.2...或者以上版本运行调试。该学生实习安全管理系统服务器的网址是：http：//14.23.103.106：1080/st2.0。用户名：test。密码：test。用户test是按学生角色登录的。

任务考核

将网络资源里的代码调试成功后连接到服务器，并实际操作相关功能。

参考文献

[1] 王雅宁. 轻松学 Android 开发 [M]. 北京：电子工业出版社，2013.
[2] 李刚. 疯狂 Android 讲义 [M]. 2 版. 北京：电子工业出版社，2013.
[3] 明日科技. Android 从入门到精通[M]. 北京：清华大学出版社，2012.
[4] 李华明. Android 游戏编程之从零开始[M]. 北京：清华大学出版社，2011.